Statistics with St

Statistics with Stata

Updated for Version 7

Lawrence C. Hamilton
University of New Hampshire

DUXBURY

THOMSON LEARNING

Australia • Canada • Mexico • Singapore • Spain • United Kingdom • United States

DUXBURY

★ ™

THOMSON LEARNING

Executive Editor, Statistics: Curt Hinrichs
Assistant Editor: Ann Day
Editorial Assistant: Katherine Brayton
Marketing Manager: Joseph Rogove
Marketing Assistant: Maria Salinas
Technology Project Manager: Burke Taft
Advertising Project Manager:
Samantha Cabaluna

Project Manager, Editorial Production:
Erica Silverstein
Print/Media Buyer: Jessica Reed
Permissions Editor: Robert Kauser
Cover Designer: Carole Lawson
Cover Image: PhotoDisc
Text and Cover Printer: Webcom Limited

Printed in Canada

2 3 4 5 6 7 06 05 04 03 02

For more information about our products, contact us at:
Thomson Learning Academic Resource Center
1-800-423-0563

For permission to use material from this text, contact us by:
Phone: 1-800-730-2214 **Fax:** 1-800-730-2215
Web: http://www.thomsonrights.com

ISBN 0-534-39654-2

Wadsworth Group/Thomson Learning
10 Davis Drive
Belmont, CA 94002-3098
USA

Asia
Thomson Learning
60 Albert Street, #15-01
Albert Complex
Singapore 189969

Australia
Nelson Thomson Learning
102 Dodds Street
South Melbourne, Victoria 3205
Australia

Canada
Nelson Thomson Learning
1120 Birchmount Road
Toronto, Ontario M1K 5G4
Canada

Europe/Middle East/Africa
Thomson Learning
Berkshire House
168–173 High Holborn
London WC1V 7AA
United Kingdom

Contents

Preface . . . . *ix*

1 Stata and Stata Resources **1**
A Typographical Note 1
An Example Stata Session 2
Stata's Documentation and Help Files 6
Searching for Information 7
Stata Corporation 7
Statalist 8
The Stata Technical Bulletin (1991–2001) and Stata Journal 9
Books Using Stata 10

2 Data Management **12**
Example Commands 12
Creating a New Dataset 14
Specifying Subsets of the Data: `in` and `if` Qualifiers 18
Generating and Replacing Variables 22
Using Functions 25
Converting between Numeric and String Formats 29
Creating New Categorical and Ordinal Variables 32
Using Explicit Subscripts with Variables 35
Importing Data from Other Programs 36
Combining Two or More Stata Files 39
Transposing, Reshaping or Collapsing Data 43
Weighting Observations 49
Creating Random Data and Random Samples 51
Writing Programs for Data Management 55
Managing Memory 56

3 Graphs **58**
Example Commands 58
Histograms 60
Scatterplots 64
Boxplots and One-Way Scatterplots 71
Pie Charts and Star Charts 73
Bar Charts 76
Symmetry and Quantile Plots 80
Quality Control Graphs 84

4   Summary Statistics and Tables **88**
    Example Commands 88
    Summary Statistics for Measurement Variables 89
    Exploratory Data Analysis 92
    Normality Tests and Transformations 94
    Frequency Tables and Two-Way Cross-Tabulations 97
    Multiple Tables and Multi-Way Cross-Tabulations 100
    Tables of Means, Medians, and other Summary Statistics 103
    Using Frequency Weights 105

5   ANOVA and Other Comparison Methods **108**
    Example Commands 108
    One-Sample Tests 110
    Two-Sample Tests 113
    One-Way Analysis of Variance (ANOVA) 115
    Two- and N-Way Analysis of Variance 118
    Analysis of Covariance (ANOCOVA) 119
    Predicted Values and Error-Bar Charts 121

6   Linear Regression Analysis **124**
    Example Commands 124
    The Regression Table 126
    Multiple Regression 128
    Predicted Values and Residuals 129
    Basic Graphs for Regression 132
    Correlations 135
    Hypothesis Tests 139
    Dummy Variables 140
    Stepwise Regression 146
    Polynomial Regression 149

7   Regression Diagnostics **152**
    Example Commands 152
    SAT Score Regression, Revisited 154
    Diagnostic Plots 156
    Diagnostic Case Statistics 160
    Multicollinearity 166

8   Fitting Curves **171**
    Example Commands 171
    Band Regression 172
    Lowess Smoothing 174
    Regression with Transformed Variables – 1 177
    Regression with Transformed Variables – 2 181
    Conditional Effect Plots 183
    Nonlinear Regression – 1 185
    Nonlinear Regression – 2 189

9 Robust Regression **193**
    Example Commands 193
    Regression with Ideal Data 194
    Y Outliers 197
    X Outliers (Leverage) 200
    Asymmetrical Error Distributions 202
    Robust Analysis of Variance 203
    Further `rreg` and `greg` Applications 208
    Robust Estimates of Variance 209

10 Logistic Regression **213**
    Example Commands 214
    Space Shuttle Data 215
    Using Logistic Regression 220
    Conditional Effect Plots 223
    Diagnostic Statistics and Plots 224
    Logistic Regression with Ordered-Category *y* 228
    Multinomial Logistic Regression 230

11 Survival Analysis and Event-Count Models **237**
    Example Commands 237
    Survival-Time Data 240
    Count-Time Data 242
    Kaplan–Meier Survivor Functions 244
    Cox Proportional Hazard Models 247
    Exponential and Weibull Regression 253
    Poisson Regression 257
    Generalized Linear Models 261

12 Principal Components and Factor Analysis **266**
    Example Commands 266
    Principal Components 267
    Rotation 269
    Factor Scores 270
    Principal Factoring 273
    Maximum-Likelihood Factoring 274

13 Time Series Analysis **277**
    Example Commands 277
    Smoothing 278
    Further Time Plot Examples 283
    Lags, Leads, and Differences 285
    Correlograms 287
    ARIMA Models 290

14 Introduction to Programming **297**
    Basic Concepts and Tools 297
    Example Program: Moving Autocorrelation 304
    Ado-File 308
    Help File 310
    Matrix Algebra 312
    Bootstrapping 317
    Monte Carlo Simulation 321

References **329**

Index **333**

Preface

Statistics with Stata is intended for students and practicing researchers, to bridge the gap between statistical textbooks and Stata's own documentation. In this intermediate role, it does not provide the detailed expositions of a proper textbook, nor does it come close to describing all of Stata's features. Instead, it demonstrates how to use Stata to accomplish some of the most common statistical tasks. Chapter topics follow conceptual themes rather than focusing on particular Stata commands, which gives *Statistics with Stata* a different structure from the *Stata Reference Manuals*. The chapter on Data Management, for example, covers a variety of procedures for creating, updating, and restructuring data files. Chapters on Summary Statistics and Tables, ANOVA and Other Comparison Methods, and Fitting Curves, among others, have similarly broad themes that encompass a number of separate techniques.

The general topics of the first six chapters (through ordinary least squares regression) roughly parallel an introductory course in applied statistics, but with digressions covering practical issues often encountered by analysts—how to aggregate data, create dummy variables, or translate ANOVA into regression, for instance. In Chapter 7 (Regression Diagnostics) and beyond, we move into the territory of advanced courses or original research. Here readers can find basic information and illustrations of how to obtain and interpret diagnostic statistics and graphs; perform robust, quantile, nonlinear, logit, ordered logit, multinomial logit, or Poisson regression; fit survival-time and event-count models; construct composite variables through factor analysis and principal components; and graph or model time series data. Stata has worked hard in recent years to maintain its state-of-the-art standing, and this effort is particu-larly apparent in the wide range of regression and model-fitting commands it now offers.

Finally, we conclude with a look at programming in Stata. Many readers will find that Stata does everything they need already, so they have no need to write original programs. For an active minority, however, programmability is one of Stata's principal attractions, and it certainly underlies Stata's currency and rapid advancement. This chapter opens the door for new users to explore Stata programming, whether for specialized data management tasks, to establish a new statistical capability, for Monte Carlo experiments, or for teaching.

Generally similar versions ("flavors") of Stata run on Windows, Macintosh, and Unix computers. Across all platforms, Stata uses the same commands, data files, and output. The versions differ in some details of screen appearance, menus, and file handling, where Stata follows the conventions native to each platform—such as \directory\filename file specifications under Windows, in contrast with the /directory/filename specifications under Unix. Rather than display all three, I employ Windows conventions, but users with other systems should find that only minor translations are needed.

Notes on the Fourth Edition

I first encountered Stata version 1 in the mid-1980s. At that time, Stata ran only on MS-DOS personal computers, but its PC orientation made it distinctly more modern than its main competitors—most of which had originated before the desktop revolution, in the 80-column punched-card Fortran environment of mainframes. Unlike mainframe statistical packages that believed each user was a stack of cards, Stata viewed the user as a conversation. Its interactive nature and integration of statistical procedures with data management and graphics supported the natural flow of analytical thought in ways that other programs did not. **graph** and **predict** soon became favorite commands. I was impressed enough by how it all fit together to start writing *Statistics with Stata*, published in 1990 for Stata version 2.

A great deal has changed about Stata since that book, in which I observed that "Stata is not a do-everything program . . . . The things it does, however, it does very well." The expansion of Stata's capabilities has been striking. This is most noticeable in the proliferation, and later in the steady rationalization, of model fitting procedures. William Gould's architecture for Stata, with its programming tools and unified syntax, has aged well, and proven able to incorporate new statistical ideas as these were developed. The formidable lists of modeling commands that begins Chapter 10, or of functions in Chapter 2, illustrate some of the ways that Stata became richer over the years. Suites of new techniques such as those for panel (**xt**), survey (**svy**), time series (**ts**), or survival time (**st**) data open worlds of possibility, as do programmable commands for nonlinear regression (**nl**) and general linear modeling (**glm**), or general procedures for maximum-likelihood estimation. Other critical expansions include the development of a matrix programming capability, and the wealth of new data-management features. Data management, with good reason, has been promoted from an incidental topic in the first *Statistics with Stata* to the longest chapter in this fourth edition.

Comparing this book about Stata 7 with its immediate predecessor, *Statistics with Stata 5*, the greatest change has been the addition of Chapter 13 on time series analysis. Internet resources deservedly get more attention in Chapter 1, as do tutorial programs and the new *Stata Journal*. Chapter 2 has fresh material about editing data, formatting variables, generating random data, and managing memory. More flexible table-making commands, and power transformations of variables, were added to Chapter 4. Chapter 5 includes a new section about graphs for ANOVA; Chapter 8 takes a sharper look at multicollinearity; Chapter 9 suggests robust and median-based analogues to ANOVA; Chapter 10 demonstrates conditional effects plotting for multinomial logit regression; and Chapter 11 gives a simple reason for considering GLM. Chapter 14 on Programming has been largely rewritten to reflect recent enhancements in matrix algebra, internal results, and output or help file displays. Other new Stata features, or improvements in old commands (including **graph** and **predict**), are scattered throughout the book. Because Stata now does so much, far beyond the scope of an introductory book, *Statistics with Stata* presents more procedures telegraphically in the "Example Commands" sections that begin most chapters, or in lists of options followed by advice that readers should "type **help whatever** " for details. Stata's online help and search features have advanced to keep pace with the program, so this is not idle advice.

Beyond the help files are Stata's web site, Internet and documentation search capabilities, user-community listserv, NetCourses, the *Stata Journal*, and of course Stata's formidable printed documentation—which contains well over 3,000 pages and is still growing. I think of this book as a gateway.

Acknowledgments

Stata's architect, William Gould, deserves credit for originating the elegant program that *Statistics with Stata* describes. Together with my editor, Curt Hinrichs, and a handful of outspoken readers, he persuaded me to begin the effort of writing a new book. Pat Branton and many others at Stata Corporation gave invaluable feedback, often on short deadlines, once this effort got underway. It would not have been possible without their assistance. James Hamilton contributed advice about time series. Leslie Hamilton helped to edit the final manuscript.

A book like this could not exist without data. To avoid endlessly recycling my own old examples, I drew on work by others including Carole Seyfrit, Sally Ward, Heather Turner, Rasmus Ole Rasmussen, Birger Poppel, Per Lyster, Erich Buch, Paul Mayewski, and Loren D. Meeker. Steve Selvin shared several of his examples from *Practical Biostatistical Methods* for Chapter 11. Data from agencies including Statistics Iceland, Statistics Greenland, Statistics Canada, the Northwest Atlantic Fisheries Organization, Greenland's Natural Resources Institute, and the Department of Fisheries and Oceans Canada can also be found among the examples. A presentation given by Brenda Topliss inspired the "gossip" programming example in Chapter 14. Other forms of support, encourage-ment, or ideas came from Fae Korsmo, Michael Ledbetter, Richard Haedrich, Cliff Brown, David Rohall, Rasmus Ole Rasmussen, James Morison, Oddmund Otterstad, James Tucker, and Cynthia M. Duncan.

Dedication

To Sarah and Dave, on their way.

Stata and Stata Resources

Stata is a full-featured statistical program for Windows, Macintosh, and Unix computers. It combines ease of use with speed, a library of pre-programmed analytical and data-management capabilities, and programmability that allows users to invent and add further capabilities as needed. Stata's consistent, intuitive command syntax simplifies both learning and use.

After introductory information, we'll begin with an example Stata session to give you a sense of the "flow" of data analysis, and of how analytical results might be used. Later chapters explain in more detail. Even without explanations, however, you can see how straightforward the commands are—**use** *filename* to retrieve dataset *filename*, **summarize** when you want summary statistics, **correlate** to get a correlation matrix, and so forth.

Stata users have available a variety of resources to help them learn about Stata and solve problems at any level of difficulty. These resources come not just from Stata Corporation, but also from an active community of users. Sections of this chapter introduce some key resources— Stata's online help and printed documentation; where to phone, fax, write, or e-mail for technical help; Stata's web site (www.stata.com), which provides many services including updates and answers to frequently asked questions; the Statalist Internet forum; and the refereed *Stata Journal*.

A Typographical Note

This book employs several typographical conventions as a visual cue to how words are used:

■ Commands typed by the user appear in a **bold Courier font**. When the whole command line is given, it starts with a period, as seen in a Stata Results window or log (output) file:

. **list** *year boats men penalty*

■ *Variable* or *file* names within these commands appear in italics to emphasize the fact that they are arbitrary and not a fixed part of the command.

■ Names of *variables* or *files* also appear in italics within the main text to distinguish them from ordinary words.

■ Items from Stata's menus are shown in an Arial font , with successive options separated by a dash. For example, we can open an existing dataset by selecting File – Open , and then finding and clicking on the name of the particular dataset. Note that some common menu actions can be accomplished either with text choices from Stata's top menu bar,

File	Edit	Prefs	Window	Help

or with the row of icons below these. For example, selecting File – Open is equivalent to clicking the leftmost icon, an opening file folder: 📂 . One could also accomplish the same thing by typing a direct command of the form

```
. use filename
```

■ Stata output as seen in the Results window is shown in a `small Courier font`. The small font allows Stata's 80-column output to fit within the margins of this book.

Thus, we show the calculation of summary statistics for a variable named *penalty* as follows:

```
. summarize penalty
```

Variable	Obs	Mean	Std. Dev.	Min	Max
penalty	10	63	59.59493	11	183

These typographic conventions exist only in this book, and not within the Stata program itself. Stata can display a variety of onscreen fonts, but it does not use italics in commands or elsewhere. Once Stata log files have been imported into a word processor, or a results table copied and pasted, you might want to format them in a Courier font, 10 point or smaller, so that columns will line up correctly.

In its commands and variable names, Stata is case sensitive. Thus, **summarize** is a command, but Summarize and SUMMARIZE are not. *Penalty* and *penalty* would be two different variables.

An Example Stata Session

As a preview showing Stata at work, this section retrieves and analyzes a previously-created dataset named *lofoten.dta*. Jentoft and Kristofferson (1989) originally published these data in an article about self-management among fishermen on Norway's arctic Lofoten Islands. There are 10 observations (years) and 5 variables, including *penalty*, a count of how many fishermen were cited each year for violating fisheries regulations.

If we might eventually want a printout of our session, the best way to prepare for this is by opening a "log file" at the start. Log files contain commands and results tables, but not graphs. To begin a log file, click the scroll-shaped Begin Log icon, 📜, and specify a name and folder for the resulting log file. Alternatively, a log file could be started by choosing File – Log – Begin from the top menu bar, or by typing a direct command such as

```
. log using monday1
```

Multiple ways of doing such things are common in Stata. Each has its own advantages, and each suits different situations or user tastes.

Log files can be created either in a special Stata format (.smcl), or in ordinary text or ASCII format (.log). A .smcl ("Stata markup and control language") file will be nicely formatted for viewing or printing within Stata. It could also contain hyperlinks that help to understand commands or error messages. .log (text) files lack such formatting, but are simpler to use if you plan later to insert or edit the output in a word processor. After selecting which type of log file you want, click **Save** . For this session, we will create a .smcl log file named *monday1.smcl*.

An existing Stata-format dataset named *lofoten.dta* will be analyzed here. To open or retrieve this dataset, we again have several options:

select **File – Open –** *lofoten.dta* using the top menu bar;

select 📂 – *lofoten.dta* ; or

type the command **use lofoten** .

Under its default Windows configuration, Stata looks for data files in folder C:\data, but we could browse through other folders in the usual way.

To see a brief description of the dataset now in memory, type

```
. describe

Contains data from C:\data\lofoten.dta
  obs:            10                          Jentoft & Kristoffersen '89
  vars:            5                          15 Dec 1996 09:53
  size:          130 (99.9% of memory free)
-------------------------------------------------------------------------
              storage  display    value
variable name   type   format     label      variable label
-------------------------------------------------------------------------
year            int    %9.0g                  Year
boats           int    %9.0g                  Number of fishing boats
men             int    %9.0g                  Number of fishermen
penalty         int    %9.0g                  Number of penalties issued
decade          byte   %9.0g      decade      Early 1970s or early 1980s
-------------------------------------------------------------------------
Sorted by: decade  year
```

Many Stata commands can be abbreviated to their first few letters. For example, we could shorten **describe** to just the letter **d** .

This dataset has only 10 observations and 5 variables, so we can easily list its contents by typing the command **list** (or just the letter **l**):

```
. list

         year    boats     men   penalty    decade
  1.     1971     1809    5281        71     1970s
  2.     1972     2017    6304       152     1970s
  3.     1973     2068    6794       183     1970s
  4.     1974     1693    5227        39     1970s
  5.     1975     1441    4077        36     1970s
  6.     1981     1540    4033        11     1980s
  7.     1982     1689    4267        15     1980s
  8.     1983     1842    4430        34     1980s
  9.     1984     1847    4622        74     1980s
 10.     1985     1365    3514        15     1980s
```

Analysis could begin with a table of means, standard deviations, minimum values, and maximum values (**summarize** or **su**):

```
. summarize
```

Variable	Obs	Mean	Std. Dev.	Min	Max
year	10	1978	5.477226	1971	1985
boats	10	1731.1	232.1328	1365	2068
men	10	4854.9	1045.577	3514	6794
penalty	10	63	59.59493	11	183
decade	10	.5	.5270463	0	1

To print results from the session so far, bring the Results window to the front by clicking on this window or on ▓ (Bring Results Window to Front), and then click 🖨 (Print).

To copy a table, commands, or other information from the Results window into a word processor, again make sure that the Results window is in front by clicking on this window or on ▓. Drag the mouse to select the results that you want, and then choose Edit – Copy Text from the menu. Finally, switch to your word processor and, at the desired insertion point, issue the appropriate "paste" command such as Edit – Paste or clicking a "clipboard" icon.

Did the number of penalties for fishing violations change over the two decades covered by these data? A table containing summary statistics for *penalty* at each value of *decade* shows that there were more penalties in the 1970s:

```
. tabulate decade, sum(penalty)
```

Early 1970s or early 1980s	Summary of Number of penalties issued		
	Mean	Std. Dev.	Freq.
1970s	96.2	67.41439	5
1980s	29.8	26.281172	5
Total	63	59.594929	10

Perhaps the number of penalties declined because fewer people were fishing in the 1980s. The number of penalties correlates strongly ($r > .8$) with the number of boats and fishermen:

```
. correlate boats men penalty
(obs=10)
```

	boats	men	penalty
boats	1.0000		
men	0.8748	1.0000	
penalty	0.8259	0.9312	1.0000

A graph might help clarify these interrelationships. Figure 1.1 shows a plot of *men* and *penalty* against *year*, produced through the **graph** command. Options after the variable list in this command specify that both *y*-variable series should be connected by line segments, with separately scaled vertical axes for each. **ylabel xlabel rlabel** tells Stata to choose round numbers when labeling the left vertical, horizontal, and right vertical axes.

```
. graph men penalty year, connect(ll) rescale ylabel xlabel rlabel
```

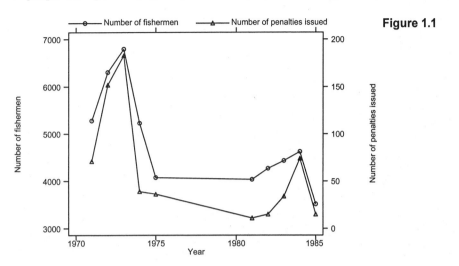

Figure 1.1

Because the years 1976 to 1980 are missing in these data, Figure 1.1 shows 1975 connected to 1981. For some purposes, we might hesitate to do this. Instead, we could either find the missing values or leave the gap unconnected by issuing a slightly more complicated set of commands.

To print this graph, click on the Graph window or on ▦ (Bring Graph Window to Front), and then click the Print icon 🖨 .

To copy the graph directly into a word processor or other document, bring the graph window to the front and select Edit – Copy Graph . Switch to your word processor, go to the desired insertion point, and issue an appropriate "paste" command such as Edit – Paste or clicking a "clipboard" icon.

To save the graph for future use, select File – Save Graph . Stata for Windows offers two formats for saving graphs:

.gph or Stata graph files, which can be retrieved by Stata to view or print later; or

.wmf or Windows metafiles, which can be imported by other programs such as word processors or slide-show designers.

Stata-format (.gph) graphs can also be saved by adding a **saving(*filename*)** option to any **graph** command. To save a graph with the filename *figure1.gph*, type

```
. graph men penalty year, connect(ll) saving(figure1)
```

Through all of the preceding analyses, the log file *monday1.smcl* has been storing our results. We could review this file to see what we have done by choosing Window – Viewer or 👁 , and then we could print the log file by choosing 🖨 (Print) . Log files close automatically at the end of a Stata session, or earlier if instructed by one of the following:

File – Log – Close ;

✧ – Close log file – OK ; or

typing the command **log close** .

Once closed, the file *monday1.smcl* could be opened again through File – View or ![eye] during a subsequent Stata session. To make an output file that can be opened easily by your word processor, create a .log format (ASCII text) log file instead of an .smcl format log file.

Stata's Documentation and Help Files

Stata documentation includes a slim *Getting Started* manual (for example, *Getting Started with Stata for Windows*), the more extensive *User's Guide*, the encyclopedic *Stata Reference Manual*, the *Stata Graphics Manual*, and the *Stata Programming Manual*. *Getting Started* helps you do just that, with the basics of installation, window management, data entry, printing, and so on. The *User's Guide* contains an extended discussion of general topics, including resources and troubleshooting. The multi-volume *Stata Reference Manual* lists all Stata commands alphabetically. Entries for each command include the full command syntax, descriptions of all available options, examples, technical notes regarding formulas and rationale, and references for further reading. Graphics and programming commands are covered in the general references, but these important topics get more detailed treatment and examples in the *Graphics Manual* and *Programming Manual*, respectively.

But when you are in the midst of a Stata session, it is often simpler to ask for onscreen help instead of consulting the manuals. Selecting Help from the top menu bar invokes a drop-down menu of further choices, including help on specific commands, general topics, online updates, the *Stata Technical Bulletin* (1991–2001), or connections to Stata's web site (www.stata.com). Alternatively, we can bring the Viewer (![eye]) to front and use its Search or Contents features to find information. We can also use the **help** command. Typing **help correlate**, for example, causes help information to appear in the Results window. Like the *Stata Reference Manual*, onscreen help provides command syntax diagrams and complete lists of options. It also includes some examples, although generally less detailed and without the technical discussions found in the *Reference Manual*. The Viewer help has several advantages over the manuals, however. It can search for keywords in the documentation or on Stata's web site. Hypertext links take you directly to related entries. Onscreen help can also include material on "unofficial" Stata programs that you have downloaded from Stata's web site or from other users.

Stata provides a number of tutorial programs that can assist you in learning to use certain features. These include

intro	An introduction to Stata
graphics	How to make graphs
tables	How to make tables
regress	Estimating regression models, including two-stage least squares (2SLS)
anova	Estimating one-, two- and N-way ANOVA and ANCOVA models
logit	Estimating maximum-likelihood logit and probit models
survival	Estimating maximum-likelihood survival models
factor	Estimating factor and principal component models
ourdata	Description of the data Stata provides

yourdata   How to input your own data into Stata

The introductory tutorial, aimed at first-time users, can be started by typing

. **tutorial intro**

Similar commands start the other tutorials. Type **tutorial contents** for a current list of available tutorials.

Searching for Information

Selecting Help – Search – Search documentation and FAQs provides a direct way to search for information either in Stata's help files, or in the web site's FAQs (frequently asked questions) and other pages. The equivalent Stata command is

. **search** *keywords*

Options available with **search** allow us to limit our search to the Stata manuals, to the *STB* or *Stata Journal*, or to the FAQs. We can ask for a search based on author names, entry identification numbers, or exact wording (no abbreviations). Typing

. **search** *median regression*

returns only the relatively few references that contain both "median" and "regression". On the other hand, typing

. **search** *median or regression*

would return a far larger set of references containing either "median" or "regression". Type **help search** for details about this command.

  Selecting Help – Search – Search net resources conducts a search not only of Stata's web site, but of other relevant Internet sites as well . The corresponding command is

. **net search** *keywords*

net search is one of many **net** commands that help to install and manage user-written programs from the Web (type **help net** to see a listing of others).

Stata Corporation

For orders, licensing, and upgrade information, you can contact Stata Corporation by e-mail at

  stata@stata.com

or visit their web site at

  http://www.stata.com

The mailing or physical address is

  Stata Corporation
  4905 Lakeway Drive
  College Station, TX 77845 USA

Telephone access includes an easy-to-remember 800 number.

telephone: 1-800-STATAPC   U.S.
           (1-800-782-8272)

           1-800-248-8272   Canada

           1-979-696-4600   International

fax: 1-979-696-4601

Online updates are free to licensed Stata users. These provide a fast and simple way to obtain the latest enhancements, bug fixes, etc. for your current version.

Technical support can be obtained by sending e-mail messages *with your Stata serial number in the subject line* to

tech_support@stata.com

Before calling or writing for technical help, though, you might want to look at www.stata.com to see whether your question is a FAQ. The site also provides product, ordering, and help information; international notes; and assorted news and announcements. Much attention is given to user support, including the following:

FAQs — Frequently asked questions and their answers. If you are puzzled by something and can't find the answer in the manuals, check here next—it might be a FAQ. Example questions range from basic—"How do I read data from my Excel file into Stata?"—to more technical queries such as "What are some of the small-sample adjustments to the sandwich estimate of variance?"

STB FTP — Easy-to-follow instructions tell you how to download free *STB* programs from the Harvard FTP (File Transfer Protocol) site.

UPDATES — Frequent minor updates or bug fixes, downloadable at no cost by licensed Stata users.

OTHER RESOURCES — Links and information including online Stata instruction (NetCourses); enhancements from the *Stata Journal* or *Stata Technical Bulletin*; a listserv (Statalist) for discussions among Stata users; a bookstore selling books about Stata and other up-to-date statistical references; downloadable datasets and programs for Stata-related books; and links to statistical web sites including Stata's competitors.

The following sections describe some of the most important user-support resources.

Statalist

Statalist provides a valuable online forum for communication among active Stata users. It is independent of Stata Corporation, although Stata programmers monitor it and often contribute to the discussion. To subscribe to Statalist, send an e-mail message to

majordomo@hsphsun2.harvard.edu

The body of this message should contain only the following words:

subscribe statalist

The list processor will acknowledge your message and send instructions for using the list, including how to post messages of your own. Any message sent to the following address goes out to all current subscribers:

```
statalist@hsphsun2.harvard.edu
```

Do *not* try to subscribe or unsubscribe by sending messages directly to the statalist address. This does not work, and your mistake goes to hundreds of subscribers. To unsubscribe from the list, write to the same majordomo address you used to subscribe:

```
majordomo@hsphsun2.harvard.edu
```

but send only the message

```
unsubscribe statalist
```

or send the equivalent message

```
signoff statalist
```

If you plan to be traveling or offline for a while, unsubscribing will keep your mailbox from filling up with Statalist messages. You can always re-subscribe.

The material on Statalist includes requests for programs, solutions, or advice, as well as answers and general discussion. Topics range from broad ("What are the advantages/ disadvantages of Stata compared with SPSS or SAS?") to specialized ("Has anyone written programs that can be used with time-varying covariates to test the reasonableness of using Cox hazard models?"). Along with the *Stata Journal* (discussed below), Statalist plays a major role in extending the capabilities both of Stata and of serious Stata users.

The *Stata Technical Bulletin* (1991–2001) and *Stata Journal*

At the time of this writing, the last issue of the *Stata Technical Bulletin* (*STB*) had just been published, and the first issue of the *Stata Journal*, scheduled for publication in fall 2001, was being assembled. For ten years, the bimonthly *STB* served as a means of distributing new commands and Stata updates, both user-written and official. Accumulated *STB* articles were published in book form each year as *Stata Technical Bulletin Reprints*. The *Reprints* and back issues of individual journals can be ordered directly from Stata Corporation. The *STB* programs themselves can be downloaded for free; select Help – STB and User-written Programs from Stata's top menu bar, or visit www.stata.com.

With the growth of the Internet, instant communication among users became possible through vehicles such as Statalist. Program files could easily be downloaded from distant sources. A bimonthly printed journal and disk no longer provided the best avenues either for communicating among users, or for distributing updates and user-written programs. To adapt to a changing world, the *STB* had to evolve into something new.

The *Stata Journal* was launched to meet this challenge and the needs of Stata's broadening user base. Like the old *STB*, the *Stata Journal* still contains articles describing new commands by users along with unofficial commands written by Stata Corporation employees. New commands are not its primary focus, however. The *Stata Journal* also contains expository articles about statistics, book reviews, and a number of interesting columns, including one by Nicholas J. Cox on effective use of the Stata programming language. Like the *STB*, the *Stata Journal* is intended for all Stata users, both novice and experienced. Unlike the *STB*, the *Stata Journal* contains peer-reviewed articles.

The *Stata Journal* is published quarterly. Subscriptions can be purchased directly from Stata Corporation by visiting www.stata.com.

Books Using Stata

A growing number of books describe Stata, or incorporate its output in their presentations. These include:

Cryer, Jonathan B. and Robert B. Miller. 1994. *Statistics for Business: Data Analysis and Modeling*, 2nd edition. Belmont, CA: Duxbury.

Davis, Duane. 2000. *Business Research for Decision Making*, 5th edition. Belmont, CA: Duxbury.

Gould, William and William Sribney. 1999. *Maximum Likelihood Estimation with Stata*. College Station, TX: Stata Press.

Hamilton, Lawrence C. 1992. *Regression with Graphics: A Second Course in Applied Statistics*. Belmont, CA: Duxbury.

Hamilton, Lawrence C. 1996. *Data Analysis for Social Scientists*. Belmont, CA: Duxbury.

Hamilton, Lawrence C. 1996. *Data Analysis for Social Scientists with StataQuest*. Belmont, CA: Duxbury.

Hardin, James and Joseph Hilbe. 2001. *Generalized Linear Models and Extensions*. College Station, TX: Stata Press.

Hosmer, David W., Jr. and Stanley Lemeshow. 1999. *Applied Survival Analysis*. New York: Wiley.

Hosmer, David W., Jr. and Stanley Lemeshow. 2000. *Applied Logistic Regression*, 2nd edition. New York: Wiley.

Howell, David C. 2002. *Statistical Methods for Psychology*, 5th edition. Belmont, CA: Duxbury.

Howell, David C. 1999. *Fundamental Statistics for the Behavioral Sciences*, 4th edition. Belmont, CA: Duxbury.

Johnston, Jack and John DiNardo. 1997. *Econometric Methods*, 4th edition. New York: McGraw-Hill.

Keller, Gerald, Brian Warrack and Henry Bartel. 2000. *Statistics for Management and Economics*, 5th edition. Belmont, CA: Duxbury.

Long, J. Scott. 1997. *Regression Models for Categorical and Limited Dependent Variables*. Thousand Oaks, CA: Sage.

Long, J. Scott and Jeremy Freese. 2001. *Regression Models for Categorical Dependent Variables using Stata*. College Station, TX: Stata Press.

Newton, H. Joseph and Jane L. Harvill. 1997. *StatConcepts: A Visual Tour of Statistical Ideas*. Pacific Grove, CA: Duxbury.

Rabe–Hesketh, Sophia and Brian Everitt. 2000. *A Handbook of Statistical Analysis Using Stata*, 2nd edition. Boca Raton, FL: Chapman & Hall.

Pagano, Marcello and Kim Gauvreau. 2000. *Principles of Biostatistics*, 2nd edition. Belmont, CA: Duxbury.

Stata Corporation. 2001. *Getting Started with Stata for Macintosh*. College Station, TX: Stata Press.

Stata Corporation. 2001. *Getting Started with Stata for Unix*. College Station, TX: Stata Press.

Stata Corporation. 2001. *Getting Started with Stata for Windows*. College Station, TX: Stata Press.

Stata Corporation. 2001. *Stata Graphics Manual, Release 7*. College Station, TX: Stata Press.

Stata Corporation. 2001. *Stata Programming Manual, Release 7*. College Station, TX: Stata Press.

Stata Corporation. 2001. *Stata Reference Manual, Release 7*, Volumes 1–4. College Station, TX: Stata Press.

Stata Corporation. 2001. *Stata User's Guide, Release 7*. College Station, TX: Stata Press.

Stine, Robert and John Fox (eds.). 1997. *Statistical Computing Environments for Social Research*. Thousand Oaks, CA: Sage.

2

Data Management

The first steps in data analysis involve organizing the raw data into a format usable by Stata. We can bring new data into Stata in several ways: type the data from the keyboard; read a text or ASCII file containing the raw data; or, using a third-party data transfer program, translate the data directly from a system file created by another spreadsheet, database, or statistical program. Once Stata has the data in memory, we can save the data in Stata format for easy retrieval and updating in the future.

Data management encompasses the initial tasks of creating a dataset, editing to correct errors, and adding internal documentation such as variable and value labels. It also encompasses many other jobs required by ongoing projects, such as adding new observations or variables; reorganizing, simplifying, or sampling from the data; separating, combing, or collapsing datasets; converting variable types; and creating new variables through algebraic or logical expressions. When data-management tasks become complex or repetitive, Stata users can write their own programs to automate the work. Although Stata is best known for its analytical capabilities, it possesses a broad range of data-management features as well. This chapter introduces some of the basics.

Example Commands

. **append using** *olddata*

Reads previously-saved dataset *olddata.dta* and adds all its observations to the data currently in memory. Subsequently typing **save** *newdata*, **replace** will save the combined dataset as *newdata.dta*.

. **browse**

Opens the spreadsheet-like Stata Browser for viewing the data. The Browser looks similar to the Editor, but it has no editing capability, so there is no risk of inadvertently changing your data. Alternatively, click .

. **browse** *boats men* **if** *year* > 1980

Opens the Browser showing only the variables *boats* and *men* for observations in which *year* is greater than 1980. This example illustrates the **if** qualifier, which can be used to focus the operation of many Stata commands.

. **compress**

Automatically converts all variables to their most efficient storage types to conserve memory and disk space. Subsequently typing the command **save, replace** will make these changes permanent.

. **drawnorm *z1 z2 z3*, n(5000)**

Creates an artificial dataset with 5,000 observations and three random variables, *z1*, *z2*, and *z3*, sampled from uncorrelated standard normal distributions. Options could specify other means, standard deviations, and correlation or covariance matrices.

. **edit**

Opens the spreadsheet-like Stata Editor where data can be entered or edited. Alternatively, choose Window – Data Editor or click ▦ .

. **edit *boats year men***

Opens the Editor with only the variables *boats*, *year*, and *men* (in that order) visible and available for editing.

. **encode *stringvar*, gen(*numvar*)**

Creates a new variable named *numvar*, with labeled numerical values based on the string (non-numeric) variable *stringvar*.

. **format *rainfall* %8.2f**

Establishes a fixed (**f**) display format for numeric variable *rainfall*: 8 columns wide, with two digits always shown after the decimal.

. **generate *newvar* = (*x* + *y*)/100**

Creates a new variable named *newvar*, equal to *x* plus *y* divided by 100.

. **generate *newvar* = uniform()**

Creates a new variable with values sampled from a uniform random distribution over the interval ranging from 0 to nearly 1, written [0,1).

. **infile *x y z* using *data.raw***

Reads an ASCII file named *data.raw* containing data on three variables we want to name *x*, *y*, and *z*. Values of those variables are separated by at least one space, and missing values are represented by a period (not a blank). Other commands exist for reading tab-delimited, comma-delimited, or fixed-column raw data (type **help infile**).

. **list *x y z* in 5/20**

Lists the *x*, *y*, and *z* values of the 5th through 20th observations, as the data are presently sorted. The **in** qualifier works in similar fashion with most other Stata commands as well.

. **merge *id* using *olddata***

Reads the previously-saved dataset *olddata.dta* and matches observations from *olddata* with observations in memory that have identical *id* values. Both *olddata* (the "using data") and the data currently in memory (the "master" data) must already be sorted in *id* order.

. **replace *oldvar* = 100 * *oldvar***

Replaces the values of *oldvar* with 100 times their previous values.

. **sample 10**

Drops all the observations in memory except for a 10% random sample. Instead of selecting a certain percentage, we could select a certain number of cases. For example, **sample 55, count** would drop all but a random sample of size $n = 55$.

. **save** *newfile*

Saves the data currently in memory, as a file named *newfile.dta*. If *newfile.dta* already exists, and you want to write over the previous version, type **save** *newfile*, **replace**. To save *newfile.dta* in a format readable by older versions of Stata, type **save** *newfile*, **old** . Alternatively, use the menus: File – Save or File – Save As .

. **set memory 24m**

(Windows or Unix systems only) Allocates 24 megabytes of memory for Stata data. The amount set could be greater or less than the current allocation. Virtual memory (disk space) is used if the request exceeds physical memory. Type **clear** to drop the current data from memory before using **set memory** .

. **sort** *x*

Sorts the data from lowest to highest values of *x*. Observations with missing *x* values appear last after sorting because Stata views missing values as very high numbers. Type **help gsort** for a more general sorting command that can arrange values in either ascending or descending order and can optionally place the missing values first.

. **tabulate** *x* **if** *y* **> 65**

Produces a frequency table for *x* using only those observations that have *y* values above 65. The **if** qualifier works similarly with most other Stata commands.

. **use** *oldfile*

Retrieves previously-saved Stata-format dataset *oldfile.dta* from disk, and places it in memory. If other data are currently in memory, and you want to discard those data without saving them, type **use** *oldfile*, **clear** . Alternatively, these tasks can be accomplished through File – Open or by clicking ⊞.

Creating a New Dataset

Data that were previously saved in Stata format can be retrieved into memory either by typing a command of the form **use** *filename*, or through menu selections. This section describes basic methods for creating a Stata-format dataset in the first place, using as our example the Canadian information (from the Federal, Provincial and Territorial Advisory Committee on Population Health, 1996) listed in Table 2.1.

Table 2.1: Data on Canada and Its Provinces

Place	1995 Pop. (1000's)	Unemployment Rate (percent)	Male Life Expectancy	Female Life Expectancy
Canada	29606.1	10.6	75.1	81.1
Newfoundland	575.4	19.6	73.9	79.8
Prince Edward Island	136.1	19.1	74.8	81.3
Nova Scotia	937.8	13.9	74.2	80.4
New Brunswick	760.1	13.8	74.8	80.6

Quebec	7334.2	13.2	74.5	81.2
Ontario	11100.3	9.3	75.5	81.1
Manitoba	1137.5	8.5	75.0	80.8
Saskatchewan	1015.6	7.0	75.2	81.8
Alberta	2747.0	8.4	75.5	81.4
British Columbia	3766.0	9.8	75.8	81.4
Yukon	30.1	—	71.3	80.4
Northwest Territories	65.8	—	70.2	78.0

The simplest way to create a dataset from Table 2.1 is through Stata's spreadsheet-like Editor, which is invoked either by clicking [icon], selecting Window – Data Editor from the top menu bar, or typing the command **edit**. Then begin typing values for each variable, in columns that Stata automatically calls *var1*, *var2*, etc. Thus, *var1* contains place names (Canada, Newfoundland, etc.); *var2*, populations; and so forth.

We can assign more descriptive variable names by double-clicking on the column headings (such as *var1*) and then typing a new name in the resulting dialog box—eight characters or fewer works best, although longer names are possible. We can also create variable labels that contain a brief description. For example, *var2* (population) might be renamed *pop*, and given the variable label "Population in 1000s, 1995".

Renaming and labeling variables can also be done outside of the Editor through the **rename** and **label variable** commands:

`. rename var2 pop`

`. label variable pop "Population in 1000s, 1995"`

Cells left empty, such as employment rates for the Yukon and Northwest Territories, will automatically be assigned Stata's missing value code, a period. At any time, we can close the Editor and then save the dataset to disk. Clicking [icon] or Window – Data Editor brings the Editor back.

If the first value entered for a variable is a number, as with population, unemployment, and life expectancy, then Stata assumes that this column is a "numerical variable" and it will thereafter permit only numerical values. Numerical values can also begin with a plus or minus sign, include decimal points, or be expressed in scientific notation. For example, we could represent Canada's population as 2.96061e+7, which means 2.96061×10^7 or about 29.6 million people. Numerical values *should not include any commas*, such as 29,606,100. If we did happen to put commas within the first value typed in a column, Stata would interpret this as a "string variable" (next paragraph) rather than as a number.

If the first value entered for a variable includes non-numerical characters, as did the place names above (or "1,000" with the comma), then Stata thereafter considers this column to be a string variable. String variable values can be almost any combination of letters, numbers, symbols, or spaces up to 80 characters long. We can thus store names, quotations, or other descriptive information. String variable values can be tabulated and counted, but do not allow the calculation of means, correlations, or most other statistics.

After typing in the information from Table 2.1 in this fashion, we close the Editor and save our data, perhaps with the name *canada0.dta*:

```
. save canada0
```

Stata automatically adds the extension .dta to any dataset name, unless we tell it to do otherwise. If we already had saved and named an earlier version of this file, it is possible to write over that with the newest version by typing

```
. save, replace
```

At this point, our new dataset looks like this:

```
. describe

Contains data from C:\data\canada0.dta
  obs:           13
  vars:           5                              11 May 2001 19:11
  size:         533 (100.0% of memory free)
-------------------------------------------------------------------------------
              storage  display    value
variable name   type   format     label     variable label
-------------------------------------------------------------------------------
var1          str21    %21s
pop           float    %9.0g                 Population in 1000s, 1995
var3          float    %9.0g
var4          float    %9.0g
var5          float    %9.0g
-------------------------------------------------------------------------------
Sorted by:

. list
                         var1        pop      var3      var4      var5
  1.                   Canada    29606.1      10.6      75.1      81.1
  2.             Newfoundland      575.4      19.6      73.9      79.8
  3.     Prince Edward Island      136.1      19.1      74.8      81.3
  4.              Nova Scotia      937.8      13.9      74.2      80.4
  5.            New Brunswick      760.1      13.8      74.8      80.6
  6.                   Quebec     7334.2      13.2      74.5      81.2
  7.                  Ontario    11100.3       9.3      75.5      81.1
  8.                 Manitoba     1137.5       8.5        75      80.8
  9.             Saskatchewan     1015.6         7      75.2      81.8
 10.                  Alberta       2747       8.4      75.5      81.4
 11.         British Columbia       3766       9.8      75.8      81.4
 12.                    Yukon       30.1         .      71.3      80.4
 13.    Northwest Territories       65.8         .      70.2        78

. summarize
    Variable |     Obs        Mean   Std. Dev.       Min        Max
-------------+------------------------------------------------------
        var1 |       0
         pop |      13    4554.769    8214.304       30.1    29606.1
        var3 |      11    12.10909    4.250048          7       19.6
        var4 |      13    74.29231    1.673052       70.2       75.8
        var5 |      13    80.71539    .9754027         78       81.8
```

Examining such output tables gives us a chance to look for errors that should be corrected. The **summarize** table, for instance, provides several numbers useful in proofreading, including the count of nonmissing observations (always 0 for string variables) and the minimum and maximum for each variable. Substantive interpretation of the summary statistics would be premature at this point, because our dataset contains one observation (Canada) that represents a combination of the other 12 provinces and territories.

The next step is to make our dataset more self-documenting. The variables could be given more descriptive names, such as the following:

```
. rename var1 place
```

```
. rename var3 unemp
```

```
. rename var4 mlife
```

```
. rename var5 flife
```

Stata also permits us to add several kinds of labels to the data. **label data** describes the dataset as a whole. For example,

```
. label data "Canadian dataset 0"
```

label variable describes an individual variable. For example,

```
. label variable place "Place name"
```

```
. label variable unemp "% 15+ population unemployed, 1995"
```

```
. label variable mlife "Male life expectancy years"
```

```
. label variable flife "Female life expectancy years"
```

By labeling data and variables, we obtain a dataset that is more self-explanatory:

```
. describe
```

```
Contains data from C:\data\canada0.dta
  obs:            13                     Canadian dataset 0
  vars:            5                     11 May 2001 19:11
  size:          533 (100.0% of memory free)
-------------------------------------------------------------------------
               storage  display   value
variable name   type    format    label     variable label
-------------------------------------------------------------------------
place          str21    %21s                 Place name
pop            float    %9.0g                Population in 1000s, 1995
unemp          float    %9.0g                % 15+ population unemployed,
                                               1995
mlife          float    %9.0g                Male life expectancy years
flife          float    %9.0g                Female life expectancy years
-------------------------------------------------------------------------
Sorted by:
```

Once labeling is completed, we should save the data to disk by using File – Save or typing

```
. save, replace
```

W can later retrieve these data any time through 📂, File – Open, or by typing

```
. use c:\data\canada0
```
```
(Canadian dataset 0)
```

We can then proceed with a new analysis. We might notice, for instance, that male and female life expectancies correlate positively with each other and also negatively with the unemployment rate. The life expectancy–unemployment rate correlation is slightly stronger for males.

```
. correlate unemp mlife flife

(obs=11)

         |   unemp    mlife    flife
---------+---------------------------
  unemp|   1.0000
  mlife|  -0.7440   1.0000
  flife|  -0.6173   0.7631   1.0000
```

The order of observations within a dataset can be changed through the **sort** command. For example, to rearrange observations from smallest to largest in population, type

```
. sort pop
```

String variables are sorted alphabetically instead of numerically. Typing the following will rearrange observations to place Alberta first, British Columbia second, and so on.

```
. sort place
```

We can control the order of variables in the data, using the **order** command. For example, we could make unemployment rate the second variable, and population last:

```
. order place unemp mlife flife pop
```

The Editor also has buttons that perform these functions. The Sort button applies to the column currently highlighted by the cursor. The << and >> buttons move the current variable to the beginning or end of the variable list, respectively. As with any other editing, these changes only become permanent if we subsequently save our data.

The Editor's Hide button does not rearrange the data, but rather makes a column temporarily invisible on the spreadsheet. This feature is convenient if, for example, we need to type in more variables and want to keep the province names or some other case identification column in view, adjacent to the "active" column where we are entering data.

We can also restrict the Editor beforehand to work only with certain variables, in a specified order, or with a specified range of values. For example,

```
. edit place mlife flife
```

or

```
. edit place unemp if pop > 100
```

The last example employs an **if** qualifier, an important tool described in the next section.

Specifying Subsets of the Data: in and if Qualifiers

Many Stata commands can be restricted to a subset of the data by adding an **in** or **if** qualifier. **in** specifies the observation numbers to which the command applies. For example, **list in 5** tells Stata to list only the 5th observation. To list the 1st through 20th observations, type

. `list in 1/20`

The letter `l` denotes the last case, and `-4` , for example, the fourth-from-last. Thus, we could list the four most populous Canadian places (which will include Canada itself) as follows:

. `sort pop`

. `list place pop in -4/l`

Note the important, although typographically subtle, distinction between `1` (number one, or first observation) and `l` (letter "el," or last observation). The `in` qualifier works in a similar way with most other analytical or data-editing commands. It always refers to the data *as presently sorted.*

The `if` qualifier also has broad applications, but it selects observations based on specific variable values. As noted, the observations in *canada0.dta* include not only 12 Canadian provinces or territories, but also Canada as a whole. For many purposes, we might want to exclude Canada from analyses involving the 12 territories and provinces. One way to do so is to restrict the analysis to only those places with populations below 20 million (20,000 thousand); that is, every place except Canada:

. `summarize if pop < 20000`

Variable	Obs	Mean	Std. Dev.	Min	Max
place	0				
pop	12	2467.158	3435.521	30.1	11100.3
unemp	10	12.26	4.44877	7	19.6
mlife	12	74.225	1.728965	70.2	75.8
flife	12	80.68333	1.0116	78	81.8

Compare this with the earlier **summarize** output to see how much has changed. The previous mean of population, for example, was grossly misleading because it counted every person twice.

The " `<` " (is less than) sign is one of six *relational operators*:

`==`	is equal to
`!=`	is not equal to (`~=` also works)
`>`	is greater than
`<`	is less than
`>=`	is greater than or equal to
`<=`	is less than or equal to

A double equals sign, " `==` ", denotes the logical test, "*Is* the value on the left side the same as the value on the right?" To Stata, a single equals sign means something different: "*Make* the value on the left side be the same as the value on the right." The single equals sign is not a relational operator and cannot be used within `if` qualifiers. Single equals signs are used instead with commands that generate new variables, or replace the values of old ones, according to algebraic expressions.

Any of these relational operators can be used to select observations based on their values for numerical variables. Only two operators, `==` and `!=`, make sense with string variables. To use string variables in an `if` qualifier, enclose the target value in double quotes. For example, we could get a summary excluding Canada (leaving in the 12 provinces and territories):

```
. summarize if place != "Canada"
```

Two or more relational operators can be combined within a single `if` expression by the use of *logical operators*. Stata's logical operators are the following:

&   and

| or (symbol is a vertical bar, not the number one or letter "el")

!   not (~ also works)

The Canadian territories (Yukon and Northwest) both have fewer than 100,000 people. To find the mean unemployment and life expectancies for the 10 Canadian provinces only, excluding both the smaller places (territories) and the largest (Canada), we could use this command:

```
. summarize unemp mlife flife if pop > 100 & pop < 20000

Variable |     Obs        Mean    Std. Dev.       Min        Max
---------+-----------------------------------------------------------
   unemp |      10       12.26     4.44877          7       19.6
   mlife |      10       74.92     .6051633      73.9       75.8
   flife |      10       80.98     .586515       79.8       81.8
```

Parentheses allow us to specify the precedence among multiple operators. For example, we might list all the places that either have unemployment below 9, *or* have life expectancies of at least 75.4 for men *and* 81.4 for women:

```
. list if unemp < 9 | (mlife >= 75.4 & flife >= 81.4)

                           place        pop    unemp    mlife    flife
  8.                    Manitoba     1137.5      8.5       75     80.8
  9.                Saskatchewan     1015.6        7     75.2     81.8
 10.                     Alberta       2747      8.4     75.5     81.4
 11.            British Columbia       3766      9.8     75.8     81.4
```

A note of caution regarding missing values: Stata shows missing values as a period, but in some operations (notably **sort** and **if**, although not in statistical calculations such as means or correlations), these same missing values are treated as if they were large positive numbers. Watch what happens if we sort places from lowest to highest unemployment rate, and then ask to see places with unemployment rates above 15%:

```
. sort unemp
```

```
. list if unemp > 15

                           place        pop    unemp    mlife    flife
 10.   Prince Edward Island        136.1     19.1     74.8     81.3
 11.            Newfoundland        575.4     19.6     73.9     79.8
 12.   Northwest Territories        65.8        .     70.2       78
 13.                  Yukon        30.1        .     71.3     80.4
```

The two places with missing unemployment rates were included among those "greater than 15." In this instance the result is obvious, but with a larger dataset we might not notice. Suppose that we were analyzing a political opinion poll. A command such as the following would tabulate the variable *vote* not only for people with ages older than 65, as intended, but also for any people whose *age* values were missing:

```
. tabulate vote if age > 65
```

Where missing values exist, we might have to deal with them explicitly as part of the **if** expression:

```
. tabulate vote if age > 65 & age != .
```

The **in** and **if** qualifiers set cases aside temporarily so that a particular command line does not apply to them. These qualifiers have no effect on the data in memory, and the next command will apply to all cases unless it too has an **in** or **if** qualifier. To drop variables from the data in memory, use the **drop** command. For example, to drop *mlife* and *flife* from memory, type

```
. drop mlife flife
```

We can drop observations from memory by using either the **in** qualifier or the **if** qualifier. Because we earlier sorted on *unemp*, the two territories occupy the 12th and 13th positions in the data. Canada itself is 6th. One way to drop these three nonprovinces employs the **in** qualifier. **drop in 12/13** means "drop the 12th through the 13th observations."

```
. list
```

	place	pop	unemp
1.	Saskatchewan	1015.6	7
2.	Alberta	2747	8.4
3.	Manitoba	1137.5	8.5
4.	Ontario	11100.3	9.3
5.	British Columbia	3766	9.8
6.	Canada	29606.1	10.6
7.	Quebec	7334.2	13.2
8.	New Brunswick	760.1	13.8
9.	Nova Scotia	937.8	13.9
10.	Prince Edward Island	136.1	19.1
11.	Newfoundland	575.4	19.6
12.	Yukon	30.1	.
13.	Northwest Territories	65.8	.

```
. drop in 12/13
(2 observations deleted)
. drop in 6
(1 observation deleted)
```

The same change could have been accomplished through an **if** qualifier, with a command that says "drop if *place* equals Canada or population is less than 100."

```
. drop if place == "Canada" | pop < 100
(3 observations deleted)
```

After dropping Canada, the territories, and the variables *mlife* and *flife*, we have the following reduced dataset:

```
. list
```

	place	pop	unemp
1.	Saskatchewan	1015.6	7
2.	Alberta	2747	8.4
3.	Manitoba	1137.5	8.5
4.	Ontario	11100.3	9.3
5.	British Columbia	3766	9.8
6.	Quebec	7334.2	13.2
7.	New Brunswick	760.1	13.8
8.	Nova Scotia	937.8	13.9
9.	Prince Edward Island	136.1	19.1
10.	Newfoundland	575.4	19.6

We can also drop selected variables or observations through the Delete button in the Editor.

Instead of telling Stata which variables or observations to drop, it sometimes is simpler to specify which to keep. The same reduced dataset could have been obtained as follows:

```
. keep place pop unemp
. keep if place != "Canada" & pop >= 100
(3 observations deleted)
```

Like any other changes to the data in memory, none of these reductions affect disk files until we save the data. At that point, we will have the option of writing over the old dataset (**save, replace**) and thus destroying it, or just saving the newly modified dataset with a new name (by choosing File – Save As , or by typing a command with the form **save newname**) so that both versions exist on disk.

Generating and Replacing Variables

The **generate** and **replace** commands allow us to create new variables or change the values of existing variables. For example, in Canada as in most industrial societies, women tend to live longer than men. To analyze regional variations in this gender gap, we might retrieve dataset *canada1.dta* and generate a new variable equal to female life expectancy (*flife*) minus male life expectancy (*mlife*). In the main part of a **generate** or **replace** statement (unlike **if** qualifiers) we use a single equals sign.

```
. use canada1, clear
(Canadian dataset 1)

. generate gap = flife - mlife

. label variable gap "Female-male gap life expectancy"

. describe
```

```
Contains data from C:\data\canada1.dta
  obs:           13                          Canadian dataset 1
  vars:           6                          11 May 2001 19:11
  size:         585 (100.0% of memory free)
-------------------------------------------------------------------------
              storage  display    value
variable name  type    format     label    variable label
-------------------------------------------------------------------------
place         str21    %21s                Place name
pop           float    %9.0g               Population in 1000s, 1995
unemp         float    %9.0g               % 15+ population unemployed,
                                              1995
mlife         float    %9.0g               Male life expectancy years
flife         float    %9.0g               Female life expectancy years
gap           float    %9.0g               Female-male gap life expectancy
-------------------------------------------------------------------------
Sorted by:
    Note:  dataset has changed since last saved
```

```
. list place flife mlife gap

                       place    flife    mlife         gap
 1.                   Canada     81.1     75.1           6
 2.             Newfoundland     79.8     73.9    5.900002
 3.    Prince Edward Island     81.3     74.8         6.5
 4.              Nova Scotia     80.4     74.2    6.200005
 5.            New Brunswick     80.6     74.8    5.799995
 6.                   Quebec     81.2     74.5    6.699997
 7.                  Ontario     81.1     75.5    5.599998
 8.                 Manitoba     80.8       75    5.800003
 9.             Saskatchewan     81.8     75.2    6.600006
10.                  Alberta     81.4     75.5    5.900002
11.         British Columbia     81.4     75.8    5.599998
12.                    Yukon     80.4     71.3    9.099998
13.    Northwest Territories       78     70.2    7.800003
```

For the province of Newfoundland, the true value of *gap* should be $79.8 - 73.9 = 5.9$ years, but the output shows this value as 5.900002 instead. Like all computer programs, Stata stores numbers in binary form, and 5.9 has no exact binary representation. The small inaccuracies that arise from approximating decimal fractions in binary are unlikely to affect statistical calculations much because calculations are done in double precision (8 bytes per number). They appear disconcerting in data lists, however. We can change the display format so that Stata shows only a rounded-off version. The following command specifies a fixed display format four numerals wide, with one digit to the right of the decimal:

```
. format gap %4.1f
```

Even when the display shows 5.9, however, a command such as the following will return no observations:

```
. list if gap == 5.9
```

This occurs because Stata believes the value does not exactly equal 5.9. (More technically, Stata stores *gap* values in single precision but does all calculations in double, and the single- and double-precision approximations of 5.9 are not identical.)

Display formats, as well as variables names and labels, can also be changed by double-clicking on a column in the Stata Editor. Fixed numeric formats such as **%4.1f** are one of three main numeric display format types. These are

%*w*.*d*g   General numeric format, where *w* specifies the total width or number of columns displayed and *d* the minimum number of digits that must follow the decimal point. Exponential notation (such as 1.00e+07, meaning 1.00×10^7 or 10 million) and shifts in the decimal-point position will be used automatically as needed, to display values in an optimal (but varying) fashion.

%*w*.*d*f   Fixed numeric format, where *w* specifies the total width or number of columns displayed and *d* the fixed number of digits that must follow the decimal point.

%*w*.*d*e   Exponential numeric format, where *w* specifes the total width or number of columns displayed and *d* the fixed number of digits that must follow the decimal point.

For example, as we saw in Table 2.1, the 1995 population of Canada was approximately 29,606,100 people, and the Yukon Territory population was 30,100. The table below shows how these two numbers appear under several different display formats:

format	Canada	Yukon
%9.0g	2.96e+07	30100
%9.1f	29606100.0	30100.0
%12.5e	2.96061e+07	3.01000e+04

Although the displayed values look different, their internal values are identical. Statistical calculations remain unaffected by display formats. Other numerical formatting options include the use of commas, right-justification, or centering within a display field. There also exist special formats for date and time series variables. Type **help format** for more information.

replace can make the same sorts of calculations as **generate**, but it changes values of an existing variable instead of creating a new variable. For example, the variable *pop* in our dataset gives population in thousands. To convert this to simple population, we just multiply (" * " means multiply) all values by 1,000:

```
. replace pop = pop * 1000
```

replace can make such wholesale changes, or it can be used with **in** or **if** qualifiers to selectively edit the data. To illustrate, suppose that we had questionnaire data with variables including *age* and year born (*born*). A command such as the following would correct one or more typos where a subject's age had been incorrectly typed as 229 instead of 29:

```
. replace age = 29 if age == 229
```

Alternatively, the following command could correct an error in the value of *age* for observation number 1453:

```
. replace age = 29 in 1453
```

For a more complicated example,

```
. replace age = 2001-born if age == . | age < 2001-born
```

This replaces values of variable *age* with 2001 minus the year of birth if *age* is missing or if the reported age is less than 2001 minus the year of birth.

generate and **replace** provide tools to create categorical variables as well. We noted earlier that our Canadian dataset includes several types of observations: 2 territories, 10 provinces, and 1 country combining them all. Although **in** and **if** qualifiers allow us to separate these, and **drop** can eliminate observations from the data, it might be most convenient to have a categorical variable that indicates the observation's "type." The following example shows one way to create such a variable. We start by generating *type* as a constant, equal to 1 for each observation. Next, we replace this with the value 2 for the Yukon and Northwest Territories, and with 3 for Canada. The final steps involve labeling new variable *type* and defining labels for values 1, 2, and 3.

```
. use canada1, clear
(Canadian dataset 1)
. generate type = 1
. replace type = 2 if place == "Yukon" | place == "Northwest
     Territories"
(2 real changes made)
. replace type = 3 if place == "Canada"
(1 real change made)
. label variable type "Province, territory or nation"
```

```
. label values type typelbl
. label define typelbl 1 "Province" 2 "Territory" 3 "Nation"
. list place flife mlife gap type

                        place    flife    mlife         gap        type
 1.                    Canada     81.1     75.1           6      Nation
 2.              Newfoundland     79.8     73.9    5.900002    Province
 3.       Prince Edward Island     81.3     74.8         6.5    Province
 4.                Nova Scotia     80.4     74.2    6.200005    Province
 5.             New Brunswick     80.6     74.8    5.799995    Province
 6.                    Quebec     81.2     74.5    6.699997    Province
 7.                   Ontario     81.1     75.5    5.599998    Province
 8.                  Manitoba     80.8       75    5.800003    Province
 9.              Saskatchewan     81.8     75.2    6.600006    Province
10.                   Alberta     81.4     75.5    5.900002    Province
11.          British Columbia     81.4     75.8    5.599998    Province
12.                     Yukon     80.4     71.3    9.099998   Territory
13. Northwest Territories         78     70.2    7.800003   Territory
```

As illustrated, labeling the values of a categorical variable requires two commands. The **label define** command specifies what labels go with what numbers. The **label values** command specifies to which variable these labels apply. One set of labels (created through one **label define** command) can apply to any number of variables (that is, be referenced in any number of **label values** commands). Value labels can have up to 80 characters, but work best if they are not too long.

generate can create new variables, and **replace** can produce new values, using any mixture of old variables, constants, random values, and expressions. For numeric variables, the following *arithmetic operators* apply:

+ add
- subtract
* multiply
/ divide
^ raise to power

Parentheses will control the order of calculation. Without them, the ordinary rules of precedence apply. Of the arithmetic operators, only addition, "+ ", works with string variables, where it concatenates two string values into one.

Although their purposes differ, **generate** and **replace** have similar syntax. Either can use any mathematically or logically feasible combination of Stata operators and **in** or **if** qualifiers. These commands can also employ Stata's broad array of special functions, introduced in the following section.

Using Functions

This section lists many of the functions available for use with **generate** or **replace**. For example, we could create a new variable named *loginc*, equal to the natural logarithm of *income*, by using the natural log function **ln** within a **generate** command:

```
. generate loginc = ln(income)
```

ln is one of Stata's *mathematical functions*. These functions are as follows:

abs(x)	Absolute value of x.
acos(x)	Arc-cosine returning radians. Because 360 degrees = 2π radians, **acos**(x)***180/_pi** gives the arc-cosine returning degrees (_pi denotes the mathematical constant π).
asin(x)	Arc-sine returning radians.
atan(x)	Arc-tangent returning radians.
comb(n,k)	Combinatorial function (number of possible combinations of n things taken k at a time).
cos(x)	Cosine of radians. To find the cosine of y degrees, type **generate y = cos(y * _pi/180)**
digamma(x)	$d\ln\Gamma(x)/dx$
exp(x)	Exponential (e to power).
ln(x)	Natural (base e) logarithm. For any other base number B, to find the base B logarithm of x, type **generate y = ln(x)/ln(B)**
lnfact(x)	Natural log of factorial. To find x factorial, type **generate y = round(exp(lnfact(x),1)**
lngamma(x)	Natural log of $\Gamma(x)$. To find $\Gamma(x)$, type **generate y = exp(lngamma(x))**
log(x)	Natural logarithm; same as **ln**(x)
log10(x)	Base 10 logarithm.
mod(x,y)	Modulus of x with respect to y.
sin(x)	Sine of radians.
sqrt(x)	Square root.
tan(x)	Tangent of radians.
trigamma(x)	$d^2\ln\Gamma(x)/dx^2$

The following *statistical functions* exist:

Binomial(n,k,π)	Probability of k or more successes in n trials, when the probability of success on a single trial is π.
binorm(h,k,ρ)	Joint cumulative probability of $\Phi(h, k, \rho)$, a bivariate normal distribution with correlation ρ, cumulative over $(-\infty, h] \times (-\infty, k]$.
chi2(df,x)	Probability that a χ^2 with df degrees of freedom is less than x.
chi2tail(df,x)	Probability that a χ^2 with df degrees of freedom is greater than x.
F(df_1,df_2,f)	Probability that an F with df_1 numerator and df_2 denominator degrees of freedom is less than f.
Ftail(df_1,df_2,f)	Probability that an F with df_1 numerator and df_2 denominator degrees of freedom is greater than f.
gammap(a,x)	Incomplete gamma function $P(a, x)$.
ibeta(a,b,x)	Incomplete beta function $I_x(a, b)$.

invbinomial(n,k,p) Inverse binomial; for $p \leq 0.5$, returns π (probability of success on a single trial) such that the probability of k or more successes in n trials equals p; for p $>$ 0.5, returns π such that the probability of k or fewer successes in n trials equals $1 - p$.

invchi2(df,p) Inverse of **chi2()**; if **chi2**$(df,x) = p$, then **invchi2**$(df,p) = x$.

invchi2tail(df,p) Inverse of **chi2tail()** ; if **chi2tail**$(df,x) = p$, then **invchi2tail**$(df,p) = x$.

invF(df_1, df_2, p) Inverse of **F()**

invFtail(df_1, df_2, p) Inverse of **Ftail()**

invgammap(a,p) Inverse of **gammap()**

invnchi2(df, λ, p) Inverse of **nchi2()**

invnorm(p) Inverse of **normprob()**

invttail(df,p) Inverse of **ttail()**.

nchi2(df, λ, x) Cumulative noncentral χ^2 distribution with df degrees of freedom and noncentrality parameter λ.

normden(z) Standard normal $N(0,1)$ density.

normden(z, σ) Normal $N(0, \sigma^2)$ density.

norm(z) Probability that a standard normal variable is less than z.

npnchi(df, x, p) Noncentrality parameter λ for noncentral χ^2.

ttail(df, t) Probability that T with df degrees of freedom is greater than t.

uniform() Pseudo-random number generator, returning values from a uniform distribution ranging from 0 to nearly 1, written [0,1).

Nothing goes inside the parentheses with **uniform()**. Optionally, we can control the pseudo-random generator's starting seed, and hence the stream of "random" numbers, by first issuing a **set seed #** command—where # could be any integer from 0 to $2^{31} - 1$ inclusive. Omitting the **set seed** command corresponds to **set seed 123456789**, which will always produce the same stream of numbers.

Stata provides more than 40 *date functions* and date-related *time series functions*. A listing can be found in Chapter 16 of the *User's Guide*, or by typing **help dates**. Below are some examples of date functions. "Elapsed date" in these functions refers to the number of days since January 1, 1960.

date$(s_1, s_2[,y])$ Returns the elapsed date corresponding to s_1. s_1 is a string variable indicating the date in virtually any format. Months can be spelled out, abbreviated to three characters, or given as numbers; years ʳᵃⁿ include or exclude the century; blanks and punctuation are allow permutation of m, d, and [##]y with their order defining month, day and year occur in s_1. ##, if specified, gives two-digit years in s_1; the default is 19y.

day(e) Returns the numeric day of the month corresponding to e, †

dow(e) Returns the numeric day of the week corresponding to e, †

mdy(m, d, y) Returns the elapsed date corresponding to m, d, and y.

| `month(e)` | Returns the numeric month corresponding to *e*, the elapsed date. |
| `year(e)` | Returns the numeric year corresponding to *e*, the elapsed date. |

Some useful *special functions* include the following:

`autocode(x,n,xmin,xmax)`	Forms categories from *x* by partitioning the interval from *xmin* to *xmax* into *n* equal-length intervals and returning the upper bound of the interval that contains *x*.
`cond(x,a,b)`	Returns *a* if *x* evaluates to "true" and *b* if *x* evaluates to "false." . `generate y = cond(inc1 > inc2, inc1, inc2)` creates the variable *y* as the maximum of *inc1* and *inc2* (assuming neither is missing).
`group(x)`	Creates a categorical variable that divides the data *as presently sorted* into *x* subsamples that are as nearly equal-sized as possible.
`int(x)`	Returns the integer obtained by truncating (dropping fractional parts of) *x*.
`max(x_1,x_2,...,x_n)`	Returns the maximum of $x_1, x_2, ..., x_n$. Missing values are ignored. For example, `max(3+2,1)` evaluates to 5.
`min(x_1,x_2,...,x_n)`	Returns the minimum of $x_1, x_2, ..., x_n$.
`recode(x,x_1,x_2,...,x_n)`	Returns missing if *x* is missing, x_1 if $x < x_1$, or x_2 if $x < x_2$, and so on.
`round(x,y)`	Returns *x* rounded to the nearest *y*.
`sign(x)`	Returns -1 if $x < 0$, 0 if $x = 0$, and $+1$ if $x > 0$ (missing if *x* is missing).
`sum(x)`	Returns the running sum of *x*, treating missing values as zero.

String functions, not described here, help to manipulate and evaluate string variables. Type **help functions** for a complete list of all Stata functions, or see the *Stata Reference Manual* and *User's Guide* for more examples and details.

Multiple functions, operators, and qualifiers can be combined in one command as needed. The functions and algebraic operators just described can also be used in another way that does not create or change any dataset variables. The **display** command performs a single calculation and shows the results onscreen. The following are some examples:

```
. display 2+3
5

. display log10(10^83)
83

. display invttail(120,.025) * 34.1/sqrt(975)
2.1622305
```

Thus, **display** works as an onscreen statistical calculator.

Unlike a calculator, **display** , **generate** , and **replace** have direct access to Stata's statistical results. For example, suppose that we summarized the unemployment rates from dataset *canada1.dta*:

```
. summarize unemp
```

Variable	Obs	Mean	Std. Dev.	Min	Max
unemp	11	12.10909	4.250048	7	19.6

After **summarize**, Stata temporarily stores the mean as a macro named r(mean).

```
. display r(mean)
12.109091
```

We could use this result to create variable *unempDEV*, defined as deviations from the mean:

```
. gen unempDEV = unemp - r(mean)
(2 missing values generated)
```

```
. summ unemp unempDEV
```

Variable	Obs	Mean	Std. Dev.	Min	Max
unemp	11	12.10909	4.250048	7	19.6
unempDEV	11	4.33e-08	4.250048	-5.109091	7.49091

Stata also provides another variable-creation command, **egen** ("extensions to **generate**"), which has its own set of functions to accomplish tasks not easily done by **generate**. These include such things as creating new variables from the sums, maxima, minima, medians, interquartile ranges, standardized values, or moving averages of existing variables or expressions. For example, the following command creates a new variable named *zscore*, equal to the standardized (mean 0, variance 1) values of *x*:

```
. egen zscore = std(x)
```

Or, the following command creates new variable *avg*, equal to the row mean of each observation's values on *x*, *y*, *z*, and *w*, ignoring any missing values.

```
. egen avg = rmean(x,y,z,w)
```

To create a new variable named *sum*, equal to the row sum of each observation's values on *x*, *y*, *z*, and *w*, treating missing values as zeroes, type

```
. egen sum = rsum(x,y,z,w)
```

The following command creates new variable *xrank*, holding ranks corresponding to values of *x*: *xrank* = 1 for the observation with highest *x*. *xrank* = 2 for the second highest, and so forth.

```
. egen xrank = rank(x)
```

Consult **help egen** for a complete list of **egen** functions, or the *Reference Manual* for further examples.

Converting between Numeric and String Formats

Dataset *canada2.dta* contains one string variable, *place*. It also has a labeled categorical variable, *type*. Both seem to have nonnumerical values.

```
. use canada2, clear
(Canadian dataset 2)
```

```
. list place type
```

	place	type
1.	Canada	Nation
2.	Newfoundland	Province
3.	Prince Edward Island	Province
4.	Nova Scotia	Province
5.	New Brunswick	Province
6.	Quebec	Province
7.	Ontario	Province
8.	Manitoba	Province
9.	Saskatchewan	Province
10.	Alberta	Province
11.	British Columbia	Province
12.	Yukon	Territory
13.	Northwest Territories	Territory

Beneath the labels, however, *type* remains a numeric variable, as we can see if we ask for the **nolabel** option:

```
. list place type, nolabel
```

	place	type
1.	Canada	3
2.	Newfoundland	1
3.	Prince Edward Island	1
4.	Nova Scotia	1
5.	New Brunswick	1
6.	Quebec	1
7.	Ontario	1
8.	Manitoba	1
9.	Saskatchewan	1
10.	Alberta	1
11.	British Columbia	1
12.	Yukon	2
13.	Northwest Territories	2

String and labeled numeric variables look similar when listed, but they behave differently when analyzed. Most statistical operations and algebraic relations are not defined for string variables, so we might want to have both string and labeled-numeric versions of the same information in our data. The **encode** command generates a labeled-numeric variable from a string variable. The number 1 is given to the alphabetically first value of the string variable, 2 to the second, and so on. In the following example, we create a labeled numeric variable named *placenum* from the string variable *place*:

```
. encode place, gen(placenum)
```

The opposite conversion is possible, too: The **decode** command generates a string variable using the values of a labeled numeric variable. Here we create string variable *typestr* from numeric variable *type*:

```
. decode type, gen(typestr)
```

When listed, the new numeric variable *placenum*, and the new string variable *typestr*, look similar to the originals:

```
. list place placenum type typestr
```

	place	placenum	type	typestr
1.	Canada	Canada	Nation	Nation
2.	Newfoundland	Newfoundland	Province	Province
3.	Prince Edward Island	Prince Edward Island	Province	Province
4.	Nova Scotia	Nova Scotia	Province	Province
5.	New Brunswick	New Brunswick	Province	Province
6.	Quebec	Quebec	Province	Province
7.	Ontario	Ontario	Province	Province
8.	Manitoba	Manitoba	Province	Province
9.	Saskatchewan	Saskatchewan	Province	Province
10.	Alberta	Alberta	Province	Province
11.	British Columbia	British Columbia	Province	Province
12.	Yukon	Yukon	Territory	Territory
13.	Northwest Territories	Northwest Territories	Territory	Territory

But with the **nolabel** option, the differences become visible. Stata views *placenum* and *type* basically as numbers.

```
. list place placenum type typestr, nolabel
```

	place	placenum	type	typestr
1.	Canada	3	3	Nation
2.	Newfoundland	6	1	Province
3.	Prince Edward Island	10	1	Province
4.	Nova Scotia	8	1	Province
5.	New Brunswick	5	1	Province
6.	Quebec	11	1	Province
7.	Ontario	9	1	Province
8.	Manitoba	4	1	Province
9.	Saskatchewan	12	1	Province
10.	Alberta	1	1	Province
11.	British Columbia	2	1	Province
12.	Yukon	13	2	Territory
13.	Northwest Territories	7	2	Territory

Statistical analyses such as finding means and standard deviations work only with basically numeric variables. For calculation purposes, the numeric variables' labels do not matter.

```
. summarize place placenum type typestr
```

Variable	Obs	Mean	Std. Dev.	Min	Max
place	0				
placenum	13	7	3.89444	1	13
type	13	1.307692	.6304252	1	3
typestr	0				

Occasionally we encounter a string variable where the values are all or mostly numbers. To convert these string values into their numerical counterparts, use the **real** function. For example, the variable *siblings* below is a string variable, although it only has one value, "4 or more," that could not be represented just as easily by a number.

```
. describe siblings
```

```
  1. siblings    str9    %9s              Number of siblings (string)
```

```
. list

        siblings
  1.           0
  2.           1
  3.           2
  4.           3
  5. 4 or more
```

```
. generate sibnum = real(siblings)
(1 missing value generated)
```

The new variable *sibnum* is numeric, with a missing value where *siblings* had "4 or more."

```
. list

        siblings    sibnum
  1.           0         0
  2.           1         1
  3.           2         2
  4.           3         3
  5. 4 or more           .
```

Creating New Categorical and Ordinal Variables

A previous section illustrated how to construct a categorical variable called *type* to distinguish among territories, provinces, and nation in our Canadian dataset. You can create categorical or ordinal variables in many other ways. This section gives a few examples.

type has three categories:

```
. tabulate type
```

Province, territory or nation	Freq.	Percent	Cum.
Province	10	76.92	76.92
Territory	2	15.38	92.31
Nation	1	7.69	100.00
Total	13	100.00	

For some purposes, we might want to re-express a multicategory variable as a set of dichotomies or "dummy variables," each coded 0 or 1. **tabulate** will create dummy variables automatically if we add the **generate** option. In the following example, this results in a set of variables called *type1*, *type2*, and *type3*, each representing one of the three categories of *type*:

```
. tabulate type, generate(type)
```

Province, territory or nation	Freq.	Percent	Cum.
Province	10	76.92	76.92
Territory	2	15.38	92.31
Nation	1	7.69	100.00
Total	13	100.00	

```
. describe

Contains data from C:\data\canada2.dta
  obs:            13                        Canadian dataset 2
  vars:           10                        13 May 2001 15:06
  size:          637 (100.0% of memory free)
-------------------------------------------------------------------------------
              storage  display   value
variable name   type   format    label    variable label
-------------------------------------------------------------------------------
place          str21   %21s               Place name
pop            float   %9.0g              Population in 1000s, 1995
unemp          float   %9.0g              % 15+ population unemployed,
                                            1995
mlife          float   %9.0g              Male life expectancy years
flife          float   %9.0g              Female life expectancy years
gap            float   %9.0g              Female-male gap life expectancy
type           byte    %9.0g    typelbl   Province, territory or nation
type1          byte    %8.0g              type==Province
type2          byte    %8.0g              type==Territory
type3          byte    %8.0g              type==Nation
-------------------------------------------------------------------------------
Sorted by:
     Note:  dataset has changed since last saved

. list place type type1-type3

                      place        type    type1    type2    type3
    1.               Canada       Nation       0        0        1
    2.         Newfoundland     Province       1        0        0
    3.  Prince Edward Island    Province       1        0        0
    4.          Nova Scotia     Province       1        0        0
    5.        New Brunswick     Province       1        0        0
    6.               Quebec     Province       1        0        0
    7.              Ontario     Province       1        0        0
    8.             Manitoba     Province       1        0        0
    9.         Saskatchewan     Province       1        0        0
   10.              Alberta     Province       1        0        0
   11.     British Columbia     Province       1        0        0
   12.                Yukon    Territory       0        1        0
   13. Northwest Territories   Territory       0        1        0
```

Re-expressing categorical information as a set of dummy variables involves no loss of information; in this example, *type1* through *type3* together tell us exactly as much as *type* itself does. Occasionally, however, analysts choose to re-express a measurement variable in categorical or ordinal form, even though this *does* result in a substantial loss of information. For example, *unemp* in *canada2.dta* gives a measure of the unemployment rate. Excluding Canada itself from the data, we see that *unemp* ranges from 7% to 19.6%, with a mean of 12.26:

```
. summarize unemp if type != 3

Variable |     Obs        Mean    Std. Dev.       Min        Max
---------+-----------------------------------------------------
   unemp |      10       12.26     4.44877          7       19.6
```

Having Canada in the data becomes a nuisance at this point, so we drop it:

```
. drop if type == 3
(1 observation deleted)
```

Two commands create a dummy variable named *unemp2* with values of 0 when unemployment is below average (12.26), 1 when unemployment is equal to or above average, and when *unemp* is missing. In reading the second command, recall that Stata's sorting and relational operators treat missing values as very large numbers.

```
. generate unemp2 = 0 if unemp < 12.26
(7 missing values generated)

. replace unemp2 = 1 if unemp >= 12.26 & unemp != .
(5 real changes made)
```

We might want to group the values of a measurement variable, thereby creating an ordered-category or ordinal variable. The **autocode** function (see "Using Functions" earlier in this chapter) provides automatic grouping of measurement variables. To create new ordinal variable *unemp3*, which groups values of *unemp* into three equal-width groups over the interval from 5 to 20, type

```
. generate unemp3 = autocode(unemp,3,5,20)
(2 missing values generated)
```

A list of the data shows how the new dummy (*unemp2*) and ordinal (*unemp3*) variables correspond to values of the original measurement variable *unemp*.

```
. list place unemp unemp2 unemp3
```

	place	unemp	unemp2	unemp3
1.	Newfoundland	19.6	1	20
2.	Prince Edward Island	19.1	1	20
3.	Nova Scotia	13.9	1	15
4.	New Brunswick	13.8	1	15
5.	Quebec	13.2	1	15
6.	Ontario	9.3	0	10
7.	Manitoba	8.5	0	10
8.	Saskatchewan	7	0	10
9.	Alberta	8.4	0	10
10.	British Columbia	9.8	0	10
11.	Yukon	.	.	.
12.	Northwest Territories	.	.	.

Both strategies just described dealt appropriately with missing values, so that Canadian places with missing values on *unemp* likewise receive missing values on the variables derived from *unemp*. Another possible approach works best if our data contain no missing values. To illustrate, we begin by dropping the Yukon and Northwest Territories:

```
. drop if unemp == .
(2 observations deleted)
```

We now can use the **group** function to create an ordinal variable not with approximately equal-width groupings, as **autocode** did, but instead with groupings of approximately equal size. We do this in two steps. First, sort the data (assuming no missing values) on the variable of interest. Second, generate a new variable using the **group(#)** function, where # indicates the number of groups desired. The example below divides our 10 Canadian provinces into 5 groups.

```
. sort unemp

. generate unemp5 = group(5)

. list place unemp unemp2 unemp3 unemp5
```

	place	unemp	unemp2	unemp3	unemp5
1.	Saskatchewan	7	0	10	1
2.	Alberta	8.4	0	10	1

3.	Manitoba	8.5	0	10	2
4.	Ontario	9.3	0	10	2
5.	British Columbia	9.8	0	10	3
6.	Quebec	13.2	1	15	3
7.	New Brunswick	13.8	1	15	4
8.	Nova Scotia	13.9	1	15	4
9.	Prince Edward Island	19.1	1	20	5
10.	Newfoundland	19.6	1	20	5

Another difference is that **autocode** assigns values equal to the upper bound of each interval, whereas **group** simply assigns 1 to the first group, 2 to the second, and so forth.

Using Explicit Subscripts with Variables

When Stata has data in memory, it also defines certain system variables that describe those data. For example, _N represents the total number of observations. _n represents the observation number: _n = 1 for the first observation, _n = 2 for the second, and so on to the last observation (_n = _N). If we issue a command such as the following, it creates a new variable, *caseID*, equal to the number of each observation as presently sorted:

. **generate** *caseID* = **_n**

Sorting the data another way will change each observation's value of _n , but its *caseID* value will remain unchanged. Thus, if we do sort the data another way, we can later return to the earlier order by typing

. **sort** *caseID*

Creating and saving unique case identification numbers in this fashion, thereby storing the order of observations at an early stage of dataset development, can be of great assistance in later data management.

We can use explicit subscripts with variable names, to specify particular observation numbers. For example, the 6th observation in dataset *canada1.dta* (if we have not dropped or re-sorted anything) is Quebec. Consequently, *pop[6]* refers to Quebec's population, 7334 thousand.

. **display** *pop[6]*
7334.2002

Similarly, *pop[12]* is the Yukon's population:

. **display** *pop[12]*
30.1

Explicit subscripting and the _n system variable have additional relevance when our data form a series. If we had the daily stock-market price of a particular stock as a variable named *price*, for instance, then either *price* or, equivalently, *price*[_n] denotes the value of the _nth observation or day. *price*[_n–1] denotes the previous day's price, and *price*[_n+1] denotes the next. Thus, we might define a new variable *difprice*, which is equal to the change in *price* since the previous day:

. **generate** *difprice* = *price* - *price[_n-1]*

Chapter 13, on time series analysis, returns to this topic.

Importing Data from Other Programs

Stata's Editor provides a simple way to enter and edit small datasets, but for larger projects we often need tools that work directly with computer files created by other programs. Such files fall into two general categories: raw-data ASCII (text) files, which can be read into Stata with the appropriate Stata commands; and system files, which must be translated to Stata format by a special third-party program before Stata can read them.

To illustrate ASCII file methods, we return to the Canadian data of Table 2.1. Suppose that, instead of typing these data into Stata's Editor, we typed them into our word processor, with at least one space between each value. String values must be in double quotes if they contain internal spaces, as does "Prince Edward Island". For other string values, quotes are optional. Word processors allow the option of saving documents as ASCII (text) files, a simpler and more universal type than the word processor's usual saved-file format. We can thus create an ASCII file named *canada.raw* that looks something like this:

```
"Canada"  29606.1  10.6  75.1  81.1
"Newfoundland"  575.4  19.6  73.9  79.8
"Prince Edward Island"  136.1  19.1  74.8  81.3
"Nova Scotia"  937.8  13.9  74.2  80.4
"New Brunswick"  760.1  13.8  74.8  80.6
"Quebec"  7334.2  13.2  74.5  81.2
"Ontario"  11100.3  9.3  75.5  81.1
"Manitoba"  1137.5  8.5  75  80.8
"Saskatchewan"  1015.6  7  75.2  81.8
"Alberta"  2747  8.4  75.5  81.4
"British Columbia"  3766  9.8  75.8  81.4
"Yukon"  30.1  .  71.3  80.4
"Northwest Territories"  65.8  .  70.2  78
```

Note the use of periods, not blanks, to indicate missing values for the Yukon and Northwest Territories. If the dataset should have five variables, then for every observation, exactly five values (including periods for missing values) must exist.

infile reads into memory an ASCII file such as *canada.raw* in which the values are separated by spaces (or commas). Its basic form is

. infile *variable-list* using *filename.raw*

With purely numeric data, the variable list could be omitted, in which case Stata assigns the names *v1*, *v2*, *v3*, and so forth. On the other hand, we might want to give each variable a distinctive name. We also need to identify string variables individually. For *canada.raw*, the **infile** command might be

. infile str30 *place pop unemp mlife flife* using canada.raw, clear
(13 observations read)

The **infile** variable list specifies variables in the order that they appear in the data file. The **clear** option drops any current data from memory before reading in the new file.

If any string variables exist, their names must each be preceded by a **str#** statement. **str30**, for example, informs Stata that the next-named variable (*place*) is a string variable with as many as 30 characters. Actually, none of the Canadian place names involve more than 21 characters, but we do not need to know that in advance. It is often easier to overestimate string variable lengths. Then, once data are in memory, use **compress** to ensure that no variable takes up more space than it needs. The **compress** command automatically changes all variables to their most memory-efficient storage type.

```
. compress
place was str30 now str21
. describe

Contains data
  obs:             13
  vars:             5
  size:           533 (100.0% of memory free)
-----------------------------------------------------------------------
              storage  display   value
variable name  type    format    label    variable label
-----------------------------------------------------------------------
place          str21   %21s
pop            float   %9.0g
unemp          float   %9.0g
mlife          float   %9.0g
flife          float   %9.0g
-----------------------------------------------------------------------
Sorted by:
```

We can now proceed to label variables and data as described earlier. At any point, the commands **save canada0** (or **save canada0, replace**) would save the new dataset in Stata format, as file *canada0.dta*. The original raw-data file, *canada.raw*, remains unchanged on disk.

If our variables have non-numeric values (for example, "male" and "female") that we want to store as labeled numeric variables, then adding the option **automatic** will accomplish this automatically. For example, we might read in raw survey data through this **infile** command:

. **infile** *gender age income vote* **using** *survey.raw*, **automatic**

Spreadsheet and database programs commonly write ASCII files that have only one observation per line, with values separated by tabs or commas. To read these files into Stata, use **insheet**. Its general syntax resembles that of **infile**, with an option telling Stata whether the data are tab- or comma-delimited. For example, assuming tab-delimited data,

. **insheet** *variable-list* **using** *filename.raw*, **tab**

Or, assuming comma-delimited data with the first row of the file containing variable names (also comma-delimited),

. **insheet** *variable-list* **using** *filename.raw*, **comma names**

With **insheet** we do not need to separately identify string variables. If we include no variable list, and do not have variable names in the file's first row, Stata automatically assigns the variable names *v1, v2, v3, ...* Errors will occur if some values in our ASCII file are not separated by tabs or commas as specified in the **insheet** command.

Raw data files created by other statistical packages can be in "fixed-column" format, where the values are not necessarily delimited at all, but do occupy predefined column positions. Both **infile** and the more specialized command **infix** permit Stata to read such files. In the command syntax itself, or in a "data dictionary" existing in a separate file or as the first part of the data file, we have to specify exactly how the columns should be read.

Here is a simple example. Data on four variables exist in an ASCII file named *nfresour.raw*:

```
198624087641691000
198725247430001044
198825138637481086
198925358964371140
1990    8615731195
1991    7930001262
```

These data concern natural resource production in Newfoundland. The four variables occupy fixed column positions: columns 1 through 4 are the years (1986...1991); columns 5–8 measure forestry production in thousands of cubic meters (2408...missing); columns 9–14 measure mine production in thousands of dollars (764,169...793,000); and columns 15–18 are the consumer price index relative to 1986 (1000...1262). Notice that in fixed-column format, unlike space or tab-delimited files, blanks indicate missing values, and the raw data contain no decimal points. To read *nfresour.raw* into Stata, we specify each variable's column position:

```
. infix year 1-4 wood 5-8 mines 9-14 CPI 15-18
     using nfresour.raw, clear
(6 observations read)

. list
```

	year	wood	mines	CPI
1.	1986	2408	764169	1000
2.	1987	2524	743000	1044
3.	1988	2513	863748	1086
4.	1989	2535	896437	1140
5.	1990	.	861573	1195
6.	1991	.	793000	1262

More complicated fixed-column formats might require a data "dictionary." Data dictionaries can be straightforward, but they offer many possible choices. Typing **help infix** or **help infile2** obtains brief outlines of these commands. For more examples and explanation, consult the *User's Guide* and *Reference Manuals*.

What if we need to export data from Stata to some other program? The **outfile** command writes ASCII files to disk. A command such as the following will create a space-delimited ASCII file named *canada6.raw*, containing whatever data were in memory:

```
. outfile using canada6
```

The **infile** , **insheet** , **infix** , and **outfile** commands just described all manipulate raw data in ASCII files. It can be faster and more convenient, however, to transfer data directly between the specialized system files saved by various spreadsheet, database, or statistical programs, without going through the intermediate step of an ASCII file. Several third-party programs perform such translations. Stat/Transfer, for example, will transfer data across many different formats including dBASE, Excel, FoxPro, Gauss, JMP, Lotus, MATLAB, Minitab, OSIRIS, Paradox, S-Plus, SAS, SPSS, SYSTAT, and Stata. It is available either through Stata Corporation or from its maker, Circle Systems:

Circle Systems
1001 Fourth Avenue Plaza, Suite 3200
Seattle, WA 98154

telephone: 206-682-3783
fax: 206-328-4788
e-mail: sales@circlesys.com

Links to this and other programs can be found on Stata's web site. Transfer programs prove indispensable for analysts working in multi-program environments or exchanging data with colleagues.

Combining Two or More Stata Files

We can combine Stata datasets in two general ways: **append** a second dataset that contains additional observations; or **merge** with a second dataset that contains new variables or values. In keeping with this chapter's Canadian theme, we will illustrate these procedures using data on Newfoundland. File *newf1.dta* records the province's population for the years 1985 to 1989.

```
. use newf1, clear
(Newfoundland 1985-89)

. describe

Contains data from C:\data\newf1.dta
  obs:            5                          Newfoundland 1985-89
  vars:           2                          13 May 2001 16:05
  size:          50 (100.0% of memory free)
-------------------------------------------------------------------------
              storage  display    value
variable name  type    format     label    variable label
-------------------------------------------------------------------------
year           int     %9.0g               Year
pop            float   %9.0g               Population
-------------------------------------------------------------------------
Sorted by:

. list

       year        pop
  1.   1985     580700
  2.   1986     580200
  3.   1987     568200
  4.   1988     568000
  5.   1989     570000
```

File *newf2.dta* has population and unemployment counts for some later years:

```
. use newf2
(Newfoundland 1990-95)

. describe

Contains data from C:\data\newf2.dta
  obs:            6                          Newfoundland 1990-95
  vars:           3                          13 May 2001 16:05
  size:          84 (100.0% of memory free)
-------------------------------------------------------------------------
              storage  display    value
variable name  type    format     label    variable label
-------------------------------------------------------------------------
year           int     %9.0g               Year
pop            float   %9.0g               Population
jobless        float   %9.0g               Number of people unemployed
-------------------------------------------------------------------------
Sorted by:
```

```
. list
```

	year	pop	jobless
1.	1990	573400	42000
2.	1991	573500	45000
3.	1992	575600	49000
4.	1993	584400	49000
5.	1994	582400	50000
6.	1995	575449	.

To combine these datasets, with *newf2.dta* already in memory, we use the **append** command:

```
. append using newf1
```

```
. list
```

	year	pop	jobless
1.	1990	573400	42000
2.	1991	573500	45000
3.	1992	575600	49000
4.	1993	584400	49000
5.	1994	582400	50000
6.	1995	575449	.
7.	1985	580700	.
8.	1986	580200	.
9.	1987	568200	.
10.	1988	568000	.
11.	1989	570000	.

Because variable *jobless* occurs in *newf2* (1990 to 1995) but not in *newf1*, its 1985 to 1989 values are missing in the combined dataset. We can now put the observations in order from earliest to latest and save these combined data as a new file, *newf3.dta*:

```
. sort year
```

```
. list
```

	year	pop	jobless
1.	1985	580700	.
2.	1986	580200	.
3.	1987	568200	.
4.	1988	568000	.
5.	1989	570000	.
6.	1990	573400	42000
7.	1991	573500	45000
8.	1992	575600	49000
9.	1993	584400	49000
10.	1994	582400	50000
11.	1995	575449	.

```
. save newf3
```

append might be compared to lengthening a sheet of paper (that is, the dataset in memory) by taping a second sheet with new observations to its bottom. **merge**, in its simplest form, corresponds to "widening" our sheet of paper by taping a second sheet to its right side, thereby adding new variables. For example, dataset *newf4.dta* contains further Newfoundland time series: the numbers of births and divorces over the years 1980 to 1994. Thus it has some observations in common with our earlier dataset *newf3.dta*, as well as one variable (*year*) in common, but it also has two new variables not present in *newf3.dta*.

```
. use newf4
(Newfoundland 1980-94)
```

```
. describe

Contains data from C:\data\newf4.dta
  obs:             15                          Newfoundland 1980-94
  vars:             3                          13 May 2001 16:06
  size:           150  (100.0% of memory free)
-------------------------------------------------------------------------------
              storage  display    value
variable name   type   format     label    variable label
-------------------------------------------------------------------------------
year            int    %9.0g               Year
births          int    %9.0g               Number of births
divorces        int    %9.0g               Number of divorces
-------------------------------------------------------------------------------
Sorted by:

. list

           year      births    divorces
  1.       1980       10332         555
  2.       1981       11310         569
  3.       1982        9173         625
  4.       1983        9630         711
  5.       1984        8560         590
  6.       1985        8080         561
  7.       1986        8320         610
  8.       1987        7656        1002
  9.       1988        7396         884
 10.       1989        7996         981
 11.       1990        7354         973
 12.       1991        6929         912
 13.       1992        6689         867
 14.       1993        6360         930
 15.       1994        6295         933
```

We want to merge *newf3* with *newf4*, matching observations according to *year* wherever possible. To accomplish this, both datasets must be sorted by the index variable (which in this example is *year*). We earlier issued a **sort year** command before saving *newf3.dta*, so we now do the same with *newf4.dta*. Then we merge the two, specifying *year* as the index variable to match.

```
. sort year

. merge year using newf3

. describe

Contains data from C:\data\newf4.dta
  obs:             16                          Newfoundland 1980-94
  vars:             6                          13 May 2001 16:06
  size:           304  (100.0% of memory free)
-------------------------------------------------------------------------------
              storage  display    value
variable name   type   format     label    variable label
-------------------------------------------------------------------------------
year            int    %9.0g               Year
births          int    %9.0g               Number of births
divorces        int    %9.0g               Number of divorces
pop             float  %9.0g               Population
jobless         float  %9.0g               Number of people unemployed
_merge          byte   %8.0g
-------------------------------------------------------------------------------
Sorted by:
     Note:  dataset has changed since last saved
```

```
. list

        year    births   divorces       pop    jobless    _merge
  1.    1980     10332        555         .          .          1
  2.    1981     11310        569         .          .          1
  3.    1982      9173        625         .          .          1
  4.    1983      9630        711         .          .          1
  5.    1984      8560        590         .          .          1
  6.    1985      8080        561     580700          .          3
  7.    1986      8320        610     580200          .          3
  8.    1987      7656       1002     568200          .          3
  9.    1988      7396        884     568000          .          3
 10.    1989      7996        981     570000          .          3
 11.    1990      7354        973     573400      42000          3
 12.    1991      6929        912     573500      45000          3
 13.    1992      6689        867     575600      49000          3
 14.    1993      6360        930     584400      49000          3
 15.    1994      6295        933     582400      50000          3
 16.    1995         .          .     575449          .          2
```

In this example, we simply used **merge** to add new variables to our data, matching observations. By default, whenever the same variables are found in both datasets, those of the "master" data (the file already in memory) are retained and those of the "using" data ignored. The **merge** command has several options, however, that override this default. A command of the following form would allow any *missing values* in the master data to be replaced by corresponding nonmissing values found in the using data (here, *newf5.dta*):

. **merge** *year* **using** *newf5***, update**

Or, a command such as the following causes *any values* from the master data to be replaced by nonmissing values from the using data, if the latter are different:

. **merge** *year* **using** *newf5***, update replace**

Suppose that the values of an index variable occur more than once in the master data; for example, suppose that the year 1990 occurs twice. Then values from the using data with *year* = 1990 are matched with each occurrence of *year* = 1990 in the master data. You can use this capability for many purposes, such as combining background data on individual patients with data on any number of separate doctor visits they made. Although **merge** makes this and many other data-management tasks straightforward, analysts should look closely at the results to be certain that the command is accomplishing what they intend.

As a diagnostic aid, **merge** automatically creates a new variable called *_merge*. Unless **update** was specified, *_merge* codes have the following meanings:

1   Observation from the master dataset only.

2   Observation from the using dataset only.

3   Observation from both master and using data (using values ignored if different).

If the **update** option was specified, *_merge* codes convey what happened:

1   Observation from the master dataset only.

2   Observation from the using dataset only.

3   Observation from both, master data agrees with using.

4   Observation from both, master data updated if missing.

5   Observation from both, master data replaced if different.

Before performing another **merge** operation, it will be necessary to discard or rename this variable. For example,

```
. drop _merge
```

Or,

```
. rename _merge _merge1
```

Transposing, Reshaping, or Collapsing Data

Long after a dataset has been created, we might discover that for some analytical purposes it has the wrong organization. Fortunately, several commands facilitate drastic restructuring of datasets. We will illustrate these using data (*growth1.dta*) on recent population growth in five eastern provinces of Canada. In these data, unlike our previous examples, province names are represented by a numerical variable with eight-character labels.

```
. use growth1, clear
(Eastern Canada growth)

. describe
```

```
Contains data from C:\data\growth1.dta
  obs:            5                        Eastern Canada growth
  vars:           5                        13 May 2001 17:05
  size:         105 (100.0% of memory free)
-------------------------------------------------------------------------------
              storage  display    value
variable name   type   format     label      variable label
-------------------------------------------------------------------------------
provinc2        byte    %8.0g      provinc2   Eastern Canadian province
grow92          float   %9.0g                 Pop. gain in 1000s, 1991-92
grow93          float   %9.0g                 Pop. gain in 1000s, 1992-93
grow94          float   %9.0g                 Pop. gain in 1000s, 1993-94
grow95          float   %9.0g                 Pop. gain in 1000s, 1994-95
-------------------------------------------------------------------------------
Sorted by:
```

```
. list
```

```
       provinc2    grow92    grow93    grow94    grow95
1.  New Brun           10       2.5       2.2       2.4
2.  Newfound          4.5        .8        -3      -5.8
3.  Nova Sco         12.1       5.8       3.5       3.9
4.    Ontario        174.9     169.1     120.9     163.9
5.     Quebec         80.6      77.4      48.5      47.1
```

In this organization, population growth for each year is stored as a separate variable. We could analyze changes in the mean or variation of population growth from year to year. On the other hand, given this organization, Stata could not readily draw a simple time plot of population growth against year, nor can Stata find the correlation between population growth in New Brunswick and Newfoundland. All the necessary information is here, but such analyses require different organizations of the data.

One simple reorganization involves transposing variables and observations. In effect, the dataset rows become its columns, and vice versa. This is accomplished by the **xpose** command. The option **clear** is required with this command, because it always clears the present

data from memory. Including the **varname** option creates an additional variable (named
_varname) in the transposed dataset, containing original variable names as strings.

. **xpose, clear varname**

. **describe**

```
Contains data
   obs:            5
   vars:           6
   size:         160  (100.0% of memory free)
------------------------------------------------------------------------
               storage  display    value
variable name   type    format     label      variable label
------------------------------------------------------------------------
v1              float   %9.0g
v2              float   %9.0g
v3              float   %9.0g
v4              float   %9.0g
v5              float   %9.0g
_varname        str8    %9s
------------------------------------------------------------------------
Sorted by:
     Note:  dataset has changed since last saved
```

. **list**

```
        v1       v2       v3       v4       v5    _varname
 1.      1        2        3        4        5    provinc2
 2.     10      4.5     12.1    174.9     80.6    grow92
 3.    2.5       .8      5.8    169.1     77.4    grow93
 4.    2.2       -3      3.5    120.9     48.5    grow94
 5.    2.4     -5.8      3.9    163.9     47.1    grow95
```

Value labels are lost along the way, so provinces in the transposed dataset are indicated only
by their numbers (1 = New Brunswick, 2 = Newfoundland, and so on). The second through last
values in each column are the population gains for that province, in thousands. Thus, variable
v1 has a province identification number (1, meaning New Brunswick) in its first row, and New
Brunswick's population growth values for 1992 to 1995 in its second through fifth rows. We
can now find correlations between population growth in different provinces, for instance, by
typing a **correlate** command with **in 2/5** (second through fifth observations only)
qualifier:

. **correlate v1-v5 in 2/5**

```
(obs=4)

         |     v1       v2       v3       v4       v5
---------+---------------------------------------------
      v1|   1.0000
      v2|   0.8058   1.0000
      v3|   0.9742   0.8978   1.0000
      v4|   0.5070   0.4803   0.6204   1.0000
      v5|   0.6526   0.9362   0.8049   0.6765   1.0000
```

The strongest correlation appears between the growth of neighboring maritime provinces New
Brunswick (*v1*) and Nova Scotia (*v3*): $r = .9742$. Newfoundland's (*v2*) growth has a much
weaker correlation with that of Ontario (*v4*): $r = .4803$.

More sophisticated restructuring is possible through the **reshape** command. This command switches datasets between two basic configurations termed "wide" and "long." Dataset *growth1.dta* is initially in wide format.

```
. use growth1, clear
(Eastern Canada growth)
. list
```

	provinc2	grow92	grow93	grow94	grow95
1.	New Brun	10	2.5	2.2	2.4
2.	Newfound	4.5	.8	-3	-5.8
3.	Nova Sco	12.1	5.8	3.5	3.9
4.	Ontario	174.9	169.1	120.9	163.9
5.	Quebec	80.6	77.4	48.5	47.1

A **reshape** command switches this to long format:

```
. reshape long grow, i(provinc2) j(year)

(note: j = 92 93 94 95)

Data                              wide   ->   long
------------------------------------------------------------------
Number of obs.                      5    ->     20
Number of variables                 5    ->      3
j variable (4 values)                    ->   year
xij variables:
            grow92 grow93 ... grow95     ->   grow
------------------------------------------------------------------

. list
```

	provinc2	year	grow
1.	New Brun	92	10
2.	New Brun	93	2.5
3.	New Brun	94	2.2
4.	New Brun	95	2.4
5.	Newfound	92	4.5
6.	Newfound	93	.8
7.	Newfound	94	-3
8.	Newfound	95	-5.8
9.	Nova Sco	92	12.1
10.	Nova Sco	93	5.8
11.	Nova Sco	94	3.5
12.	Nova Sco	95	3.9
13.	Ontario	92	174.9
14.	Ontario	93	169.1
15.	Ontario	94	120.9
16.	Ontario	95	163.9
17.	Quebec	92	80.6
18.	Quebec	93	77.4
19.	Quebec	94	48.5
20.	Quebec	95	47.1

```
. label data "Eastern Canadian growth--long"

. label variable grow "Population growth in 1000s"
```

```
. save growth2
file C:\data\growth2.dta saved
```

The **reshape** command above began by stating that we want to put the dataset in **long** form. Next, it named the new variable to be created, *grow*. The **i(provinc2)** option specified the observation identifier, or the variable whose unique values denote logical observations. In this example, each province forms a logical observation. The **j(year)** option specifies the sub-observation identifier, or the variable whose unique values (within each logical observation) denote sub-observations. Here, the sub-observations are years within each province.

Figure 2.1 shows a possible use for the long-format dataset. With one **graph** command, we can now produce a time plot comparing the population gains in New Brunswick, Newfoundland, and Nova Scotia (observations for which *provinc2* < 4). The **graph** command below calls for a plot of *grow* (as *y*-axis variable) against *year* (*x* axis) if *province2* < 4, with the *y*-axis and *x*-axis values labeled neatly, a horizontal line at *y* = 0 (zero population growth), data points connected by line segments, and separate plots for each value of *provinc2*. Chapter 3 gives more details concerning the versatile **graph** command.

```
. graph grow year if provinc2 < 4, ylabel xlabel yline(0)
     connect(1) by(provinc2)
```

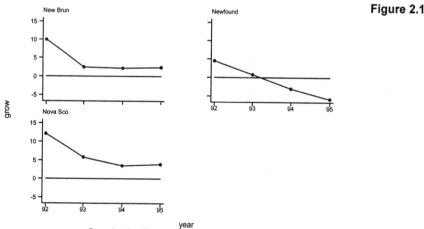

Graphs by Eastern Canadian province

Declines in their fisheries during the early 1990s caused economic hardship for the three provinces. Growth slowed dramatically in New Brunswick and Nova Scotia, while Newfoundland (the most fisheries-dependent province) actually lost population.

reshape works equally well in reverse, to switch data from "long" to "wide" format. Dataset *growth3.dta* serves as an example of long format.

```
. use growth3, clear
(Eastern Canadian growth--long)
```

```
. list
        provinc2        grow        year
  1.  New Brun           10          92
  2.  New Brun          2.5          93
  3.  New Brun          2.2          94
  4.  New Brun          2.4          95
  5.  Newfound          4.5          92
  6.  Newfound           .8          93
  7.  Newfound           -3          94
  8.  Newfound         -5.8          95
  9.  Nova Sco         12.1          92
 10.  Nova Sco          5.8          93
 11.  Nova Sco          3.5          94
 12.  Nova Sco          3.9          95
 13.   Ontario        174.9          92
 14.   Ontario        169.1          93
 15.   Ontario        120.9          94
 16.   Ontario        163.9          95
 17.    Quebec         80.6          92
 18.    Quebec         77.4          93
 19.    Quebec         48.5          94
 20.    Quebec         47.1          95
```

To convert this to wide format, we use **reshape wide** :

```
. reshape wide grow, i(provinc2) j(year)

(note: j = 92 93 94 95)

Data                                  long   ->   wide
-------------------------------------------------------------------------
Number of obs.                          20   ->      5
Number of variables                      3   ->      5
j variable (4 values)                 year   ->   (dropped)
xij variables:
                                      grow   ->   grow92 grow93 ... grow95
-------------------------------------------------------------------------

. list
        provinc2     grow92      grow93      grow94      grow95
  1.  New Brun           10         2.5         2.2         2.4
  2.  Newfound          4.5          .8          -3        -5.8
  3.  Nova Sco         12.1         5.8         3.5         3.9
  4.   Ontario        174.9       169.1       120.9       163.9
  5.    Quebec         80.6        77.4        48.5        47.1
```

Notice that we have recreated the organization of dataset *growth1.dta*.

Another important tool for restructuring datasets is the **collapse** command, which creates an aggregated dataset of statistics (for example, means, medians, or sums). The long *growth3* dataset has four observations for each province:

```
. use growth3, clear
(Eastern Canadian growth--long)

. list
        provinc2        grow        year
  1.  New Brun           10          92
  2.  New Brun          2.5          93
  3.  New Brun          2.2          94
  4.  New Brun          2.4          95
  5.  Newfound          4.5          92
```

```
 6.  Newfound        .8        93
 7.  Newfound        -3        94
 8.  Newfound      -5.8        95
 9.  Nova Sco      12.1        92
10.  Nova Sco       5.8        93
11.  Nova Sco       3.5        94
12.  Nova Sco       3.9        95
13.   Ontario     174.9        92
14.   Ontario     169.1        93
15.   Ontario     120.9        94
16.   Ontario     163.9        95
17.    Quebec      80.6        92
18.    Quebec      77.4        93
19.    Quebec      48.5        94
20.    Quebec      47.1        95
```

We might want to aggregate the different years into a mean growth rate for each province. In the collapsed dataset, each observation will correspond to one value of the **by ( )** variable, that is, one province.

```
. collapse (mean) grow, by(provinc2)

. list
```

```
      provinc2       grow
 1.  New Brun       4.275
 2.  Newfound   -.8750001
 3.  Nova Sco       6.325
 4.   Ontario       157.2
 5.    Quebec        63.4
```

For a slightly more complicated example, suppose we had a dataset similar to *growth3.dta* but also containing the variables *births*, *deaths*, and *income*. We want an aggregate dataset with each province's total numbers of births and deaths over these years, the mean income (to be named *meaninc*), and the median income (to be named *medinc*). If we do not specify a new variable name, as with *grow* in the previous example or *births* and *deaths*, the collapsed variable takes on the same name as the old variable.

```
. collapse (sum) births deaths (mean) meaninc = income
      (median) medinc = income, by(provinc2)
```

collapse can create variables based on the following summary statistics:

mean	Means (the default; used if the type of statistic is not specified)
sd	Standard deviations
sum	Sums
rawsum	Sums ignoring optionally specified weight
count	Number of nonmissing observations
max	Maximums
min	Minimums
median	Medians
p1	1st percentile
p2	2nd percentile (and so forth to **p99**)
iqr	Interquartile range

Weighting Observations

Stata understands four types of weighting:

aweight　Analytical weights, used in weighted least squares (WLS) regression and similar procedures.

fweight　Frequency weights, counting the number of duplicated observations. Frequency weights must be integers.

iweight　Importance weights, however you define "importance."

pweight　Probability or sampling weights, equal to the inverse of the probability that an observation is included due to sampling strategy.

Researchers sometimes speak of "weighted data." This might mean that the original sampling scheme selected observations in a deliberately disproportionate way, as reflected by weights equal to 1/(*probability of selection*). Appropriate use of **pweight** can compensate for disproportionate sampling in certain analyses. On the other hand, "weighted data" might mean something different—an aggregate dataset, perhaps constructed from a frequency table or cross-tabulation, with one or more variables indicating how many times a particular value or combination of values occurred. In that case, we need **fweight**.

Not all types of weighting have been defined for all types of analyses. We cannot, for example, use **pweight** with the **tabulate** command. Using weights in any analysis requires a clear understanding of what we want weighting to accomplish in that particular analysis. The weights themselves can be any variable in the dataset.

The following small dataset (*nfschool.dta*), containing results from a survey of 1,381 rural Newfoundland high school students, illustrates a simple application of frequency weighting.

```
. describe

Contains data from C:\data\nfschool.dta
  obs:             6                    Newf.school/univer.(Seyfrit 93)
  vars:            3                    13 May 2001 19:02
  size:           48 (99.6% of memory free)
-------------------------------------------------------------------------
              storage  display    value
variable name  type    format     label    variable label
-------------------------------------------------------------------------
univers       byte     %8.0g      yes      Expect to attend university?
year          byte     %8.0g               What year of school now?
count         int      %8.0g               observed frequency
-------------------------------------------------------------------------
Sorted by:

. list

      univers    year    count
1.        no      10      210
2.        no      11      260
3.        no      12      274
4.       yes      10      224
5.       yes      11      235
6.       yes      12      178
```

At first glance, the dataset seems to contain only 6 observations, and when we cross-tabulate whether students expect to attend a university (*univers*) by their current year in high school (*year*), we get a table with one observation per cell.

```
. tabulate univers year
```

```
Expect to  | What year of school now?
attend     |
university?|      10           11          12 |     Total
-----------+--------------------------------+----------
        no |       1            1           1 |         3
       yes |       1            1           1 |         3
-----------+--------------------------------+----------
     Total |       2            2           2 |         6
```

To understand these data, we need to apply frequency weights. The variable *count* gives frequencies: 210 of these students are tenth graders who said they did not expect to attend a university, 260 are eleventh graders who said no, and so on. Specifying **[fweight = count]** obtains a cross-tabulation showing responses of all 1,381 students.

```
. tabulate univers year [fweight = count]
```

```
Expect to  | What year of school now?
attend     |
university?|      10           11          12 |     Total
-----------+--------------------------------+----------
        no |     210          260         274 |       744
       yes |     224          235         178 |       637
-----------+--------------------------------+----------
     Total |     434          495         452 |      1381
```

Carrying the analysis further, we might add options asking for a table with column percentages (**col**), no cell frequencies (**nof**), and a χ^2 test of independence (**chi2**). This reveals that the percentage of students expecting to go to college declines with each year of high school.

```
. tabulate univers year [fw = count], col nof chi2
```

```
Expect to  | What year of school now?
attend     |
university?|      10           11          12 |     Total
-----------+--------------------------------+----------
        no |   48.39        52.53       60.62 |     53.87
       yes |   51.61        47.47       39.38 |     46.13
-----------+--------------------------------+----------
     Total |  100.00       100.00      100.00 |    100.00
```

$$\text{Pearson chi2(2)} = 13.8967 \quad Pr = 0.001$$

Survey data often reflect complex sampling designs, based on one or more of the following:

disproportionate sampling — for example, oversampling particular subpopulations, in order to get enough cases to draw conclusions about them.

clustering — for example, selecting voting precincts at random, and then sampling individuals within the selected precincts.

stratification — for example, dividing precincts into "urban" and "rural" strata, and then sampling precincts and/or individuals within each stratum.

Complex sampling designs require specialized analytical tools. **pweights** and Stata's ordinary analytical commands do not suffice.

Stata's procedures for complex survey data include special tabulation, means, regression, logit, probit, tobit, and Poisson regression commands. Before applying these commands, users must first set up their data by identifying variables that indicate the strata, PSUs (primary sampling units) or clusters, finite population correction, and probability weights. This is accomplished through **svyset** commands. For example:

```
. svyset strata urb_rur
. svyset psu precinct
. svyset fpc finite
. svyset pweight invPsel
```

For each observation in this example, the value of variable *urb_rur* identifies the strata, *precinct* identifies the cluster, *finite* gives the finite population correction, and *invPsel* gives the probability weight or inverse of the probability of selection. After the data have been **svyset** and saved, the survey analytical procedures (**svyreg** , **svytab** , **svylogit** , etc.) are relatively straightforward. See Chapter 30 of the *User's Guide* for an overview of survey methods. The *Reference Manual* describes specific commands, which all begin with the letters **svy** . For online guidance, type **help svy** and follow the links to particular commands.

Creating Random Data and Random Samples

The pseudo-random number function **uniform()** lies at the heart of Stata's ability to generate random data or to sample randomly from the data at hand. The *User's Guide* provides a technical description of this 32-bit pseudo-random generator. If we presently have data in memory, then a command such as the following creates a new variable named *randnum*, having apparently random 16-digit values over the interval [0,1) for each case in the data.

```
. generate randnum = uniform()
```

Alternatively, we might create a random dataset from scratch. Suppose we want to start a new dataset containing 10 random values. We first clear any other data from memory (if they were valuable, **save** them first). Next, set the number of observations desired for the new dataset. Explicitly setting the seed number makes it possible to later reproduce the same "random" results. Finally, we generate our random variable.

```
. clear
. set obs 10
obs was 0, now 10
. set seed 12345
. generate randnum = uniform()
. list

        randnum
  1.    .309106
  2.    .6852276
  3.    .1277815
  4.    .5617244
  5.    .3134516
```

```
  6.    .5047374
  7.    .7232868
  8.    .4176817
  9.    .6768828
 10.    .3657581
```

In combination with Stata's algebraic, statistical, and special functions, **uniform()** can simulate values sampled from a variety of theoretical distributions. If we want *newvar* sampled from a uniform distribution over [0,428) instead of the usual [0,1), we type

```
. generate newvar = 428 * uniform()
```

These will still be 16-digit values. Perhaps we want only integers from 1 to 428 (inclusive):

```
. generate newvar = 1 + int(428 * uniform())
```

To simulate 1,000 rolls of a six-sided die, type

```
. clear
```

```
. set obs 1000
obs was 0, now 1000
```

```
. generate roll = 1 + int(6 * uniform())
```

```
. tabulate roll
```

die	Freq.	Percent	Cum.
1	171	17.10	17.10
2	164	16.40	33.50
3	150	15.00	48.50
4	170	17.00	65.50
5	169	16.90	82.40
6	176	17.60	100.00
Total	1000	100.00	

We might theoretically expect 16.67% ones, 16.67% twos, and so on, but in any one sample like these 1,000 "rolls," the observed percentages will vary randomly around their expected values.

To simulate 1,000 rolls of a pair of six-sided dice, type

```
. generate dice = 2 + int(6 * uniform()) + int(6 * uniform())
```

```
. tabulate dice
```

dice	Freq.	Percent	Cum.
2	26	2.60	2.60
3	62	6.20	8.80
4	78	7.80	16.60
5	120	12.00	28.60
6	153	15.30	43.90
7	149	14.90	58.80
8	146	14.60	73.40
9	96	9.60	83.00
10	88	8.80	91.80
11	53	5.30	97.10
12	29	2.90	100.00
Total	1000	100.00	

We can use _n to begin an artificial dataset as well. The following commands create a new 5,000-observation dataset with one variable named *index*, containing values from 1 to 5,000.

```
. set obs 5000
obs was 0, now 5000

. generate index = _n

. summarize
```

Variable	Obs	Mean	Std. Dev.	Min	Max
index	5000	2500.5	1443.52	1	5000

It is possible to generate variables from a normal (Gaussian) distribution using `uniform()`. The following example creates a dataset with 2,000 observations and 2 variables, z from an $N(0,1)$ population, and x from $N(500,75)$.

```
. clear

. set obs 2000
obs was 0, now 2000

. generate z = invnorm(uniform())

. generate x = 500 + 75*invnorm(uniform())
```

The actual sample means and standard deviations differ slightly from their theoretical values:

```
. summarize
```

Variable	Obs	Mean	Std. Dev.	Min	Max
z	2000	.0375032	1.026784	-3.536209	4.038878
x	2000	503.322	75.68551	244.3384	743.1377

If z follows a normal distribution, $w = e^z$ follows a lognormal distribution. To form a lognormal variable w based upon a standard normal z,

```
. generate w = exp(invnorm(uniform()))
```

To form a lognormal variable w based on an $N(100,15)$ distribution,

```
. generate w = exp(100 + 15*invnorm(uniform()))
```

Taking logarithms, of course, normalizes a lognormal variable.

To simulate y values drawn randomly from an exponential distribution with mean and standard deviation $\mu = \sigma = 3$,

```
. generate y = -3 * ln(uniform())
```

For other means and standard deviations, substitute other values for 3.

$X1$ follows a χ^2 distribution with one degree of freedom, which is the same as a squared standard normal:

```
. generate X1 = (invnorm(uniform())^2
```

By similar logic, $X2$ follows a χ^2 with two degrees of freedom:

```
. generate X2 = (invnorm(uniform()))^2 + (invnorm(uniform()))^2
```

Other statistical distributions, including t and F, can be simulated along the same lines. In addition, programs have been written for Stata to generate random samples following distributions such as binomial, Poisson, gamma, and inverse Gaussian.

Although `invnorm(uniform())` can be adjusted to yield normal variates with particular correlations, a much easier way to do this is through the **drawnorm** command. To generate 5,000 observations from N(0,1), type

```
. clear
. drawnorm z, n(5000)
. summ
```

```
    Variable |     Obs        Mean    Std. Dev.       Min        Max
-------------+-------------------------------------------------------
           z |    5000   -.0005951    1.019788   -4.518918   3.923464
```

Below, we will create three further variables. Variable $x1$ is from an N(0,1) population, variable $x2$ is from N(100,15), and $x3$ is from N(500,75). Furthermore, we define these variables to have the following population correlations:

	x1	x2	x3
x1	1.0	0.4	−0.8
x2	0.4	1.0	0.0
x3	−0.8	0.0	1.0

The procedure for creating such data requires first defining the correlation matrix *C,* and then using *C* in the **drawnorm** command:

```
. mat C = (1, .4, -.8 \ .4, 1, 0 \ -.8, 0, 1)
. drawnorm x1 x2 x3, means(0,100,500) sds(1,15,75) corr(C)
. summarize x1-x3
```

```
    Variable |     Obs        Mean    Std. Dev.       Min        Max
-------------+-------------------------------------------------------
          x1 |    5000    .0024364    1.01648    -3.478467   3.598916
          x2 |    5000    100.1826    14.91325    46.13897   150.7634
          x3 |    5000    500.7747    76.93925   211.5596    769.6074
```

```
. correlate x1-x3
(obs=5000)
             |      x1        x2        x3
-------------+---------------------------------
          x1 |   1.0000
          x2 |   0.3951    1.0000
          x3 |  -0.8134   -0.0072    1.0000
```

Compare the sample variables' correlations and means with the theoretical values given earlier. Random data generated in this fashion can be viewed as samples drawn from theoretical populations. We should not expect the samples to have exactly the theoretical population parameters (in this example, an $x3$ mean of 500, $x1$–$x2$ correlation of 0.4, $x1$–$x3$ correlation of −.8, and so forth).

The command **sample** makes unobtrusive use of **uniform**'s random generator to obtain random samples of the data in memory. For example, to discard all but a 10% random sample of the original data, type

```
. sample 10
```

When we add an **in** or **if** qualifier, **sample** applies only to those observations meeting our criteria. For example,

```
. sample 10 if age < 26
```

would leave us with a 10% sample of those observations with *age* less than 26, plus 100% of the original observations with *age* ≥ 26.

We could also select random samples of a particular size. To discard all but 90 randomly-selected observations from the dataset in memory, type

```
. sample 90, count
```

The sections in Chapter 14 on bootstrapping and Monte Carlo simulations provide further examples of random sampling and random variable generation.

Writing Programs for Data Management

Data management on larger projects often involves repetitive or error-prone tasks that are best handled by writing specialized Stata programs. Chapter 14 addresses programming in more detail, but you can begin by writing simple programs that consist of nothing more than a sequence of Stata commands, typed and saved as an ASCII file. ASCII files can be created using your favorite word processor or text editor, which should offer "ASCII text file" among its options under File – Save As. An even easier way to create such text files is through Stata's Do-file Editor, which is brought up by clicking Window – Do-file Editor or the icon 🔧.

For example, using the Do-file Editor we might create a file named *canada.do* (which contains the commands to read in a raw data file named *canada.raw*), then label the dataset and its variables, compress it, and save it in Stata format. The commands in this file are identical to those seen earlier when we went through the example step by step.

```
infile str30 place pop unemp mlife flife using canada.raw
label data "Canadian dataset 1"
label variable pop "Population in 1000s, 1995"
label variable unemp "% 15+ population unemployed, 1995"
label variable mlife "Male life expectancy years"
label variable flife "Female life expectancy years"
compress
save canada1, replace
```

Once this *canada.do* file has been written and saved, simply typing the following command causes Stata to read the file and run each command in turn:

```
. do canada
```

Such batch-mode programs, termed "do-files," are usually saved with a .do extension. More elaborate programs (defined by do-files or "automatic do" files) can be stored in memory, and can call other programs in turn—creating new Stata commands and opening worlds of possibility for adventurous analysts.

Stata ordinarily interprets the end of a command line as the end of that command. This is reasonable onscreen, where the line can be arbitrarily long, but does not work as well when we are typing commands in a text file. One way to avoid line-length problems is through the

#delimit command, which can set some other character as the end-of-command delimiter. In the following example, we make a semicolon the delimiter; then type two long commands that do not end until a semicolon appears; and then finally reset the delimiter to its usual value, a carriage return (cr):

```
#delimit ;
infile str30 place pop unemp mlife flife births deaths
    marriage medinc mededuc using newcan.raw;
order place pop births deaths marriage medinc mededuc
    unemp mlife flife;
#delimit cr
```

Stata normally pauses each time the Results window becomes full of information, and waits to proceed until we press any key (or 🔘). Instead of pausing, we can ask Stata to continue scrolling until the output is complete. Typed in the Command window or as part of a program, the command

```
. set more off
```

calls for continuous scrolling. This is convenient if our program produces much screen output that we don't want to see, or if it is writing to a log file that we will examine later. Typing

```
. set more on
```

returns to the usual mode of waiting for keyboard input before scrolling.

Managing Memory

When we **use** or File – Open a dataset, Stata reads the disk file and loads it into memory. Loading the data into memory permits rapid analysis, but it is only possible if the dataset can fit within the amount of memory currently allocated to Stata. If we try to open a too-large dataset, we get an error message saying "no room to add more observations."

```
. use C:\data\labrador.dta
(Scientific surveys off N. Newfoundland)
no room to add more observations
r(901);
```

Small Stata allocates 400 kilobytes to data, and this limit cannot be changed. Intercooled Stata is flexible, however. Its default allocation is 1 megabyte, but the *Getting Started* manuals for Windows, Macintosh, or Unix computers describe how to allocate more memory automatically on Intercooled Stata startup.

Windows and Unix users also can reallocate Intercooled Stata's memory on the fly, through the **set memory** command. In the example given above, *labrador.dta* is a 4.8 megabyte dataset that would not fit into Stata's 1-megabyte default allocation. Encountering this problem, we can easily surmount it by asking for more memory. It is necessary to first clear any existing data from memory. We then ask for 12 megabytes of memory, which is more than enough to accommodate *labrador.dta*.

```
. clear
```

```
. set memory 12m
(12288k)
```

```
. use C:data\labrador.dta
(Scientific surveys off N. Newfoundland)

. describe

Contains data from C:\data\labrador.dta
  obs:         55,578                          Scientific surveys off N.
                                                 Newfoundland
  vars:            15                          2 Nov 1999 13:49
  size:     5,168,754 (58.9% of memory free)
-------------------------------------------------------------------------------
              storage  display      value
variable name type     format       label      variable label
-------------------------------------------------------------------------------
trip          int      %8.0g
set           int      %8.0g
year          byte     %4.0g
nafo_div      str2     %2s
mean_dep      int      %8.0g
temp_sur      int      %8.0g
temp_fs_      int      %8.0g
lat           float    %9.0g
long          float    %9.0g
number        long     %9.0g
weight        long     %9.0g
station       int      %8.0g
latin         str31    %31s
common        str23    %23s
id            float    %9.0g
-------------------------------------------------------------------------------
Sorted by:
```

Dataset *labrador.dta* contains 55,578 observations from scientific surveys of fish populations in the Labrador Sea, conducted over the years 1978 to 1993. When we **describe** the data (above), Stata reports "58.9% of memory free," meaning not 58.9% of the computer's total resources, but 58.9% of the approximately 12 megabytes we allocated for Stata data. It is often advisable to ask for more memory than our data actually require. Many statistical and data-management operations consume additional memory, in part because they temporarily create new variables as they work.

It is possible to **set memory** to values higher than the computer's available physical memory. In that case, Stata uses "virtual memory," which is really disk storage. Although virtual memory allows bypassing hardware limitations, it can be terribly slow. If you regularly work with datasets that push the limits of your computer, you might soon conclude that it is time to buy more memory.

Type **help limits** to see a list of limitations in Stata, not only on dataset size but also other dimensions including matrix size, command lengths, lengths of names, and numbers of variables in commands. Some of these limitations can be adjusted by the user.

Graphs

The **graph** command forms the core of Stata's graphics. It offers seven basic styles:

hist	histograms (default if only one variable is listed)
oneway	one-way scatterplots
twoway	two-way scatterplots (default if two or more variables are listed)
matrix	scatterplot matrix
box	boxplots
bar	bar charts
star	star charts
pie	pie charts

With most of these styles we can accept simple default versions or choose among many further options that control details of their appearance.

In addition to **graph** , Stata has numerous specialized graphing commands, including symmetry plots, quantile plots, and quality control charts. Other graphs used in connection with particular statistical models are described in later chapters such as those about regression diagnostics, nonlinear regression, survival analysis, and time series.

Example Commands

. **graph y**
Draws basic unadorned histogram of variable *y* (same as **graph y, hist**).

. **graph y, bin(9) norm ylabel xlabel border**
Draws histogram of *y* with 9 vertical bars (default is 5), normal curve, automatically-selected *y* and *x* axis values labeled, and a border drawn around.

. **sort x**
. **graph y, by(x) ylabel xlabel total**
In one figure, draws separate histograms of *y* for each value of *x*. Also includes a "total" histogram showing distribution of *y* across all values of *x*.

. **graph y x**
Displays a basic two-variable scatterplot of *y* against *x*. Equivalent to **graph y x, twoway** . Add the option **border** to include a border.

. `graph y x, oneway twoway box ylabel xlabel`

Constructs scatterplot of *y* versus *x*, with automatically labeled axes. In the margins, shows one-way scatterplots and boxplots of *y* and *x*.

. `graph y x, rbox symbol([country]) psize(130)`

Constructs scatterplot of *y* versus *x*, containing a rangefinder box (simplified boxplots) showing both variables' medians, interquartile ranges, and "fences." Uses the values or labels of variable *country* (which could be the country names) as plotting symbols. Makes these plotting symbols 130% of their normal size.

. `sort x2`
. `graph y x1, by(x2)`

In one figure, draws separate *y* versus *x1* scatterplots for each value of *x2*.

. `graph y x1 [iweight = x2]`

Draws a scatterplot of *y* versus *x1*, with symbols sized in proportion to *x2*.

. `graph y time, connect(1) sort ylabel xlabel(1970,1980,1990,2000)`

Plots *y* against *time*. Connects the points with line segments. Labels the *y* axis automatically, but labels the *x* axis at points 1970, 1980, 1990, and 2000. Such labeling could be accomplished compactly with the option `xlabel(1970,1980 to 2000)`.

. `graph y1 y2 time, connect(ll) sort ylabel xlabel rlabel rescale`

Plots *y1* and *y2* against *time*, connects with line segments, and rescales the right-hand axis to match the units of *y2*.

. `graph y y time, connect(ls) bands(15) symbol(pi) ylabel yreverse`
  `xlabel(1900,1910 to 2000)`

Draws time plot showing both raw data and a smoothed curve. Connects data points (plus-sign symbols) with line segments. Connects the (*time,y*) cross-medians of 15 vertical (*time*) bands (invisible) with a smooth cubic-spline curve. Labels the *y* axis automatically, but with values reversed to read from high (bottom) to low (top). Labels the *x* axis specifically at points 1900, 1910, 1920, ... , 2000.

. `graph y1 y2 y3, box ylabel`

Constructs boxplots of variables *y1*, *y2*, and *y3*, with labeled vertical axis.

. `sort x`
. `graph y, box by(x) ylabel`

Constructs boxplots of *y* for each value of *x*.

. `graph w x y z, pie`

Draws one pie chart with slices indicating the relative amounts of variables *w*, *x*, *y*, and *z*. The variables must have similar units.

. `graph w x y z, star`

Draws a star chart for each observation in the data, showing that observation's relative values on variables *w*, *x*, *y*, and *z*. The variables need not have similar units.

. `graph w x y, bar`

Shows the sums of variables *w*, *x*, and *y* as side-by-side bars in a bar chart.

. `sort z`
. `graph w x y, bar by(z) means`

Constructs bar chart showing the means of *w*, *x*, and *y* for each value of *z*.

. **qnorm y**

Draws a quantile-normal plot (normal probability plot) showing quantiles of *y* versus corresponding quantiles of a normal distribution.

. **rchart x1 x2 x3 x4 x5, connect(1)**

Constructs a quality-control R chart graphing the range of values represented by variables *x1 − x5*.

At the time this book was printed, Stata's graphics features were in the process of being revised. All of the commands described in this chapter will continue to work, but there might be additional features available to you. Type **help graph** to read about the full capabilities now available.

Histograms

Typing simply **graph *varname*** produces a histogram, Stata's default one-variable display. For examples we turn to *states.dta*, which contains selected environment and education measures on the 50 U.S. states plus the District of Columbia (data originally from the League of Conservation Voters 1991; National Center for Education Statistics 1992, 1993; World Resources Institute 1993).

. **use *states***
(U.S. states data 1990-91)

. **describe**

```
Contains data from c:\data\states.dta
  obs:            51                          U.S. states data 1990-91
  vars:           21                          24 Jun 2001 13:04
  size:         4,080 (99.9% of memory free)
-------------------------------------------------------------------------
              storage  display    value
variable name  type    format     label    variable label
-------------------------------------------------------------------------
state          str20   %20s                 State
region         byte    %9.0g      region    Geographical region
pop            float   %9.0g                1990 population
area           float   %9.0g                Land area, square miles
density        float   %7.2f                People per square mile
metro          float   %5.1f                Metropolitan area population, %
waste          float   %5.2f                Per capita solid waste, tons
energy         int     %8.0g                Per capita energy consumed, Btu
miles          int     %8.0g                Per capita miles driven/year
toxic          float   %5.2f                Per capita toxics released, lbs
green          float   %5.2f                Per capita greenhouse gas, tons
house          byte    %8.0g                House '91 environ. voting, %
senate         byte    %8.0g                Senate '91 environ. voting, %
csat           int     %9.0g                Mean composite SAT score
vsat           int     %8.0g                Mean verbal SAT score
msat           int     %8.0g                Mean math SAT score
percent        byte    %9.0g                % HS graduates taking SAT
expense        int     %9.0g                Per pupil expenditures prim&sec
income         long    %10.0g               Median household income
high           float   %9.0g                % over 25 w/HS diploma
college        float   %9.0g                % over 25 w/bachelor's degree +
-------------------------------------------------------------------------
Sorted by:  state
```

Figure 3.1 shows a simple histogram of *college*, the percentage of a state's over-25 population with a bachelor's degree or higher. It was produced by the following command:

```
. graph college, title(Figure 3.1)
```

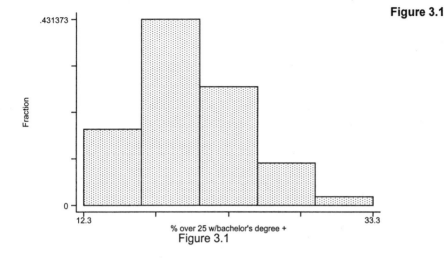

Figure 3.1

Figure 3.1

Menu options allow us to print whatever graph is onscreen, save it to disk, or cut and paste it into another program such as a word processor.

Figure 3.2 illustrates two ways to enhance a graph:

1. **ylabel** and **xlabel** options ask for round-number labeling on the *y* and x axes (compare with the default labeling seen in Figure 3.1).

2. We can add up to eight titles, two each at the top, bottom, left, and right. The first bottom title (called **title**) appears in a larger font.

```
. graph college, ylabel xlabel title(Figure 3.2)
      b2(this is bottom 2) l1(this is left 1) l2(this is left 2)
      t1(this is top 1) t2(this is top 2) r1(this is right 1)
      r2(this is right 2)
```

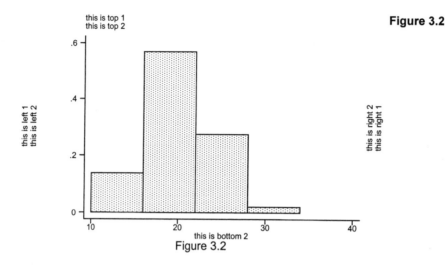

Figure 3.2

Some further **graph** options include:

bin(#)   Draws the histogram with # vertical bars. The default is **bin(5)** .

noaxis   Suppresses the lines for *x* and *y* axes.

norm   Draws a normal curve over the histogram, based on sample mean and standard deviation. We can specify other parameters; **norm(100,15)**, for example, would draw a normal curve with $\mu = 100$ and $\sigma = 15$.

ylabel   Uses automatically-chosen round numbers to label the *y* axis. We can ask for certain labels specifically; **ylabel(0,2,4,6)** calls for labels at *y* = 0, 2, 4, and 6. The same thing is accomplished by **ylabel(0,2 to 6)**.

xlabel   Same as **ylabel** , applied to the *x* axis.

ytick( )   Places *y*-axis tick marks at the specified locations. **ytick(1,3,5,7)**, or **ytick(1,3 to 7)**, calls for ticks at *y* = 1, 3, 5, and 7.

yline( )   Draws horizontal lines across the graph at specified heights. **yline(0,5)** calls for lines at *y* = 0 and at *y* = 5.

xtick( )   Same as **ytick( )**, applied to the *x* axis.

xline( )   Same as **yline( )**, applied to the *x* axis.

freq   Label the histogram's vertical axis with frequencies rather than proportions.

Figure 3.3 illustrates these options.

```
. graph college, norm border xlabel(12,16 to 32) xtick(14,18 to 34)
    bin(11) freq ylabel(0,2 to 12) ytick(1,3 to 13)
```

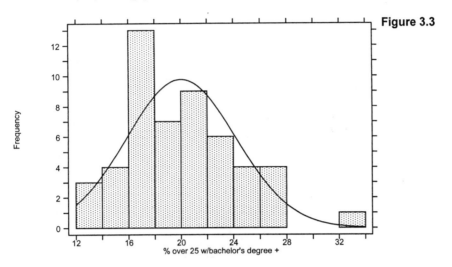

Figure 3.3

Suppose we want to see how the distribution of *college* varies by *region*. The **by** option obtains a separate histogram for each value of the categorical variable *region*. Before using **by** , we must first sort the data. Figure 3.4 shows the result.

```
. sort region
. graph college, by(region) ylabel xlabel bin(4)
```

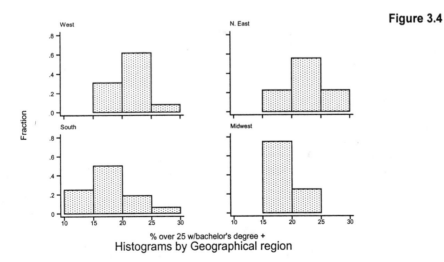

Figure 3.4

Histograms by Geographical region

Figure 3.5 contains a similar set of graphs, overlaid by normal curves based not on the sample mean and standard deviation (defaults), but on the more outlier-resistant median and pseudo-standard deviation (obtained via **lv college** ; see Chapter 4). We also include a **total** plot, showing the whole-sample distribution.

. graph *college*, by(*region*) ylabel xlabel bin(4) total
    norm(20.02,4.17)

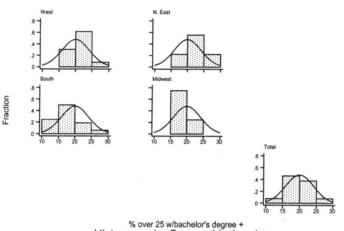

Figure 3.5

% over 25 w/bachelor's degree +
Histograms by Geographical region

These examples illustrate some general points regarding the **graph** command. Its syntax has the form

. graph *variable-list*, option1 option2 ...

We can list the desired options in any order, after the single comma. Except for **freq** , **bin()**, and **norm**, the options illustrated in Figures 3.2 through 3.5 work with other styles of graphs besides histograms. For example, **ylabel** also labels *y*-axis values in a scatterplot or a boxplot. **by(varname)** can produce a display containing separate one or two-way scatterplots, boxplots, bar charts, or pie charts for each value of *varname*.

A good way to get exactly the graph you want is to build it in steps. Begin with a simple version, view that, and then add options a few at a time. Build upon your previous commands retrieved from the Review window. Save or print only your final version.

Stata's histograms are limited to a maximum of 50 bins, or vertical bars. For a way to graph frequency distributions with more than 50 bins, see **help spikeplot** .

Scatterplots

Scatterplots are Stata's most versatile style of graph. These can range from simple *y* versus *x* plots, to graphs with many different "*y*" variables, plotted by different symbols and connected in a variety of ways. If **graph** is followed by two or more variable names, Stata assumes that we want a scatterplot unless told otherwise. The last-named variable in the command defines the horizontal or *x* axis, while all other variables are plotted against the vertical or *y* axis.

For example, again using the *states.dta* dataset, we could plot *waste* (per capita solid wastes) against *metro* (percent population in metropolitan areas), with the result shown in Figure 3.6. Each point in Figure 3.6 represents one of the 50 U.S. states (or Washington DC).

```
. graph waste metro
```

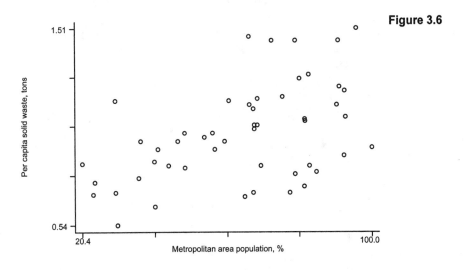

Figure 3.6

As with histograms, we can add options to control titles or the labeling of axes. Scatterplots also permit "importance weighting" (**iweight**), in which the area of each plotting symbol is made proportional to a weight variable. For example, weighting the symbols by *pop* (state population) and labeling the axes yields Figure 3.7.

```
. graph waste metro [iweight = pop], ylabel xlabel
```

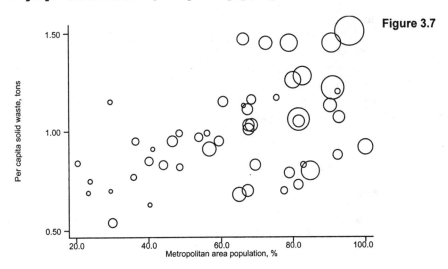

Figure 3.7

Other graph styles, with the exception of star plots, also support weighting. The **graph** command treats importance, frequency, and analytical weights (**iweight**, **fweight**, and **aweight**) the same. Weights affect the size of graphing symbols and also, where appropriate, the calculation of summary statistics that underlie the graph.

Two-variable scatterplots can have boxplots or oneway scatterplots in their margins, as seen in Figure 3.8. The options **twoway oneway box** invoke all three styles at once. Marginal plots display the individual distributions of the *y* and *x* variable, at the same time the scatterplot shows their joint distribution. Marginal boxplots (or the related **rbox** option) assist in noticing outliers and skewness.

. **graph waste metro, twoway oneway box ylabel xlabel**

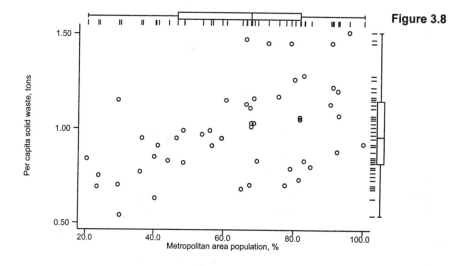

Figure 3.8

The **matrix** option creates a scatterplot matrix, containing small plots with all pairwise combinations of the variables listed (Figure 3.9). The **half** option with **matrix** displays only the lower triangular half of a scatterplot matrix. If our variable list includes one "dependent" variable and a number of "independent" variables, it makes sense to list the dependent variable last—as done here with *waste*. Then the bottom row of the matrix consists of a series of dependent-versus-independent variable plots. The option **label** asks Stata to find round-number labels for the axes; **ylabel** and **xlabel** do not work with scatterplot matrices.

. graph *metro pop waste*, matrix half label symbol(p)

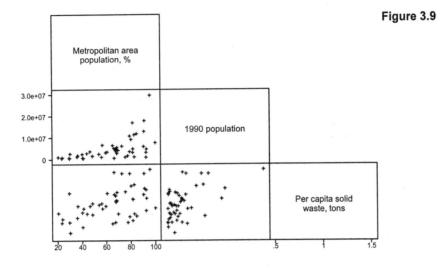

Figure 3.9

Figure 3.9 also employed a **symbol(p)** option, which caused data points to be plotted as small plus signs (+) instead of as the default large circles seen in Figure 3.8. Possible plotting symbols are

symbol(O)	Large circle (default for first-listed *y* variable)
symbol(T)	Large triangle (default for second-listed *y* variable)
symbol(S)	Large square
symbol(o)	Small circle
symbol(p)	Small plus sign
symbol(d)	Small diamond
symbol(x)	x
symbol(.)	Dot
symbol(I)	Invisible
symbol([_n])	Observation number (note inside square brackets)
symbol([varname])	Values of variable *varname* (note square brackets)

We can specify a different plotting symbol for each *y* variable. For example,

symbol(Sd[varname]) requests the first-listed *y* plotted as large squares, the second as small diamonds, and the third using values of *varname*, which could be numbers, labels, or strings. When we omit the **symbol( )** option, **graph** assigns symbols automatically.

The **connect( )** option allows for connecting points in various ways:

connect(1)   Connects the points with line segments. Unless the data are already in order by *x*, you might want to specify **connect(1) sort**.

connect(L)   Connects with line segments, so long as *x* keeps ascending.

connect(m)   Connects cross-medians of several vertical bands. The **bands( )** option specifies how many bands. For example **connect(m) bands(6)** calls for dividing the data into six equal-width bands. The default is **bands(200)**.

connect(s)   Connects cross-medians using smooth curves (cubic splines). As with **connect(m)**, we can also specify **bands( )**. We can also specify **density( )**, the number of points to be calculated on the cubic spline. The default is **density(5)**.

connect(.)   Does not connect points. This is helpful if we have two or more *y* variables. For example, **connect(.1)** does not connect the first *y* variable, but connects the second with line segments.

connect(J)   Connects in steps, like stairs.

connect(| |)   Connects a pair of variables with vertical bars (as in high-low plots).

connect(II)   Same as **connect(| |)**, but with bars capped like capital I's (as in error-bar plots).

Note that the similar-appearing lower-case letter **1**, vertical line **|**, and capital letter **I** mean quite different things to **graph**.

To illustrate the **symbol( )** and **connect( )** options, we turn to time series data on Newfoundland's Northern Cod (*cod.dta*). The Northern Cod, among the world's richest fishery resources, collapsed due to overfishing in 1992.

```
. use cod
(Newfoundland's Northern Cod, 1960-1997)

. describe
```

```
Contains data from c:\data\cod.dta
  obs:            38                          Newfoundland's Northern Cod,
                                                1960-1997
  vars:            5                          24 Jun 2001 17:23
  size:          684 (99.8% of memory free)
-------------------------------------------------------------------------
                storage  display    value
variable name    type    format     label     variable label
-------------------------------------------------------------------------
year             int     %8.0g                Year
total            float   %8.0g                Northern Cod landings, 1000 tons
canada           int     %8.0g                Canadian Northern Cod landings,
                                                1000 tons
TAC              int     %8.0g                Total Allowable Catch, 1000 tons
biomass          float   %9.0g                Estimated Northern Cod biomass,
                                                1000 tons
-------------------------------------------------------------------------
Sorted by:  year
```

A simple time plot showing Canadian and total landings can be constructed by graphing both variables against *year*, and connecting them with line segments (Figure 3.10). The option **connect(11)** means "connect the first-listed *y* variable with straight line segments, and connect the second-listed *y* variable also with straight line segments." The options **symbol(Op)** means "plot the first-listed *y* variable with large circles, and the second-listed *y* variable with small plus signs." In this graph, we see that Canadian landings comprised most of the total after 1976 when Canada, like other coastal nations, extended its jurisdiction out to 200 miles.

. **graph** *total canada year*, **symbol(Op) connect(ll) ylabel xlabel**

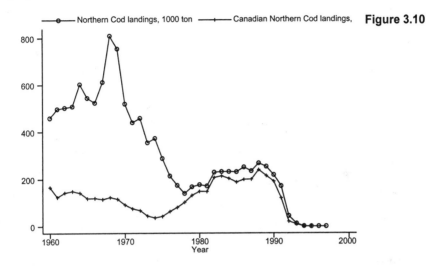

Figure 3.10

An alternative way to distinguish between variables in a scatterplot or time plot involves changing line patterns from solid to dashes or dots. For example, we could use invisible plotting symbols for both variables (**symbol(ii)**), but connect *total* with dashes, and *canada* with dots (**connect(l[_]l[.])**) as shown in Figure 3.11.

. **graph** *total canada year*, **symbol(ii) connect(l[_]l[.]) ylabel**
    xlabel border

We have several choices for line patterns:

l	solid line (letter "el", default)
_	long dash (underscore)
-	short dash (hyphen)
.	dot (period)
#	space

These line-pattern choices are given in square brackets, after specifying how we want data points connected, within the **connect( )** option. They can be combined creatively. For example, a dot-dot-dash pattern would be **connect(l[.._])**.

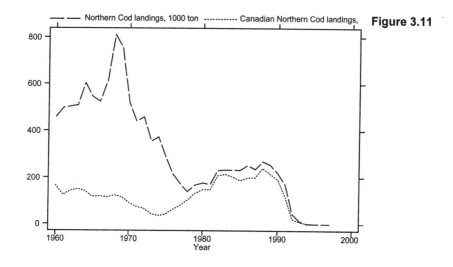

Figure 3.12 adds the Total Allowable Catch, connected in steps (**connect (J)**), show-ing the legal limits on catches after 1972. Note also the *x*-axis labeling at five-year intervals, and the horizontal lines at 200 (thousand) ton intervals.

```
. graph total canada TAC year, symbol(iii) connect(1[_]1[.]J) ylabel
    xlabel(1960 ,1965 to 1995) l1(Thousands of metric tons)
    border yline(0,200 to 600)
```

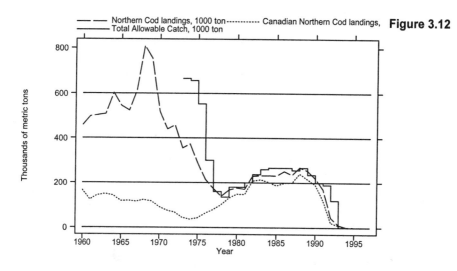

This section illustrated just a few of the many possible variations on scatterplots. Others appear in later chapters, notably those on regression and time series methods. Type **help graph** for a summary of options, which include controls over such elements as graph colors, size of text and plotting symbols, and the thickness of printed lines.

Boxplots and One-Way Scatterplots

Boxplots convey information about center, spread, symmetry, and outliers at a glance. To obtain a single boxplot, type a command of the form

```
. graph y, box
```

If several different variables have roughly similar scales, we can visually compare their distributions through commands of the form

```
. graph w x y z, box
```

One of the most common applications for boxplots involves comparing the distribution of one variable across categories of a second. Figure 3.13 compares the distribution of *college* across states of four U.S. regions, from dataset *states.dta*. Because Washington DC has no value for *region* in these data, it gets left out of the graph.

```
. sort region
```

```
. graph college, box by(region) ylabel yline(19.1)
```

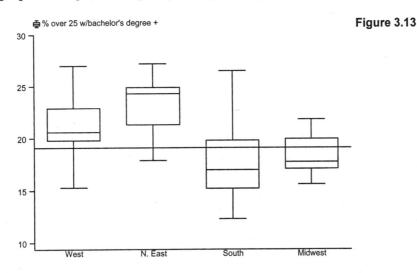

Figure 3.13

The **yline(19.1)** option drew a horizontal line at $y = 19.1$, the 50-state median of *college*. This median was obtained by typing

```
. summarize college if region != ., detail
```

Chapter 4 describes the **summarize, detail** command. The **if region != .** qualifier above restricted our analysis to observations that have nonmissing values of *region*; that is, to every place except Washington DC.

The box in a boxplot extends from approximate first to third quartiles, a distance called the interquartile range (IQR). Outliers, defined as observations more than 1.5IQR beyond the first or third quartile, are plotted individually in a boxplot—usually as large circles, although we can control what symbol gets used for outliers by specifying the **symbol( )** option. No outliers appear among the four distributions in Figure 3.13. Stata's boxplots define quartiles in the same manner as **summarize, detail**. This is not the same approximation used to calculate "fourths" for letter-value displays, **lv** (Chapter 4). See Frigge, Hoaglin, and Iglewicz (1989) and Hamilton (1992b) for more about quartile approximations and their role in identifying outliers.

One-way scatterplots, Stata's simplest style of graph, also display variable distributions. They work best for continuous variables that have few or no tied values. Figure 3.14 shows a one-way scatterplot of *college*.

```
. graph college, oneway
```

Figure 3.14

We earlier saw one-way scatterplots combined with boxplots and a two-way scatterplot (Figure 3.8). Combining just boxplots and one-way scatterplots also makes a useful display for comparing distributions of several different variables.

```
. graph w x y z, oneway box
```

Alternatively, we could compare the distribution of one variable across categories of a second. Figure 3.15 does this, giving a slightly different view of the same information presented in Figure 3.13. The percentage of adults with a college degree tends to be highest among Northeastern states and lower in the South.

```
. sort region
. graph college, oneway box by(region)
```

Figure 3.15

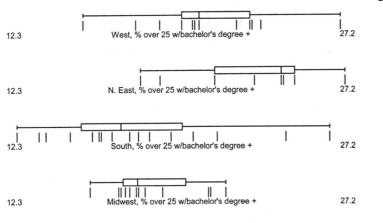

Pie Charts and Star Charts

Pie charts are popular tools for "presentation graphics," although they have less value in analytical work. Stata's basic pie chart command has the form

```
. graph w x y z, pie
```

where the variables *w*, *x*, *y*, and *z* all measure quantities of something, in similar units (for example, all are in dollars, hours, or people).

Dataset *AKethnic.dta*, on the ethnic composition of Alaska's population, provides an illustration. Alaska's indigenous Native population divides into three broad cultural/linguistic groups: Aleut, Indian (including Athabaska, Tlingit, and Haida), and Eskimo (Yupik and Inupiat). The variables *aleut*, *indian*, *eskimo*, and *nonnativ* are population counts for each group, taken from the 1990 U.S. Census. This dataset contains only three observations, representing three types or sizes of communities: cities of 10,000 people or more; towns of 1,000 to 10,000; and villages with fewer than 1,000 people.

```
Contains data from C:\data\AKethnic.dta
  obs:            3                      Alaska ethnicity 1990
 vars:            7                      25 Jun 2001 21:06
 size:           63 (100.0% of memory free)
-------------------------------------------------------------------------
               storage  display   value
variable name    type   format    label     variable label
-------------------------------------------------------------------------
comtype         byte    %8.0g     popcat    Community type (size)
pop             float   %9.0g               Population
n               int     %8.0g               number of communities
aleut           int     %8.0g               Aleuts
indian          int     %8.0g               Indians
eskimo          int     %8.0g               Eskimos
nonnativ        float   %9.0g               Non-Natives
-------------------------------------------------------------------------
Sorted by:
```

The majority of the state's population is non-Native, as clearly seen in a pie chart (Figure 3.16). Like bar charts and many other graphics, pie charts look better in color.

```
. graph aleut indian eskimo nonnativ, pie
```

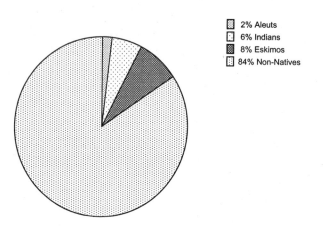

Figure 3.16

□ 2% Aleuts
□ 6% Indians
▨ 8% Eskimos
▨ 84% Non-Natives

Non-Natives are the largest group in Figure 3.16, but if we draw separate pies for each type of community, new details emerge (Figure 3.17). Although Natives are only a small fraction of the population in Alaska cities, they constitute the majority among those living in villages. This gives Alaska villages a different character from Alaska cities.

```
. sort comtype
```

```
. graph aleut indian eskimo nonnativ, pie by(comtype) total
```

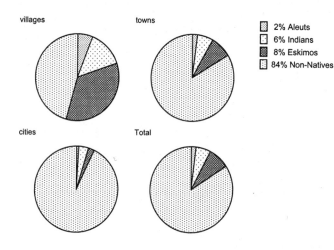

Figure 3.17

□ 2% Aleuts
□ 6% Indians
▨ 8% Eskimos
▨ 84% Non-Natives

In contrast to the ubiquitous pie chart, star charts are relatively uncommon. They belong to a family of methods (that includes "Chernof faces") meant to compare observations visually on a number of variables at once. A basic star chart command has the form

```
. graph w x y z, star label(name)
```

Unlike pie charts, star charts do not require that the variables be in similar units. **by(*varname*)** options do not work with star charts. Instead, we automatically get one star for each observation. Values of the variable specified in the **label( )** option can provide the labels for each star.

We could use star charts to compare information about a number of technically similar (and now outdated) microcomputers, found in file *micro.dta*.

```
Contains data from C:\data\micro.dta
  obs:            51                          PC Magazine's benchmark tests
  vars:            9                          25 Jun 2001 21:30
  size:        1,887 (100.0% of memory free)
-------------------------------------------------------------------------------
              storage  display   value
variable name   type   format    label     variable label
-------------------------------------------------------------------------------
make            byte    %8.0g     mk        Manufacturer
model           float   %9.0g     md        Model
chip            float   %9.0g               CPU chip type
disk            float   %9.0g               DOS disk access time
clock           float   %9.0g               Clock speed in MHz
nop             float   %9.0g               No operation loop
mix             float   %9.0g               CPU instruction mix
fp              float   %9.0g               Floating point test
ram             float   %9.0g               RAM read/write test
-------------------------------------------------------------------------------
Sorted by:  chip
```

The variables *disk*, *clock*, *nop*, *mix*, *fp*, and *ram* measure different aspects of computing speed. Most are indices, and higher values mean a slower computer. Variable *make* stores the computer brand names. Figure 3.18 shows a star chart for the 17 computers that used an 80386 chip and do not have missing values on the memory read/write test (**if *chip* == 80386 & *ram* != .**). Before drawing the graph, we **set textsize 140**, which asks that all text in our graph appears at 140% of its normal size to make it more readable. Afterwards, we return to the usual setting by **set textsize 100**.

```
. set textsize 140
. graph disk clock nop mix fp ram if chip == 80386 & ram != .,
     star label(make)
. set textsize 100
```

The **set textsize #** command applies to other Stata graph styles as well. It remains in effect for the remainder of the session, or until another **set textsize** command is typed.

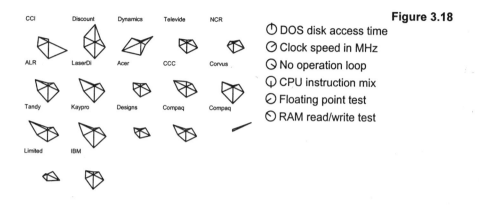

Figure 3.18

A long line segment in the *disk* direction (straight up) in Figure 3.18 indicates that the computer has a high value on this variable. That is, its disk access is slow. The same holds for the other five variables: the slower the computer, the longer the corresponding line segment. We cannot read the variables' actual values from a star chart. Instead we scan for general patterns. For example, the second Compaq and the Limited stand out as fast machines overall; the Discount and Laser appear generally slow. Several computers have quite similar performance profiles, such as the Laser and Kaypro, or the Televideo and Acer. Note that we compare these 17 computers on six variables at once—the chart contains a lot of information.

Bar Charts

Bar charts provide simple yet versatile tools for depicting the distributions of categorical variables, or summarizing measurement variables across categories. The height of the bars can represent either sums (default) or means (**means** option).

Returning to the Alaska ethnicity dataset *AKethnic.dta*, we can display the same information as the earlier pie charts but this time in bar chart form. Figure 3.19 shows the state's overall ethnic composition.

. graph *aleut indian eskimo nonnativ*, ylabel bar

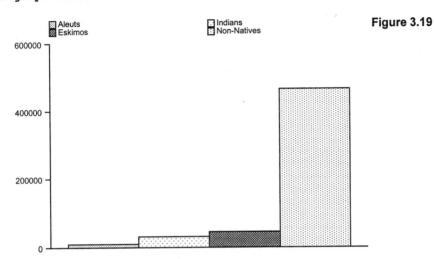

Figure 3.19

Figure 3.20 breaks the population down by community type (*comtype*), and stacks bars for different ethnic groups on top of each other. Without the **stack** option, we would have four side-by-side bars for the ethnic population of each community type. The pie charts in Figure 3.17 depict the relative sizes (percentages) of different ethnic groups, whereas the bar charts in Figure 3.20 depict their absolute sizes (sums). Consequently, Figure 3.20 reveals something that Figure 3.17 could not: the majority of Alaska's Eskimo (Yupik and Inupiat) population lives in villages.

. sort *comtype*

. graph *aleut indian eskimo nonnativ*, ylabel bar stack by(*comtype*)

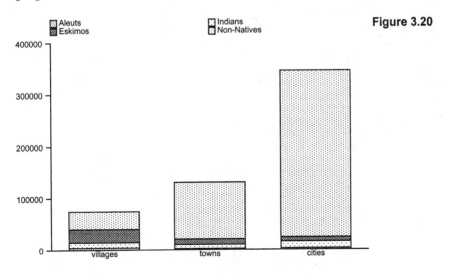

Figure 3.20

Bar charts can display means instead of sums. Dataset *student1.dta* contains some results from a survey of students in one undergraduate statistics class. The variables include each student's gender, age, political party preference, and score on a recent quiz.

```
Contains data from C:\data\student1.dta
  obs:            43                        Statistics students survey
  vars:            6                        26 Jun 2001 10:20
  size:          473 (99.8% of memory free)
-------------------------------------------------------------------------
              storage  display    value
variable name  type    format     label    variable label
-------------------------------------------------------------------------
gender         byte    %8.0g      gender    Gender
age            byte    %8.0g                Age
party          byte    %11.0g     party     Political party preference
quiz           byte    %8.0g                Quiz score on readings
grade          byte    %8.0g      grade     Expected grade in statistics
msat           int     %9.0g                Mathematics SAT score
-------------------------------------------------------------------------
Sorted by:
```

The Democrats in this class had slightly higher mean quiz scores, as seen in Figure 3.21. The **means** option calls for bars showing means, instead of the default sums.

```
. sort party

. graph quiz, ylabel bar means by(party)
```

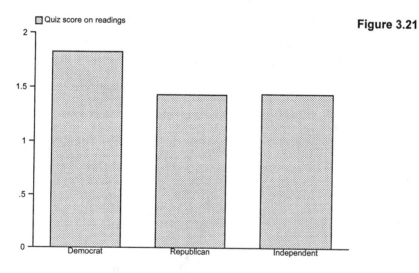

Figure 3.21

Stata's bar charts work most naturally with measurement variables such as quiz scores, but they can also show categorical-variable distributions. This class contained 17 Democrats, 7 Republicans, and 16 Independents:

```
. tabulate party

  Political |
      party |
 preference |      Freq.      Percent         Cum.
------------+-----------------------------------------
    Democrat |         17        42.50        42.50
  Republican |          7        17.50        60.00
 Independent |         16        40.00       100.00
------------+-----------------------------------------
       Total |         40       100.00
```

party is a labeled numerical variable, with values 1 for "Democrat," 2 for "Republican," and 3 for "Independent." If we type **graph *party*, bar**, we would see the sums of these codes; **graph *party*, bar means** would display the means. Neither choice makes sense, given the arbitrary nature of *party*'s 1, 2, 3 codes. Graphing the distribution of a categorical variable requires an extra step:

1. Create a new "variable" (if one does not already exist) equal to 1 for every observation in the data by typing

    . generate *count* = 1

2. Draw a bar chart for the sums of this new "variable," by values of the categorical variable of interest. Because *count* equals 1 for each observation, the sums of *count* within each category equal the frequency of that category, as shown in Figure 3.22.

    . sort *party*

    . graph *count*, ylabel bar by(*party*)

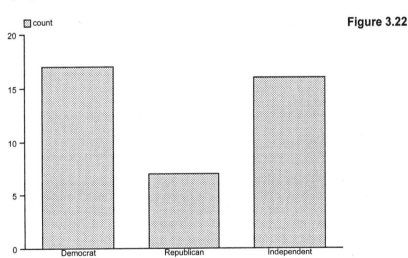

Figure 3.22

A similar trick allows us to graph the relation between two categorical variables such as *party* and *gender* in one bar chart. *gender* is a dummy variable, coded 0 for male and 1 for female. For this graph we will want a new indicator coded 0 for male and 100 for female. Also, for reasons that will soon be apparent, we label this new indicator "Percent female" although really it is just an indicator with values of 0 or 100. We then use the new variable *female* as the basis for our bar chart. The result, Figure 3.23, shows the percent female within each category of political party preference.

```
. generate female=0 if gender == 0
(31 missing values generated)
. replace female = 100 if gender == 1
(31 real changes made)
. label variable female "Percent female"
. graph female, bar by(party) means ylabel yline(72)
```

Note that we used the **means** option in Figure 3.23. Means are needed here because the mean of a {0,100} dichotomy really does equal the percentage coded 100, that is, the percent female. (Similarly, the mean of a {0,1} dummy variable equals the proportion coded 1.) The **yline(72)** option marks the overall percentage: 72% of these students are female.

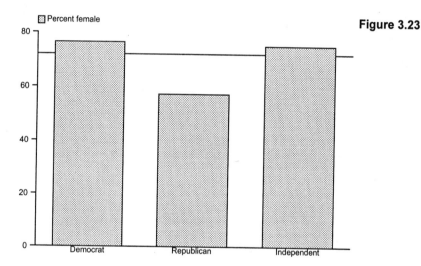

Figure 3.23

Symmetry and Quantile Plots

Boxplots and histograms summarize measurement variable distributions, hiding individual data points to clarify overall patterns. Symmetry and quantile plots, on the other hand, include points for every observation in a distribution. They are harder to read than summary graphs, but convey more detailed information.

A histogram of per-capita energy consumption in the 50 U.S. states (from *states.dta*) appears in Figure 3.24. The skewed distribution includes a handful of very high-consumption states, which happen to be oil producers. With the **freq** option, bar heights indicate frequencies. For example, 17 of these states had *energy* values between 200 and 300.

. **graph** *energy*, **xlabel ylabel bin(8) norm freq**

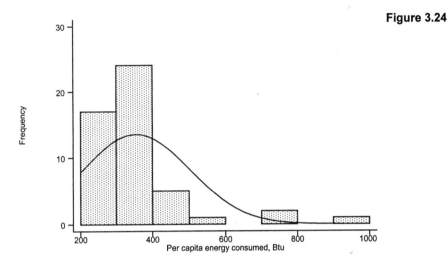

Figure 3.24

Figure 3.25 depicts this distribution as a symmetry plot. It plots the distance of the ith observation above the median (vertical) against the distance of the ith observation below the median. All points would lie on the diagonal line if this distribution were symmetrical. Instead, we see that distances above the median grow steadily larger than corresponding distances below the median, a symptom of positive skew. Unlike Figure 3.24, Figure 3.25 also reveals that the energy-consumption distribution is approximately symmetrical near its center.

. **symplot** *energy*, **xlabel ylabel**

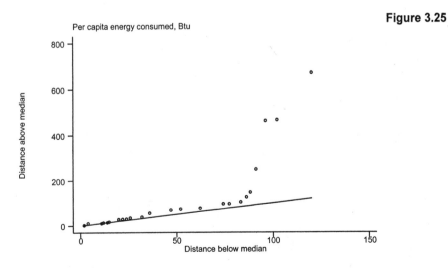

Figure 3.25

Quantiles are values below which a certain fraction of the data lie. For example, a .3 quantile is that value higher than 30% of the data. If we sort n observations in ascending order, the *i*th value forms the $(i - .5)/n$ quantile. The following commands would calculate quantiles of variable *energy*:

```
. drop if energy == .
. sort energy
. generate quant = (_n - .5)/_N
```

As mentioned in Chapter 2, _n and _N are Stata system variables, always unobtrusively present when there are data in memory. _n represents the current observation number , and _N the total number of observations.

Quantile plots automatically calculate what fraction of the observations lie below each data value, and display the results graphically as in Figure 3.26. Quantile plots provide a graphic reference for someone who does not have the original data at hand. From well-labeled quantile plots, we can estimate order statistics such as median (.5 quantile) or quartiles (.25 and .75 quantiles). The IQR equals the rise between .25 and .75 quantiles. We could also read a quantile plot to estimate the fraction of observations falling below a given value.

```
. quantile energy, ylabel
```

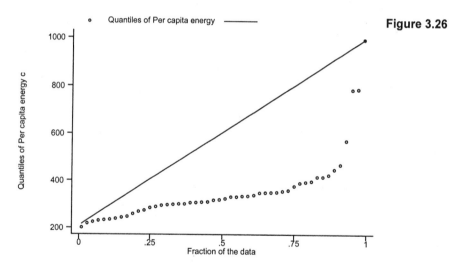

Figure 3.26

Quantile-normal plots, also called normal probability plots, compare quantiles of a variable's distribution with quantiles of a theoretical normal distribution having the same mean and standard deviation. They allow visual inspection for departures from normality in every part of a distribution, which can help guide decisions regarding normality assumptions and efforts to find a normalizing transformation. Figure 3.27, a quantile-normal plot of *energy*, confirms the severe positive skew that we had already observed. The **grid** option causes the plot to include a set of lines marking the .05, .10, .25 (first quartile), .50 (median), .75 (third quartile), .90, and .95 quantiles of both distributions. The .05, .50, and .95 quantile values are printed along the top and right-hand axes.

. **qnorm** *energy*, **ylabel xlabel grid**

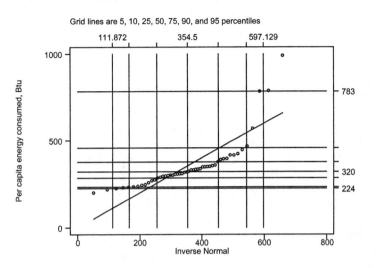

Figure 3.27

Quantile-quantile plots resemble quantile-normal plots, but they compare quantiles (ordered data points) of two empirical distributions instead of comparing one empirical distribution with a theoretical normal distribution. Figure 3.28 shows a quantile-quantile plot of the mean math SAT score versus the mean verbal SAT score in 50 states and the District of Columbia. If the two distributions were identical, we would see points along the diagonal line. Instead, data points form a straight line roughly parallel to the diagonal, indicating that the two variables have different means but similar shapes and standard deviations.

. **qqplot** *msat vsat*, **ylabel xlabel**

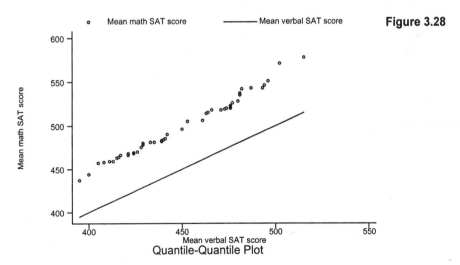

Figure 3.28

Regression with Graphics (Hamilton 1992a) includes an introduction to reading quantile-based plots. Chambers et al. (1983) provide more details. Related Stata commands include **pnorm** (standard normal probability plot), **pchi** (chi-squared probability plot), and **qchi** (quantile – chi-squared plot).

Quality Control Graphs

You can use quality control charts to monitor output from a repetitive process such as industrial production. Stata offers four basic types: c chart, p chart, R chart, and $\bar{x}$ chart. A fifth type, called Shewhart after the inventor of these methods, consists of vertically-aligned $\bar{x}$ and R charts. Iman (1994) provides a brief introduction to R and $\bar{x}$ charts, including the tables used in calculating their control limits. The *Reference Manual* gives the command details and formulas used by Stata. Basic outlines of these commands are as follows:

. **cchart** *defects unit*

Constructs a c chart with the number of nonconformities or defects (*defects*) graphed against the unit number (*unit*). Upper and lower control limits, based on the assumption that number of nonconformities per unit follows a Poisson distribution, appear as horizontal lines in the chart. Observations with values outside these limits are said to be "out of control."

. **pchart** *rejects unit ssize*

Constructs a p chart with the proportion of items rejected (*rejects / ssize*) graphed against the unit number (*unit*). Upper and lower control limit lines derive from a normal approximation, taking sample size (*ssize*) into account. If *ssize* varies across units, the control limits will vary too, unless we add the option **stabilize** .

. **rchart** *x1 x2 x3 x4 x5*, **connect(1)**

Constructs an R (range) chart using the replicated measurements in variables *x1* through *x5*—that is, in this example, five replications per sample. Graph the range within each sample against the sample number, and (optionally) connect successive ranges with line segments. Horizontal lines indicate the mean range and control limits. Control limits are estimated from the sample size if the process standard deviation is unknown. When σ is known we can include this information in the command. For example, assuming σ = 10,

. **rchart** *x1 x2 x3 x4 x5*, **connect(1) std(10)**

. **xchart** *x1 x2 x3 x4 x5*, **connect(1)**

Constructs an $\bar{x}$ (mean) chart using the replicated measurements in variables *x1* through *x5*. Graphs the mean within each sample against the sample number and connect successive means with line segments. The mean range is estimated from the mean of sample means and control limits from sample size, unless we override these defaults. For example, if we know that the process actually has μ = 50 and σ =10,

. **xchart** *x1 x2 x3 x4 x5*, **connect(1) mean(50) std(10)**

Alternatively, we could specify particular upper and lower control limits:

. **xchart** *x1 x2 x3 x4 x5*, **connect(1) mean(50) lower(40)**
      upper(60)

. **shewhart** *x1 x2 x3 x4 x5*, **mean(50) std(10)**

In one figure, vertically align an $\bar{x}$ chart with an R chart.

To illustrate a p chart, we turn to the quality inspection data in *quality1.dta*.

```
Contains data from C:\data\quality1.dta
  obs:            16                      Quality control example 1
  vars:            3                      26 Jun 2001 19:58
  size:          112 (99.8% of memory free)
-------------------------------------------------------------------------
              storage  display   value
variable name  type    format    label    variable label
-------------------------------------------------------------------------
day            byte    %9.0g              Day sampled
ssize          byte    %9.0g              Number of units sampled
rejects        byte    %9.0g              Number of units rejected
-------------------------------------------------------------------------
Sorted by:

. list in 1/5

          day      ssize    rejects
   1.      58        53        10
   2.       7        53        12
   3.      26        52        12
   4.      21        52        10
   5.       6        51        10
```

Note that sample size varies from unit to unit, and that the units (days) are not in order. **pchart** handles these complications automatically, creating the graph with changing control limits seen in Figure 3.29. (For constant control limits despite changing sample sizes, add the **stabilize** option.)

`. pchart rejects day ssize, ylabel xlabel`

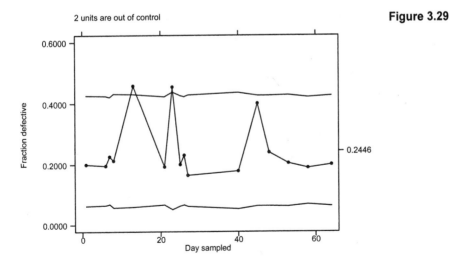

Figure 3.29

Dataset *quality2.dta*, borrowed from Iman (1994:662), serves to illustrate **rchart** and **xchart** . Variables *x1* through *x4* represent repeated measurements from an industrial production process; 25 units with four replications each form the dataset.

```
Contains data from C:\data\quality2.dta
  obs:              25                        Quality control (Iman 1994:662)
  vars:              4                        26 Jun 2001 20:04
  size:             500 (99.8% of memory free)
-------------------------------------------------------------------------------
              storage  display   value
variable name  type    format    label       variable label
-------------------------------------------------------------------------------
x1            float    %9.0g
x2            float    %9.0g
x3            float    %9.0g
x4            float    %9.0g
-------------------------------------------------------------------------------
Sorted by:
```

. **list in 1/5**

```
        x1        x2        x3        x4
1.     4.6         2         4       3.6
2.     6.7       3.8       5.1       4.7
3.     4.6       4.3       4.5       3.9
4.     4.9         6       4.8       5.7
5.     7.6       6.9       2.5       4.7
```

Figure 3.30, an R chart, graphs variation in the process range over the 25 units. **rchart** informs us that one unit's range is "out of control."

. **rchart x1 x2 x3 x4, connect(l)**

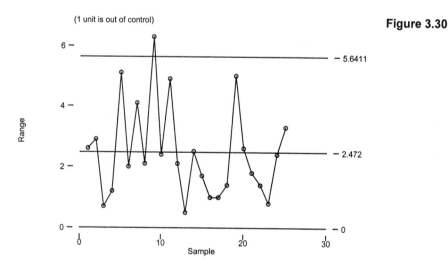

Figure 3.31, an $\bar{x}$ chart, shows variation in the process mean. None of these 25 means falls outside the control limits.

. xchart *x1 x2 x3 x4*, connect(l)

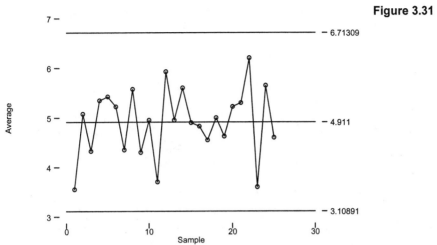

Figure 3.31

4

Summary Statistics and Tables

The **summarize** command obtains descriptive statistics such as medians, means, and standard deviations of measurement variables. Similar summary statistics are available with the more flexible command **tabstat**. **tabulate** obtains frequency distribution tables, cross-tabulations, and assorted tests or measures of association for categorical and ordinal variables. **tabulate** can also construct one- or two-way tables of means and standard deviations. A more general table-making command, **table**, produces as many as seven-way tables in which the cells contain statistics such as frequencies, sums, means, or medians.

This chapter also describes further one-variable procedures including normality tests, transformations, and displays for exploratory data analysis (EDA). Stata also provides many tables of particular interest to epidemiologists. These are not described in this chapter, but can be viewed by typing **help epitab**. Selvin (1996) introduces the topic.

Example Commands

. **summarize y1 y2 y3**
   Calculates simple summary statistics (means, standard deviations, and numbers of observations) for the variables listed.

. **summarize y1 y2 y3, detail**
   Obtains detailed summary statistics including percentiles, median, mean, standard deviation, variance, skewness, and kurtosis.

. **summarize y1 if x1 > 3 & x2 != .**
   Finds summary statistics for $y1$ using only those observations for which variable $x1$ is greater than 3 and $x2$ is not missing.

. **summarize y1 [fweight = w], detail**
   Calculates detailed summary statistics for $y1$ using the frequency weights in variable w.

. **tabstat y1, stats(mean sd skewness kurtosis n)**
   Calculates only the specified summary statistics for variable $y1$.

. **tabstat y1, stats(min p5 p25 p50 p75 p95 max) by(x1)**
   Calculates the specified summary statistics (minimum, 5th percentile, 25th percentile, etc.) for measurement variable $y1$, within categories of $x1$.

   ate x1
   ays a frequency distribution table for all nonmissing values of variable $x1$.

. **tab1** *x1 x2 x3 x4*
Displays a series of frequency distribution tables, one for each of the variables listed.

. **tabulate** *x1 x2*
Displays a two-variable cross-tabulation with *x1* as the row variable, and *x2* as the columns.

. **tabulate** *x1 x2*, **chi2 nof column**
Produces a cross-tabulation and Pearson χ^2 test of independence. Does not show cell frequencies, but instead gives the column percentages in each cell.

. **tabulate** *x1 x2*, **missing row all**
Produces a cross-tabulation that includes missing values in the table and in the calculation of percentages. Calculates "all" available statistics (Pearson and likelihood χ^2, Cramer's *V*, Goodman and Kruskal's gamma, and Kendall's τ_b).

. **tab2** *x1 x2 x3 x4*
Performs all possible two-way cross-tabulations of the listed variables.

. **tabulate** *x1*, **summ(y)**
Produces a one-way table showing the mean, standard deviation, and frequency of *y* values within each category of *x1*.

. **tabulate** *x1 x2*, **summ(y) means**
Produces a two-way table showing the mean of *y* at each combination of *x1* and *x2* values.

. **by** *x3*, **sort:** **tabulate** *x1 x2*, **exact**
Creates a three-way cross-tabulation, with subtables for *x1* (row) by *x2* (column) at each value of *x3*. Calculates Fisher's exact test for each subtable. **by** *varname*, **sort:** works as a prefix for almost any Stata command where it makes sense. The **sort** option is unnecessary if the data already are sorted on *varname*.

. **table** *y x1 x3*, **by(x4 x5) contents(freq)**
Creates a five-way cross-tabulation, of *y1* (row) by *x2* (column) by *x3* (supercolumn), by *x4* (superrow 1) by *x5* (superrow 2). Cells contain frequencies.

. **table** *x1 x2*, **contents(mean y1 median y2)**
Creates a two-way table of *x1* (row) by *x2* (column). Cells contain the mean of *y1* and the median of *y2*.

Summary Statistics for Measurement Variables

Dataset *VTtown.dta* contains information from residents of a town in Vermont. A survey was conducted soon after routine state testing had detected trace amounts of toxic chemicals in the town's water supply. Higher concentrations were found in several private wells and near the public schools. Worried citizens held meetings to discuss possible solutions to this problem.

```
Contains data from C:\data\VTtown.dta
  obs:             153                        VT town survey (Hamilton 1985)
  vars:              7                        30 Jul 2001 07:30
  size:          1,683 (99.9% of memory free)
-------------------------------------------------------------------------------
                 storage  display    value
variable name    type     format     label      variable label
-------------------------------------------------------------------------------
gender           byte     %8.0g      sexlbl     Respondent's gender
lived            byte     %8.0g                 Years lived in town
kids             byte     %8.0g      kidlbl     Have children <19 in town?
educ             byte     %8.0g                 Highest year school completed
meetings         byte     %8.0g      kidlbl     Attended meetings on pollution
contam           byte     %8.0g      contamlb   Believe own property/water
                                                  contaminated
school           byte     %8.0g      close      School closing opinion
-------------------------------------------------------------------------------
Sorted by:
```

To find the mean and standard deviation of the variable *lived* (years the respondent had lived in town), type

. **summarize** *lived*

```
    Variable |      Obs        Mean    Std. Dev.       Min        Max
-------------+--------------------------------------------------------
       lived |      153    19.26797    16.95466          1         81
```

This table also gives the number of nonmissing observations and the variable's minimum and maximum values. If we had simply typed **summarize** with no variable list, we would obtain means and standard deviations for every numerical variable in the dataset.

To see more detailed summary statistics, type

. **summarize lived, detail**

```
                        Years lived in town
-------------------------------------------------------------
        Percentiles     Smallest
 1%           1              1
 5%           2              1
10%           3              1      Obs                  153
25%           5              1      Sum of Wgt.          153

50%          15                     Mean            19.26797
                        Largest     Std. Dev.       16.95466
75%          29             65
90%          42             65      Variance        287.4606
95%          55             68      Skewness        1.208804
99%          68             81      Kurtosis        4.025642
```

This **summarize, detail** output includes basic statistics plus the following:

Percentiles: Notably the first quartile (25th percentile), median (50th percentile), and third quartile (75th percentile). Because many samples do not divide evenly into quarters or other standard fractions, these percentiles are approximations.

Four smallest and four largest values, where outliers might show up.

Sum of weights: Stata understands four types of weights: analytical weights (**aweight**), frequency weights (**fweight**), importance weights (**iweight**), and sampling weights (**pweight**). Different procedures allow, and make sense with, different kinds of weights. **summarize** , for example, permits only **aweight** or **fweight** . Type **help weights** for explanations.

Variance: Standard deviation squared (more properly, standard deviation equals the square root of variance).

Skewness: The direction and degree of asymmetry. A perfectly symmetrical distribution has skewness = 0. Positive skew (heavier right tail) results in skewness > 0; negative skew (heavier left tail) results in skewness < 0.

Kurtosis: Tail weight. A normal (Gaussian) distribution is symmetrical and has kurtosis = 3. If a symmetrical distribution has heavier-than-normal tails (that is, is sharply peaked), it will have kurtosis > 3. Kurtosis < 3 indicates lighter-than-normal tails.

The **tabstat** command provides a more flexible alternative to **summarize** . We can specify just which summary statistics we want to see. For example,

```
. tabstat lived, stats(mean range skewness)

    variable |      mean     range  skewness
-------------+------------------------------
       lived |  19.26797        80  1.208804
-------------------------------------------
```

With a **by(varname)** option, **tabstat** constructs a table containing summary statistics for each value of *varname*. The following example contains means, standard deviations, medians, interquartile ranges, and number of nonmissing observations of *lived*, for each category of *gender*. The means and medians both indicate that the women in this sample had lived in town for fewer years than the men. Note that the median column is labeled "p50", meaning 50th percentile.

```
. tabstat lived, stats(mean sd median iqr n) by(gender)

Summary for variables: lived
     by categories of: gender (Respondent's gender)

gender |      mean        sd       p50       iqr         N
-------+--------------------------------------------------
  male |  23.48333  19.69125      19.5        28        60
female |  16.54839  14.39468        13        19        93
-------+--------------------------------------------------
 Total |  19.26797  16.95466        15        24       153
----------------------------------------------------------
```

Statistics available for the **stats()** option of **tabstat** include:

mean	Mean
count	Count of nonmissing observations
n	Same as **count**
sum	Sum
max	Maximum
min	Minimum
range	Range = max – min
sd	Standard deviation

`var`	Variance
`semean`	Standard error of mean = sd/sqrt(n)
`skewness`	Skewness
`kurtosis`	Kurtosis
`median`	Median (same as `p50`)
`p1`	1st percentile (similarly, `p5` , `p10` , `p25` , `p50` , `p75` , `p95` , or `p99`)
`iqr`	Interquartile range = p75 – p25
`q`	Quartiles; equivalent to specifying `p25 p50 p75`

Further `tabstat` options give control over the table layout and labeling. Type `help tabstat` to see a complete list.

The statistics produced by **summarize** or **tabstat** describe the sample at hand. We might also want to make inferences about the population, for example, by constructing a 99% confidence interval for the mean of *lived*:

```
. ci lived, level(99)
```

Variable	Obs	Mean	Std. Err.	[99% Conf. Interval]
lived	153	19.26797	1.370703	15.69241 22.84354

Based on this sample, we could be 99% confident that the population mean lies somewhere in the interval from 15.69 to 22.84 years. Here we used a **level( )** option to specify a 99% confidence interval. If we omit this option, **ci** defaults to a 95% confidence interval.

Other options allow **ci** to calculate exact confidence intervals for variables that follow binomial or Poisson distributions. A related command, **cii**, calculates normal, binomial, or Poisson confidence intervals directly from summary statistics, such as we might encounter in a published article. It does not require the raw data. Type **help ci** for details.

Exploratory Data Analysis

Statistician John Tukey invented a toolkit of methods for analyzing data in an exploratory and skeptical way without making unneeded assumptions (see Tukey 1977; also Hoaglin, Mosteller, and Tukey 1983, 1985). Boxplots, introduced in Chapter 3, are one of Tukey's best-known innovations. Another is the stem-and-leaf display, in which initial digits form the "stems" and following digits for each observation make up the "leaves."

```
. stem lived
```

```
Stem-and-leaf plot for lived (Years lived in town)

  0* | 111111122222333333344444444
  0. | 5555555555566666666777889999
  1* | 0000001122223333334
  1. | 55555567788899
  2* | 0000001111112224444
  2. | 56778899
  3* | 00000124
  3. | 5555666789
  4* | 0012
  4. | 59
  5* | 00134
  5. | 556
  6* |
  6. | 5558
  7* |
  7. |
  8* | 1
```

stem automatically chose a double-stem version here, in which 1* denotes first digits
of 1and second digits of 0–4 (that is, respondents who had lived in town 10–14 years). 1.
denotes first digits of 1 and second digits of 5 to 9 (15–19 years). We can control the number
of lines per initial digit with the **lines ( )** option. For example, a five-stem version in which
the 1* stem hold leaves of 0–1, 1t leaves of 2–3, 1f leaves of 4–5, 1s leaves of 6–7,
and 1. leaves of 8–9 could be obtained by typing

. **stem** *lived*, **lines(5)**

Type **help stem** for information about other options.

Letter-value displays (**lv**) use order statistics to dissect a distribution.

. **lv** *lived*

```
  #     153              Years lived in town
                   -------------------------------------
  M     77   |                   15               |    spread   pseudosigma
  F     39   |         5         17         29  |      24       17.9731
  E     20   |         3         21         39  |      36       15.86391
  D     10.5 |         2         27         52  |      50       16.62351
  C     5.5  |         1       30.75       60.5 |     59.5      16.26523
  B     3    |         1         33         65  |      64       15.15955
  A     2    |         1        34.5        68  |      67       14.59762
  Z     1.5  |         1       37.75       74.5 |     73.5      15.14113
        1    |         1         41         81  |      80       15.32737
             |                                    |
             |                                    |   # below     # above
inner fence  |        -31                   65  |      0           5
outer fence  |        -67                  101  |      0           0
```

M denotes the median, and F the "fourths" (quartiles, using a different approximation than the
quartile approximation used by **summarize, detail** and **tabsum**). E, D, C,...
denote cutoff points such that roughly 1/8, 1/16, 1/32, . . . of the distribution remains outside
in the tails. The second column of numbers gives the "depth," or distance from nearest extreme,
for each letter value. Within the center box, the middle column gives "midsummaries," which
are averages of the two letter values. If midsummaries drift away from the median, as they do
for *lived*, this tells us that the distribution becomes progressively more skewed as we move
farther out into the tails. The "spreads" are differences between pairs of letter values. For
instance, the spread between F 's equals the approximate interquartile range. Finally,

"pseudosigmas" in the right-hand column estimate what the standard deviation should be if these letter values described a Gaussian population. The F pseudosigma, sometimes called a "pseudo standard deviation" (*PSD*), provides a simple and outlier-resistant check for approximate normality in symmetrical distributions:

1.   Comparing mean with median diagnoses overall skew:

mean > median	positive skew
mean = median	symmetry
mean < median	negative skew

2.   If the mean and median are similar, indicating symmetry, then a comparison between standard deviation and *PSD* helps to evaluate tail normality:

standard deviation > *PSD*	heavier-than-normal tails
standard deviation = *PSD*	normal tails
standard deviation < *PSD*	lighter-than-normal tails

Let F_1 and F_3 denote 1st and 3rd fourths (approximate 25th and 75th percentiles). Then the interquartile range, *IQR*, equals $F_3 - F_1$, and $PSD = IQR / 1.349$.

lv also identifies mild and severe outliers.   We call an *x* value a "mild outlier" when it lies outside the inner fence, but not outside the outer fence:

$$F_1 - 3IQR \le x < F_1 - 1.5IQR \quad \text{or} \quad F_3 + 1.5IQR < x \le F_3 + 3IQR$$

x is a "severe outlier" if it lies outside the outer fence:

$$x < F_1 - 3IQR \quad \text{or} \quad x > F_3 + 3IQR$$

lv gives these cutoffs and the number of outliers of each type. Severe outliers, values beyond the outer fences, occur sparsely (about two per million) in normal populations. Monte Carlo simulations suggest that the presence of any severe outliers in samples of $n = 15$ to about 20,000 should be sufficient evidence to reject a normality hypothesis at $\alpha = .05$ (Hamilton 1992b). Severe outliers create problems for many statistical techniques.

summarize, **stem**, and **lv** all confirm that *lived* has a positively skewed sample distribution, not at all resembling a theoretical normal curve. The next section introduces more formal normality tests, and transformation methods that can reduce a variable's skew.

Normality Tests and Transformations

Many statistical procedures work best when applied to variables that follow normal distributions. The preceding section described exploratory methods to check for approximate normality, extending the graphical tools presented in Chapter 3. A skewness–kurtosis test, making use of the skewness and kurtosis statistics shown by **summarize, detail**, can more formally evaluate the null hypothesis that the sample at hand came from a normally-distributed population.

```
. sktest lived

              Skewness/Kurtosis tests for Normality
                                              ------- joint ------
     Variable |  Pr(Skewness)   Pr(Kurtosis)  adj chi2(2)    Prob>chi2
------------+-------------------------------------------------------
        lived |     0.000          0.028          24.79        0.0000
```

sktest here rejects normality: *lived* appears significantly nonnormal in skewness ($P =$ .000), kurtosis ($P = .028$), and in both statistics considered jointly ($P = .0000$). Stata rounds off displayed probabilities to three or four decimals; "0.0000" really means $P < .00005$.

Other normality or log-normality tests include Shapiro–Wilk W (**swilk**) and Shapiro–Francia W' (**sfrancia**) methods. Type **help sktest** to see the options.

Nonlinear transformations such as square roots and logarithms are often employed to change distributions' shapes, with the aim of making skewed distributions more symmetrical and perhaps more nearly normal. Transformations might also help linearize relations between variables (Chapter 8). Table 4.1 shows a progression called the "ladder of powers" (Tukey 1977) that provides guidance for choosing transformations to change distributional shape. The variable *lived* exhibits mild positive skew, so its square root might be more symmetrical. We could create a new variable equal to the square root of *lived* by typing

. **generate srlived = lived^.5**

Instead of **lived^.5**, we could equally well have written **sqrt(lived)** .

Logarithms are another transformation that can reduce positive skew. To generate a new variable equal to the natural (base *e*) logarithm of *lived*, type

. **generate loglived = ln(lived)**

In the ladder of powers and related transformation schemes such as Box–Cox, logarithms take the place of a "0" power. Their effect on distribution shape is intermediate between .5 (square root) and −.5 (reciprocal root) transformations.

Table 4.1: Ladder of Powers

Transformation	Formula	Effect
cube	**new = old ^3**	reduce severe negative skew
square	**new = old ^2**	reduce mild negative skew
raw	**old**	no change (raw data)
square root	**new = old ^.5**	reduce mild positive skew
$\log_e$ (or $\log_{10}$)	**new = ln(old)** **new = log10(old)**	reduce positive skew
negative reciprocal root	**new = -(old ^-.5)**	reduce severe positive skew
negative reciprocal	**new = -(old ^-1)**	reduce very severe positive skew
negative reciprocal square	**new = -(old ^-2)**	"
negative reciprocal cube	**new = -(old ^-3)**	"

When raising to a power less than zero, we take negatives of the result in order to preserve the original order—the highest value of *old* becomes transformed into the highest value of *new*, and so forth. When *old* itself contains negative or zero values, it is necessary to add a constant before transformation. For example, if *arrests* measures the number of times a person has been arrested (0 for many people), then a suitable log transformation could be

. **generate larrests = ln(arrests + 1)**

The **ladder** command combines the ladder of powers with **sktest** tests for normality. It tries each power on the ladder, and reports whether the result is significantly nonnormal. This can be illustrated using the severely skewed variable *energy*, per capita energy consumption, from *states.dta*.

. **ladder energy**

Transformation	formula	chi2(2)	P(chi2)
cube	energy^3	53.74	0.000
square	energy^2	45.53	0.000
raw	energy	33.25	0.000
square-root	sqrt(energy)	25.03	0.000
log	log(energy)	15.88	0.000
reciprocal root	1/sqrt(energy)	7.36	0.025
reciprocal	1/energy	1.32	0.517
reciprocal square	1/(energy^2)	4.13	0.127
reciprocal cube	1/(energy^3)	11.56	0.003

It appears that the reciprocal transformation, 1/*energy* (or *energy*$^{-1}$), most closely resembles a normal distribution. Most of the other transformations (including the raw data) are significantly nonnormal. Figure 4.1 (produced by the **gladder** command) visually supports this conclusion by comparing histograms of each transformation to normal curves.

. **gladder energy, ylabel xlabel**

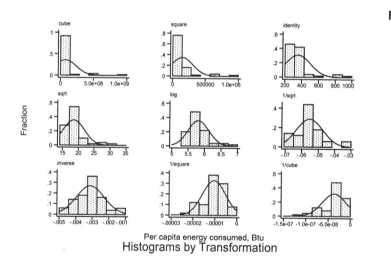

Figure 4.1

Histograms by Transformation

Figure 4.2 shows a corresponding set of quantile-normal plots for these ladder of powers transformations, obtained by the **qladder** command. With **qladder** (unlike **gladder**) we can and often will want to increase the scale of text labeling the small graphs. The default for **qladder** is **scale(1.25)**, but Figure 4.2 employs the larger **scale(3)**.

. **qladder energy, scale(3)**

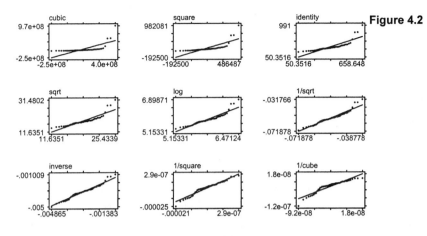

Per capita energy consumed, Btu
Quantile-Normal Plots by Transformation

An alternative technique called Box–Cox transformation offers finer gradations among transformations and automates the choice among them (easier for the analyst, but not always a good thing). The command **bcskew0** finds a value of λ (lambda) for the Box–Cox transformations

$$y^{(\lambda)} = \{y^{\lambda} - 1\} / \lambda \qquad \lambda > 0 \text{ or } \lambda < 0$$

or

$$y^{(\lambda)} = \ln(y) \qquad \lambda = 0$$

such that $y^{(\lambda)}$ has approximately 0 skewness. Applying this to *energy*, we obtain the transformed variable *benergy*:

. **bcskew0 *benergy* = *energy*, level(95)**

```
        Transform |        L    [95% Conf. Interval]      Skewness
------------------+-----------------------------------------------
  (energy^L-1)/L  |  -1.246052   -2.052503  -.6163383     .000281
(1 missing value generated)
```

That is, *benergy* = (*energy*$^{-1.246}$ – 1)/(–1.246) is the transformation that comes closest to symmetry (as defined by the skewness statistic). The Box–Cox parameter $\lambda = -1.246$ is not far from our ladder-of-powers choice, the –1 power. The confidence interval for λ,

$$-2.0525 < \lambda < -.6163$$

allows us to reject some other possibilities including logarithms ($\lambda = 0$) or square roots ($\lambda = .5$). Chapter 8 describes a Box–Cox approach to regression modeling.

Frequency Tables and Two-Way Cross-Tabulations

The methods described above apply to measurement variables. Categorical variables require other approaches, such as tabulation. Returning to the survey data in *VTtown.dta*, we could find the percentage of respondents who attending meetings concerning the pollution problem by tabulating the categorical variable *meetings*:

```
. tabulate meetings
```

Attended meetings on pollution	Freq.	Percent	Cum.
no	106	69.28	69.28
yes	47	30.72	100.00
Total	153	100.00	

tabulate can produce frequency distributions for variables that have many values—up to 500 with Small Stata or 3,000 with Intercooled Stata. To construct a manageable frequency distribution table, however, you might first want to group values by applying **generate** with its **recode** or **autocode** options (see Chapter 2 or **help generate**).

tabulate followed by two variable names creates a two-way cross-tabulation. For example, here is a cross-tabulation of *meetings* by *kids* (whether respondent has children under 19 living in town):

```
. tabulate meetings kids
```

Attended meetings on pollution	Have children <19 in town? no	yes	Total
no	52	54	106
yes	11	36	47
Total	63	90	153

The first-named variable forms the rows, and the second forms columns in the resulting table. We see that only 11 of these 153 people were non-parents who attended the meetings.

tabulate has a number of options that are useful with frequency tables:

cell   Shows total percentages for each cell.

chi2   Pearson χ^2 (chi-squared) test of hypothesis that row and column variables are independent.

column   Shows column percentages for each cell.

exact   Fisher's exact test of the independence hypothesis. Superior to **chi2** if the table contains thin cells with low expected frequencies. Often too slow to be practical in large tables, however.

gamma   Goodman and Kruskal's γ (gamma), with its asymptotic standard error (ASE). Measures association between ordinal variables, based on the number of concordant and discordant pairs (ignoring ties). $-1 \le \gamma \le 1$.

generate(new)   Creates a set of dummy variables named *new1*, *new2*, and so on to represent the values of the tabulated variable.

lrchi2   Likelihood-ratio χ^2 test of independence hypothesis. Not obtainable if table contains any empty cells.

missing   Includes "missing" as one row and/or column of the table.

nolabel   Shows numerical values rather than value labels of labeled numeric variables.

row	Shows row percentages for each cell.
nofreq	Does not show cell frequencies.
taub	Kendall's τ_b (tau-b), with its asymptotic standard error (ASE). Measures association between ordinal variables. **taub** is similar to **gamma**, but uses a correction for ties. $-1 \le \tau_b \le 1$.
v	Cramer's V, a measure of association for nominal variables. In 2×2 tables, $-1 \le V \le 1$. In larger tables, $0 \le V \le 1$.
all	Equivalent to the options **chi2 lrchi2 gamma taub V**. Not all of these options will be equally appropriate for a given table. **gamma** and **taub** assume that both variables have ordered categories, whereas **chi2**, **lrchi2**, and **V** do not.

To get the column percentages (because the column variable, *kids*, is the independent variable) and a χ^2 test for the cross-tabulation of *meetings* by *kids*, type

```
. tabulate meetings kids, column chi2

Attended |
meetings | Have children <19 in
      on |        town?
pollution |        no        yes |      Total
----------+----------------------+----------
      no |        52         54 |        106
         |     82.54      60.00 |      69.28
----------+----------------------+----------
     yes |        11         36 |         47
         |     17.46      40.00 |      30.72
----------+----------------------+----------
   Total |        63         90 |        153
         |    100.00     100.00 |     100.00

         Pearson chi2(1) =    8.8464   Pr = 0.003
```

Forty percent of the respondents with children attended meetings, compared with about 17% of the respondents without children. This association is statistically significant ($P = .003$).

Occasionally we might need to re-analyze a published table, or without retrieving the original raw data. A special command, **tabi** ("immediate" tabulation), accomplishes this. Type the cell frequencies on the command line, with table rows separated by " \ ". For illustration, here is how **tabi** could reproduce the previous χ^2 analysis, given only the four cell frequencies:

```
. tabi 52 54 \ 11 36, column chi2
```

```
             |          col
         row |         1          2 |     Total
-------------+----------------------+----------
           1 |        52         54 |       106
             |     82.54      60.00 |     69.28
-------------+----------------------+----------
           2 |        11         36 |        47
             |     17.46      40.00 |     30.72
-------------+----------------------+----------
       Total |        63         90 |       153
             |    100.00     100.00 |    100.00

        Pearson chi2(1) =    8.8464    Pr = 0.003
```

Unlike **tabulate**, **tabi** does not require or refer to any data in memory. By adding the **replace** option, however, we can ask **tabi** to replace whatever data are in memory with the new cross-tabulation. Statistical options (**chi2**, **exact**, **nofreq**, and so forth) work the same with **tabi** as they do with **tabulate**.

Multiple Tables and Multi-Way Cross-Tabulations

With surveys and other large datasets, we sometimes need frequency distributions for many different variables. Instead of asking for each table separately, for example by typing **tabulate** *meetings*, then **tabulate** *gender*, and finally **tabulate** *kids*, we could simply use another specialized command, **tab1**:

```
. tab1 meetings gender kids
```

Or, to produce one-way frequency tables for each variable from *gender* through *kids* in this dataset (the maximum is 30 variables at one time), type

```
. tab1 gender-school
```

Similarly, **tab2** creates multiple two-way tables. For example, the following command cross-tabulates every two-way combination of the listed variables:

```
. tab2 meetings gender kids
```

tab1 and **tab2** offer the same options as **tabulate**.

Returning to the ordinary **tabulate** command: with a **by** prefix, we can use this to form multi-way contingency tables. Here is a three-way cross-tabulation of *meetings* by *kids* by *contam* (respondent believes his or her own property or water contaminated), with χ^2 tests for the independence of *meetings* and *kids* within each level of *contam*:

```
. by contam, sort: tabulate meetings kids, nofreq col chi2
```

```
-> contam = no

Attended |
meetings | Have children <19 in
      on |       town?
pollution |      no      yes |    Total
----------+----------------------+----------
      no |   91.30    68.75 |    78.18
     yes |    8.70    31.25 |    21.82
----------+----------------------+----------
   Total |  100.00   100.00 |   100.00

        Pearson chi2(1) =   7.9814   Pr = 0.005
```

```
-> contam = yes

Attended |
meetings | Have children <19 in
      on |       town?
pollution |      no      yes |    Total
----------+----------------------+----------
      no |   58.82    38.46 |    46.51
     yes |   41.18    61.54 |    53.49
----------+----------------------+----------
   Total |  100.00   100.00 |   100.00

        Pearson chi2(1) =   1.7131   Pr = 0.191
```

Parents were more likely to attend meetings, among both the contaminated and uncontaminated groups. Only among the larger uncontaminated group is this "parenthood effect" statistically significant, however. As multi-way tables separate the data into smaller subsamples, the size of these subsamples has noticeable effects on significance-test outcomes.

This approach can be extended to tabulations of greater complexity. For example, to get a four-way cross-tabulation of *gender* by *contam* by *meetings* by *kids,* with χ^2 tests for each *meetings* by *kids* subtable (results not shown), type the command

```
. by gender contam, sort:  tabulate meetings kids, column chi2
```

A better way to produce multi-way tables, if we do not need percentages or statistical tests, is through Stata's general table-making command, **table**. This versatile command has many options, only a few of which are illustrated here. To construct a simple frequency table of *meetings*, type

```
. table meetings, contents(freq)
```

```
----------------------
Attended  |
meetings  |
on        |
pollution |     Freq.
----------+-----------
      no |      106
     yes |       47
----------------------
```

For a two-way frequency table or cross-tabulation, type

```
. table meetings kids, contents(freq)
```

```
---------------------
           |    Have
Attended   |  children
meetings   |  <19 in
on         |   town?
pollution  |  no    yes
-----------+-----------
       no  |  52    54
      yes  |  11    36
---------------------
```

If we specify a third categorical variable, it forms the "supercolumns" of a three-way table:

```
. table meetings kids contam, contents(freq)
```

```
------------------------------------
           |       Believe own
           |      property/water
Attended   |   contaminated and Have
meetings   |    children <19 in town?
on         |  --- no ---    --- yes --
pollution  |  no     yes    no    yes
-----------+------------------------
       no  |  42     44     10     10
      yes  |   4     20      7     16
------------------------------------
```

More complicated tables require the **by ( )** option, which allows up to four "supperrow" variables. **table** thus can produce up to seven-way tables: one row, one column, one supercolumn, and up to four superrows. Here is a four-way example:

```
. table meetings kids contam, contents(freq) by(gender)
```

```
------------------------------------
Responden |
t's       |
gender    |
and       |        Believe own
Attended  |       property/water
meetings  |    contaminated and Have
on        |     children <19 in town?
pollution |  --- no ---    --- yes --
          |  no     yes    no    yes
----------+------------------------
male      |
      no  |  18     18      3      3
     yes  |   2      7      3      6
----------+------------------------
female    |
      no  |  24     26      7      7
     yes  |   2     13      4     10
------------------------------------
```

The **contents( )** option of **table** specifies what statistics the table's cells contain.

contents(freq)	Frequency
contents(mean *varname*)	Mean of *varname*
contents(sd *varname*)	Standard deviation of *varname*
contents(sum *varname*)	Sum of *varname*

`contents(rawsum varname)`	Sums ignoring optionally specified weight
`contents(count varname)`	Count of nonmissing observations of *varname*
`contents(n varname)`	Same as `count`
`contents(max varname)`	Maximum of *varname*
`contents(min varname)`	Minimum of *varname*
`contents(median varname)`	Median of *varname*
`contents(iqr varname)`	Interquartile range (IQR) of *varname*
`contents(p1 varname)`	1st percentile of *varname*
`contents(p2 varname)`	2nd percentile of *varname* (so forth to `p99`)

The next section illustrates several more of these options.

Tables of Means, Medians, and Other Summary Statistics

`tabulate` readily produces tables of means and standard deviations within categories of the tabulated variable. For example, to form a one-way table with means of *lived* within each category of *meetings*, type

`. tabulate meetings, summ(lived)`

```
  Attended |
meetings on |     Summary of Years lived in town
  pollution |        Mean    Std. Dev.        Freq.
------------+----------------------------------------
         no |   21.509434    17.743809          106
        yes |   14.212766    13.911109           47
------------+----------------------------------------
      Total |   19.267974    16.954663          153
```

Meetings attenders appear to be relative newcomers, averaging 14.2 years in town, compared with 21.5 years for those who did not attend.

We can also use **`tabulate`** to form a two-way table of means by typing

`. tabulate meetings kids, sum(lived) means`

```
                      Means of Years lived in town

  Attended |
  meetings |    Have children <19
        on |        in town?
  pollution |       no          yes |      Total
------------+--------------------------+----------
         no |  28.307692    14.962963 |  21.509434
        yes |  23.363636    11.416667 |  14.212766
------------+--------------------------+----------
      Total |  27.444444    13.544444 |  19.267974
```

Both parents and nonparents among the meeting attenders tend to have lived fewer years in town, so the newcomer/oldtimer division noticed in the previous table is not a spurious reflection of the fact that parents with young children were more likely to attend.

The **means** option used above called for a table containing only means. Otherwise we get a bulkier table with means, standard deviations, and frequencies in each cell. Chapter 5 describes statistical tests for hypotheses about subgroup means.

Although it performs no tests, **table** nicely builds up to seven-way tables containing means, standard deviations, sums, medians, or other statistics (see the option list in previous section). Here is a one-way table showing means of *lived* within categories of *meetings*:

```
. table meetings, contents(mean lived)

-------------------------------
Attended  |
meetings  |
on        |
pollution | mean(lived)
----------+--------------------
      no  |    21.5094
     yes  |    14.2128
-------------------------------
```

A two-way table of means is a straightforward extension:

```
. table meetings kids, contents(mean lived)

----------------------------------
Attended  |
meetings  |Have children <19
on        |    in town?
pollution |    no        yes
----------+-----------------------
      no  | 28.3077    14.963
     yes  | 23.3636    11.4167
----------------------------------
```

Table cells can contain more than one statistic. Suppose we wanted a two-way table with both means and medians of the variable *lived*:

```
. table meetings kids, contents(mean lived median lived)

----------------------------------
Attended  |
meetings  |Have children <19
on        |    in town?
pollution |    no        yes
----------+-----------------------
      no  | 28.3077    14.963
          |    27.5      12.5
          |
     yes  | 23.3636    11.4167
          |      21         6
----------------------------------
```

The medians in the table above confirm our earlier finding based on means: the meeting attenders, both parents and nonparents, tended to have lived fewer years in town than their non-attending counterparts. Medians within each cell are less than the means, reflecting the positive skew (means pulled up by a few long-time residents) of the variable *lived*.

The cell contents shown by **table** could be means, medians, sums, or other summary statistics for two or more different variables.

Using Frequency Weights

`summarize` , `tabulate` , `table` , and related commands can be used with frequency weights that indicate the number of replicated observations. For example, file *sextab2.dta* contains results from a British survey of sexual behavior (Johnson et al. 1992). It apparently has 48 observations:

```
Contains data from C:\data\sextab2.dta
  obs:            48                         British sex survey (Johnson 92)
  vars:            4                         29 Jun 2001 11:26
  size:          432 (99.9% of memory free)
-------------------------------------------------------------------------------
              storage  display   value
variable name  type    format    label     variable label
-------------------------------------------------------------------------------
age            byte    %8.0g     age       Age
gender         byte    %8.0g     gender    Gender
lifepart       byte    %8.0g     partners  # heterosex partners lifetime
count          int     %8.0g               Number of individuals
-------------------------------------------------------------------------------
Sorted by:  age  lifepart  gender
```

One variable, *count*, indicates the number of individuals with each combination of characteristics, so this small dataset actually contains information from over 18,000 respondents. For example, 405 respondents were male, ages 16 to 24, and reported having no heterosexual partners so far in their lives.

```
. list in 1/5

        age    gender  lifepart    count
1.    16-24      male      none      405
2.    16-24    female      none      465
3.    16-24      male       one      323
4.    16-24    female       one      606
5.    16-24      male       two      194
```

We use *count* as a frequency weight to create a cross-tabulation of *lifepart* by *gender*:

```
. tabulate lifepart gender [fw = count]

        # |
heterosex |
partners  |         Gender
lifetime  |    male      female |     Total
----------+----------------------+----------
    none  |     544         586 |      1130
     one  |    1734        4146 |      5880
     two  |     887        1777 |      2664
     3-4  |    1542        1908 |      3450
     5-9  |    1630        1364 |      2994
     10+  |    2048         708 |      2756
----------+----------------------+----------
   Total  |    8385       10489 |     18874
```

The usual `tabulate` options work as expected with frequency weights. Here is the same table showing column percentages instead of frequencies:

```
. tabulate lifepart gender [fweight = count], column nof
```

```
        # |
heterosex |
 partners |        Gender
 lifetime |     male     female |      Total
----------+---------------------+----------
     none |     6.49       5.59 |       5.99
      one |    20.68      39.53 |      31.15
      two |    10.58      16.94 |      14.11
      3-4 |    18.39      18.19 |      18.28
      5-9 |    19.44      13.00 |      15.86
      10+ |    24.42       6.75 |      14.60
----------+---------------------+----------
    Total |   100.00     100.00 |     100.00
```

Other types of weights such as probability or analytical weights do not work as well with **tabulate** because their meanings are unclear regarding the command's principal options.

A different application of frequency weights can be demonstrated with **summarize**. File *college1.dta* contains information on a random sample consisting of 11 U.S. colleges, drawn from *Barron's Compact Guide to Colleges* (1992).

```
Contains data from C:\data\college1.dta
  obs:           11                          Colleges sample 1 (Barron's 92)
  vars:           5                          29 Jun 2001 11:40
  size:         429 (99.9% of memory free)
-------------------------------------------------------------------------------
              storage  display   value
variable name  type    format    label    variable label
-------------------------------------------------------------------------------
school         str28    %28s               College or university
enroll         int      %8.0g              Full-time students 1991
pctmale        byte     %8.0g              Percent male 1991
msat           int      %8.0g              Average math SAT
vsat           int      %8.0g              Average verbal SAT
-------------------------------------------------------------------------------
Sorted by:
```

The variables include *msat*, the mean math Scholastic Aptitude Test score at each of the 11 schools:

```
. list school enroll msat
```

```
                           school    enroll     msat
  1.           Brown University      5550      680
  2.                U. Scranton      3821      554
  3.  U. North Carolina/Asheville    2035      540
  4.           Claremont College      849      660
  5.           DePaul University     6197      547
  6.     Thomas Aquinas College       201      570
  7.            Davidson College     1543      640
  8.        U. Michigan/Dearborn     3541      485
  9.        Mass. College of Art      961      482
 10.             Oberlin College     2765      640
 11.         American University     5228      587
```

We can easily find the mean *msat* value among these 11 schools by typing

```
. summarize msat

    Variable |      Obs        Mean    Std. Dev.       Min        Max
-------------+--------------------------------------------------------
        msat |       11    580.4545    67.63189        482        680
```

This summary table gives each school's mean math SAT score the same weight. DePaul University, however, has 30 times as many students as Thomas Aquinas College. To take the different enrollments into account we could weight by *enroll*,

```
. summarize msat [fweight = enroll]

    Variable |      Obs        Mean    Std. Dev.       Min        Max
-------------+--------------------------------------------------------
        msat |    32691    583.064    63.10665        482        680
```

Typing

```
. summarize msat [freq = enroll]
```

would accomplish the same thing.

The enrollment-weighted mean, unlike the unweighted mean, is equivalent to the mean for the 32,691 students at these colleges (assuming they all took the SAT). Note, however, that we could not say the same thing about the standard deviation, minimum, or maximum. Apart from the mean, most individual-level statistics cannot be calculated simply by weighting data that already are aggregated. Thus, we need to use weights with caution. They might make sense in the context of one particular analysis, but seldom do so for the dataset as a whole, when many different kinds of analyses are needed.

5

ANOVA and Other Comparison Methods

Analysis of variance (ANOVA) encompasses a set of methods for testing hypotheses about differences between means. Its applications range from simple analyses where we compare the means of y across categories of x, to more complicated situations with multiple categorical and measurement x variables. t tests for hypotheses regarding a single mean (one-sample) or a pair of means (two-sample) correspond to elementary forms of ANOVA.

Rank-based "nonparametric" tests, including sign, Mann–Whitney, and Kruskal–Wallis, take a different approach to comparing distributions. These tests make weaker assumptions about measurement, distribution shape, and spread. Consequently, they remain valid under a wider range of conditions than ANOVA and its "parametric" relatives. Careful analysts sometimes use parametric and nonparametric tests together, checking to see whether both point toward the same conclusions. Further troubleshooting is called for when parametric and nonparametric results disagree.

anova is the first of Stata's model-fitting commands to be introduced in this book. Like the others, it has considerable flexibility and can fit a wide variety of models. **anova** can estimate one-way and N-way ANOVA or analysis of covariance (ANCOVA) for balanced and unbalanced designs, including designs with missing cells. It can also estimate factorial, nested, mixed, or repeated-measures designs. One follow-up command, **predict**, calculates predicted values, several types of residuals, and assorted standard errors and diagnostic statistics after **anova**. Another follow-up command, **test**, obtains tests of user-specified hypotheses. Both **predict** and **test** work similarly with other Stata model-fitting commands, such as **regress** (Chapter 6).

Example Commands

. `anova y x1 x2`

   Performs two-way ANOVA, testing for differences among the means of y across categories of $x1$ and $x2$.

. `anova y x1 x2 x1*x2`

   Performs a two-way factorial ANOVA, including both the main and interaction ($x1*x2$) effects of categorical variables $x1$ and $x2$.

. `anova y x1 x2 x3 x1*x2 x1*x3 x2*x3 x1*x2*x3`

Performs a three-way factorial ANOVA, including the three-way interaction *x1*x2*x3*, as well as all two-way interactions and main effect.

. `anova reading curriculum / teacher|curriculum`

Fits a nested model to test the effects of three types of curriculum on students' reading ability (*reading*). *teacher* is nested within *curriculum* (`teacher|curriculum`) because several different teachers were assigned to each curriculum. The *Reference Manual* provides other nested ANOVA examples, including a split-plot design.

. `anova headache subject medication, repeated(medication)`

Fits a repeated-measures ANOVA model to test the effects of three types of headache medication (*medication*) on the severity of subjects' headaches (*headache*). The sample consists of 20 subjects who report suffering from frequent headaches. Each subject tried each of the three medications at separate times during the study.

. `anova y x1 x2 x3 x4 x2*x3, continuous(x3 x4) regress`

Performs analysis of covariance (ANCOVA) with four independent variables, two of them (*x1* and *x2*) categorical and two of them (*x3* and *x4*) measurements. Includes the *x2*x3* interaction, and shows results in the form of a regression table instead of the default ANOVA table.

. `kwallis y, by(x)`

Performs a Kruskal–Wallis test of the null hypothesis that *y* has identical distributions across the *k* categories of *x* ($k > 2$).

. `oneway y x`

Performs a one-way analysis of variance (ANOVA), testing for differences among the means of *y* across categories of *x*. The same analysis, with a different output table, is produced by `anova y x`.

. `oneway y x, tabulate scheffe`

Performs one-way ANOVA, including a table of sample means and Scheffé multiple-comparison tests in the output.

. `ranksum y, by(x)`

Performs a Wilcoxon rank-sum test (also known as a Mann–Whitney *U* test) of the null hypothesis that *y* has identical rank distributions for both categories of dichotomous variable *x*. If we can assume that both rank distributions possess the same shape, this amounts to a test for whether the two medians of *y* are equal.

. `serrbar ymean se x, scale(2)`

Constructs a standard-error-bar plot from a dataset of means. Variable *ymean* holds the group means of *y*; *se* the standard errors; and *x* the values of categorical variable *x*. `scale(2)` asks for bars extending to ±2 standard errors around each mean (default is ±1 standard error).

. `signrank y1 = y2`

Performs a Wilcoxon matched-pairs signed-rank test for the equality of the rank distributions of *y1* and *y2*. We could test whether the median of *y1* differs from a constant such as 23.4 by typing the command `signrank y1 = 23.4` .

. **signtest y1 = y2**

   Tests the equality of the medians of *y1* and *y2* (assuming matched data; that is, both variables measured on the same sample of observations). Typing **signtest y1 = 5** would perform a sign test of the null hypothesis that the median of *y1* equals 5.

. **ttest y = 5**

   Performs a one-sample *t* test of the null hypothesis that the population mean of *y* equals 5.

. **ttest y1 = y2**

   Performs a one-sample (paired difference) *t* test of the null hypothesis that the population mean of *y1* equals that of *y2*. The default form of this command assumes that the data are paired. With unpaired data (*y1* and *y2* are measured from two independent samples), add the option **unpaired**.

. **ttest y, by(x) unequal**

   Performs a two-sample *t* test of the null hypothesis that the population mean of *y* is the same for both categories of variable *x*. Does not assume that the populations have equal variances. (Without the **unequal** option, **ttest** does assume equal variances.)

One-Sample Tests

One-sample *t* tests have two seemingly different applications:

1. Testing whether a sample mean $\bar{y}$ differs significantly from an hypothesized value μ_0.
2. Testing whether the means of y_1 and y_2, two variables measured over the same set of observations, differ significantly from each other. This is equivalent to testing whether the mean of a "difference score" variable created by subtracting y_1 from y_2 equals zero.

We use essentially the same formulas for either application, although the second starts with information on two variables instead of one.

   The data in *writing.dta* were collected to evaluate a college writing course based on word processing (Nash and Schwartz 1987). Measures such as the number of sentences completed in timed writing were collected both before and after students took the course. The researchers wanted to know whether the post-course measures showed improvement.

```
Contains data from C:\data\writing.dta
  obs:            24                          Nash and Schwartz (1987)
  vars:            9                          29 Jun 2001 13:47
  size:          312 (99.6% of memory free)
-------------------------------------------------------------------------------
              storage  display    value
variable name   type   format     label      variable label
-------------------------------------------------------------------------------
id             byte    %8.0g      slbl       Student ID
preS           byte    %8.0g                 # of sentences (pre-test)
preP           byte    %8.0g                 # of paragraphs (pre-test)
preC           byte    %8.0g                 Coherence scale 0-2 (pre-test)
preE           byte    %8.0g                 Evidence scale 0-6 (pre-test)
postS          byte    %8.0g                 # of sentences (post-test)
postP          byte    %8.0g                 # of paragraphs (post-test)
postC          byte    %8.0g                 Coherence scale 0-2 (post-test)
postE          byte    %8.0g                 Evidence scale 0-6 (post-test)
-------------------------------------------------------------------------------
Sorted by:
```

Suppose that we knew that students in previous years were able to complete an average of 10 sentences. Before examining whether the students in *writing.dta* improved during the course, we might want to learn whether at the start of the course they were essentially like earlier students—that is, whether their pre-test (*preS*) mean differs significantly from the mean of previous students (10). To see a one-sample *t* test of $H_0:\mu = 10$, type

`. ttest preS = 10`

```
One-sample t test

-------------------------------------------------------------------------------
Variable |    Obs       Mean    Std. Err.    Std. Dev.    [95% Conf. Interval]
---------+---------------------------------------------------------------------
   preS |     24    10.79167    .9402034     4.606037     8.846708    12.73663
-------------------------------------------------------------------------------
Degrees of freedom: 23

                          Ho: mean(preS) = 10

   Ha: mean < 10              Ha: mean ~= 10              Ha: mean > 10
     t =   0.8420               t =   0.8420               t =   0.8420
  P < t =   0.7958          P > |t| =   0.4084          P > t =   0.2042
```

The notation `P > t` means "the probability of a greater value of *t*"—that is, the one-tail test probability. The two-tail probability of a greater absolute *t* appears as `P > |t| = .4084`. Because this probability is high, we have no reason to reject $H_0:\mu = 10$. Note that **ttest** automatically provides a 95% confidence interval for the mean. We could get a different confidence interval, such as 99%, by adding a **level(99)** option to this command.

A nonparametric counterpart, the sign test, employs the binomial distribution to test hypotheses about single medians. For example, we could test whether the median of *preS* equals 10. **signtest** gives us no reason to reject that null hypothesis either.

`. signtest preS = 10`

```
Sign test

       sign |   observed    expected
------------+-----------------------
   positive |      12          11
   negative |      10          11
       zero |       2           2
------------+-----------------------
        all |      24          24

One-sided tests:
  Ho: median of preS - 10 = 0 vs.
  Ha: median of preS - 10 > 0
      Pr(#positive >= 12) =
        Binomial(n = 22, x >= 12, p = 0.5) =   0.4159

  Ho: median of preS - 10 = 0 vs.
  Ha: median of preS - 10 < 0
      Pr(#negative >= 10) =
        Binomial(n = 22, x >= 10, p = 0.5) =   0.7383

Two-sided test:
  Ho: median of preS - 10 = 0 vs.
  Ha: median of preS - 10 ~= 0
      Pr(#positive >= 12 or #negative >= 12) =
        min(1, 2*Binomial(n = 22, x >= 12, p = 0.5)) =   0.8318
```

Like **ttest**, **signtest** includes right-tail, left-tail, and two-tail probabilities. Unlike the symmetrical *t* distributions used by **ttest**, however, the binomial distributions used by **signtest** have different left- and right-tail probabilities. In this example, only the two-tail probability matters because we were testing whether the *writing.dta* students "differ" from their predecessors.

Next, we can test for improvement during the course by testing the null hypothesis that the mean number of sentences completed before and after the course (that is, the means of *preS* and *postS*) are equal. The **ttest** command accomplishes this as well, finding a significant improvement.

```
. ttest postS = preS

Paired t test
```

Variable	Obs	Mean	Std. Err.	Std. Dev.	[95% Conf. Interval]	
postS	24	26.375	1.693779	8.297787	22.87115	29.87885
preS	24	10.79167	.9402034	4.606037	8.846708	12.73663
diff	24	15.58333	1.383019	6.775382	12.72234	18.44433

```
                    Ho: mean(postS - preS) = mean(diff) = 0

 Ha: mean(diff) < 0          Ha: mean(diff) ~= 0           Ha: mean(diff) > 0
     t =  11.2676                t =  11.2676                  t =  11.2676
 P < t =   1.0000            P > |t| =   0.0000            P > t =   0.0000
```

Because we expect "improvement," not just "difference" between the *preS* and *postS* means, a one-tail test is appropriate. The displayed one-tail probability rounds off four decimal places to zero ("0.0000" really means $P < .00005$). Students' mean sentence completion does significantly improve. Based on this sample, we are 95% confident that it improves by between 12.7 and 18.4 sentences.

t tests assume that variables follow a normal distribution. This assumption is usually not critical because the tests are moderately robust. When nonnormality involves severe outliers, however, or occurs in small samples, we might be safer turning to medians instead of means and employing a nonparametric test that does not assume normality. The Wilcoxon signed-rank test, for example, assumes only that the distributions are symmetrical and continuous. Applying a signed-rank test to these data yields essentially the same conclusion as **ttest**, that students' sentence completion significantly improved. Because both tests agree on this conclusion, we can assert it with more assurance.

```
. signrank postS = preS

Wilcoxon signed-rank test
```

sign	obs	sum ranks	expected
positive	24	300	150
negative	0	0	150
zero	0	0	0
all	24	300	300

```
unadjusted variance      1225.00
adjustment for ties        -1.63
adjustment for zeros        0.00
                         ----------
adjusted variance        1223.38

Ho: postS = preS
             z =    4.289
    Prob > |z| =   0.0000
```

Two-Sample Tests

The remainder of this chapter draws examples from a survey of college undergraduates from Ward and Ault (1990) (*student2.dta*).

```
Contains data from C:\data\student2.dta
  obs:           243                    Student survey (Ward & Ault 1990)
  vars:           19                    29 Jun 2001 13:57
  size:        6,561 (99.3% of memory free)
-------------------------------------------------------------------------
              storage  display    value
variable name   type   format     label    variable label
-------------------------------------------------------------------------
id              int     %8.0g               Student ID
year            byte    %8.0g      year     Year in college
age             byte    %8.0g               Age at last birthday
gender          byte    %9.0g      s        Gender (male)
major           byte    %8.0g               Student major
relig           byte    %8.0g      v4       Religious preference
drink           byte    %9.0g               33-point drinking scale
gpa             float   %9.0g               Grade Point Average
grades          byte    %8.0g      grades   Guessed grades this semester
belong          byte    %8.0g      belong   Belong to fraternity/sorority
live            byte    %8.0g      v10      Where do you live?
miles           byte    %8.0g               How many miles from campus?
study           byte    %8.0g               Avg. hours/week studying
athlete         byte    %8.0g      yes      Are you a varsity athlete?
employed        byte    %8.0g      yes      Are you employed?
allnight        byte    %8.0g      allnight How often study all night?
ditch           byte    %8.0g      times    How many class/month ditched?
hsdrink         byte    %9.0g               High school drinking scale
aggress         byte    %9.0g               Aggressive behavior scale
-------------------------------------------------------------------------
Sorted by:  id
```

About 19% of these students belong to a fraternity or sorority:

```
. tabulate belong

   Belong to |
  fraternity/ |
    sorority |     Freq.     Percent        Cum.
------------+-----------------------------------
      member |        47       19.34       19.34
   nonmember |       196       80.66      100.00
------------+-----------------------------------
       Total |       243      100.00
```

Another variable, *drink*, measures how often and heavily a student drinks alcohol, on a 33-point scale. Campus rumors might lead one to suspect that fraternity/sorority members tend to differ from other students in their drinking behavior. Boxplots comparing the *drink* values reported by fraternity and sorority members and nonmembers (Figure 5.1) appear consistent with these rumors.

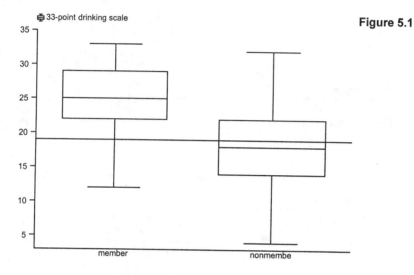

Figure 5.1

The **ttest** command, used earlier for one-sample and paired-difference tests, can perform two-sample tests as well. In this application its general syntax is **ttest** *measurement*, **by(***categorical***)**. For example,

. **ttest** *drink*, **by(***belong***)**

Two-sample t test with equal variances

Group	Obs	Mean	Std. Err.	Std. Dev.	[95% Conf. Interval]	
member	47	24.7234	.7124518	4.884323	23.28931	26.1575
nonmembe	196	17.7602	.4575013	6.405018	16.85792	18.66249
combined	243	19.107	.431224	6.722117	18.25756	19.95643
diff		6.9632	.9978608		4.997558	8.928842

Degrees of freedom: 241

                  Ho: mean(member) - mean(nonmembe) = diff = 0

Ha: diff < 0	Ha: diff ~= 0	Ha: diff > 0
t = 6.9781	t = 6.9781	t = 6.9781
P < t = 1.0000	P > |t| = 0.0000	P > t = 0.0000

As the output notes, this *t* test rests on an equal-variances assumption. But the fraternity and sorority members' sample standard deviation appears somewhat lower—they are more alike than nonmembers in their reported drinking behavior. To perform a similar test without assuming equal variances, add the option **unequal**:

```
. ttest drink, by(belong) unequal

Two-sample t test with unequal variances

---------------------------------------------------------------------------
   Group |      Obs        Mean    Std. Err.   Std. Dev.   [95% Conf. Interval]
---------+-----------------------------------------------------------------
  member |       47     24.7234    .7124518    4.884323    23.28931     26.1575
nonmembe |      196     17.7602    .4575013    6.405018    16.85792    18.66249
---------+-----------------------------------------------------------------
combined |      243      19.107    .431224     6.722117    18.25756    19.95643
---------+-----------------------------------------------------------------
    diff |               6.9632    .8466965                5.280627    8.645773
---------------------------------------------------------------------------
Satterthwaite's degrees of freedom:    88.22

          Ho: mean(member) - mean(nonmembe) = diff = 0

   Ha: diff < 0                Ha: diff ~= 0                Ha: diff > 0
      t =   8.2240                t =   8.2240                t =   8.2240
  P < t =   1.0000           P > |t| =   0.0000           P > t =   0.0000
```

Adjusting for unequal variances does not alter our basic conclusion that members and nonmembers are significantly different. We can further check this conclusion by trying a nonparametric Mann–Whitney U test, also known as a Wilcoxon rank-sum test. Assuming that the rank distributions have similar shape, the rank-sum test here indicates that we can reject the null hypothesis of equal population medians.

```
. ranksum drink, by(belong)

Two-sample Wilcoxon rank-sum (Mann-Whitney) test

      belong |     obs    rank sum    expected
-------------+----------------------------------
      member |      47        8535        5734
   nonmember |     196       21111       23912
-------------+----------------------------------
    combined |     243       29646       29646

unadjusted variance     187310.67
adjustment for ties       -472.30
                        ----------
adjusted variance       186838.36

Ho: drink(belong==member) = drink(belong==nonmember)
          z =    6.480
  Prob > |z| =   0.0000
```

One-Way Analysis of Variance (ANOVA)

Analysis of variance (ANOVA) provides another way, more general than t tests, to test for differences among means. The simplest type, one-way ANOVA, tests whether the means of y differ across categories of x. For example,

. **oneway** *drink belong*, **tabulate**

```
    Belong to |
 fraternity/  | Summary of 33-point drinking scale
   sorority   |      Mean   Std. Dev.        Freq.
--------------+-----------------------------------
     member   |  24.723404   4.8843233           47
  nonmember   |  17.760204   6.4050179          196
--------------+-----------------------------------
      Total   |  19.106996   6.7221166          243
```

```
                       Analysis of Variance
     Source              SS        df      MS            F      Prob > F
-----------------------------------------------------------------------
Between groups      1838.08426      1  1838.08426      48.69    0.0000
Within groups       9097.13385    241  37.7474433
-----------------------------------------------------------------------
      Total         10935.2181    242  45.1868517
```

Bartlett's test for equal variances: chi2(1) =   4.8378 Prob>chi2 = 0.028

The **tabulate** option produces a table of means and standard deviations in addition to the analysis of variance table itself. One-way ANOVA with a dichotomous x variable is equivalent to a two-sample t test, and its F statistic equals the corresponding t statistic squared. **oneway** offers more options and processes faster, but it lacks **ttest**'s **unequal** option for abandoning the equal-variances assumption.

oneway formally tests the equal-variances assumption, using Bartlett's χ^2. A low Bartlett's probability implies that ANOVA's equal-variance assumption is implausible, in which case we should not trust the ANOVA F test results. In the **oneway** *drink belong* example above, Bartlett's $P = .028$ casts doubt on the ANOVA finding of a significant difference between means.

ANOVA's real value lies not in two-sample comparisons but, rather, in more complicated comparisons of three or more means. For example, we could test whether mean drinking behavior varies by year in college. Figure 5.2 shows these four distributions as boxplots. The following command tests whether the four means are equal.

. **oneway** *drink year*, **tabulate scheffe**

```
   Year in |  Summary of 33-point drinking scale
   college |      Mean   Std. Dev.        Freq.
-----------+-----------------------------------
  Freshman |    18.975   6.9226033           40
 Sophomore | 21.169231   6.5444853           65
    Junior | 19.453333   6.2866081           75
    Senior | 16.650794   6.6409257           63
-----------+-----------------------------------
     Total | 19.106996   6.7221166          243
```

```
                       Analysis of Variance
     Source              SS        df      MS            F      Prob > F
-----------------------------------------------------------------------
Between groups       666.200518     3  222.066839       5.17    0.0018
Within groups       10269.0176    239  42.9666008
-----------------------------------------------------------------------
      Total         10935.2181    242  45.1868517
```

Bartlett's test for equal variances: chi2(3) =   0.5103 Prob>chi2 = 0.917

```
               Comparison of 33-point drinking scale by Year in college
                                    (Scheffe)
Row Mean-|
Col Mean |    Freshman    Sophomor      Junior
---------+-----------------------------------------
Sophomor |    2.19423
         |      0.429
         |
  Junior |    .478333     -1.7159
         |      0.987       0.498
         |
  Senior |   -2.32421     -4.51844    -2.80254
         |      0.382        0.002       0.103
```

We can reject the hypothesis of equal means ($P = .0018$), but not the hypothesis of equal variances ($P = .917$). The latter (supported by the boxplots of Figure 5.2) is "good news" regarding the ANOVA's validity.

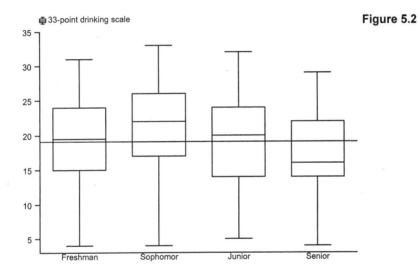

Figure 5.2

The **scheffe** option (Scheffé multiple-comparison test) produces a table showing the differences between each pair of means. The freshman mean equals 18.975 and the sophomore mean equals 21.16923, so the sophomore–freshman difference is $21.16923 - 18.975 = 2.19423$, not statistically distinguishable from zero ($P = .429$). Of the six contrasts in this table, only the senior–sophomore difference, $16.6508 - 21.1692 = -4.5184$, is significant ($P = .002$). Thus, our overall conclusion that these four groups' means are not the same arises mainly from the contrast between seniors (the lightest drinkers) and sophomores (the heaviest).

oneway offers three multiple-comparison options: **scheffe**, **bonferroni**, or **sidak** (see the *Reference Manual* for definitions). The Scheffé test remains valid under a wider variety of conditions, although it is sometimes less sensitive.

The Kruskal–Wallis test (**kwallis**), a K-sample generalization of the two-sample rank-sum test, provides a nonparametric alternative to one-way ANOVA. It tests the null hypothesis of equal population medians.

```
. kwallis drink, by(year)

Test: Equality of populations (Kruskal-Wallis test)

       year         _Obs    _RankSum
   Freshman          40      4914.00
  Sophomore          65      9341.50
     Junior          75      9300.50
     Senior          63      6090.00

chi-squared =     14.453 with 3 d.f.
probability =      0.0023

chi-squared with ties =    14.490 with 3 d.f.
probability =      0.0023
```

Here, the **kwallis** results ($P = .0023$) agree with our **oneway** findings of significant differences in *drink* by year in college. Kruskal–Wallis is generally safer than ANOVA if we have reason to doubt ANOVA's equal-variances or normality assumptions, or if we suspect problems caused by outliers. **kwallis**, like **ranksum**, makes the weaker assumption of similar-shaped distributions within each group. In principle, **ranksum** and **kwallis** should produce similar results when applied to two-sample comparisons, but in practice this is true only if the data contain no ties. **ranksum** incorporates an exact method for dealing with ties, which makes it preferable for two-sample problems.

Two- and *N*-Way Analysis of Variance

One-way ANOVA examines how the means of one measurement variable *y* vary across categories of a second variable *x*. *N*-way ANOVA generalizes this approach to deal with two or more categorical *x* variables. For example, we might consider how drinking behavior varies not only by fraternity or sorority membership, but also by gender. We start by examining a two-way table of means:

```
. table belong gender, contents(mean drink) row col

-----------------------------------------
Belong to |
fraternit |
y/sororit |        Gender (male)
y         |   Female     Male      Total
----------+------------------------------
   member | 22.44444  26.13793   24.7234
nonmember | 16.51724  19.5625    17.7602
          |
    Total | 17.31343  21.31193   19.107
-----------------------------------------
```

It appears that in this sample, males drink more than females and members drink more than nonmembers. The member–nonmember difference appears similar among males and females. Stata's *N*-way ANOVA command, **anova**, can test for significant differences among these means attributable to belonging to a fraternity or sorority, gender, or the interaction of belonging and gender (written *belong*gender*):

. **anova** *drink belong gender belong*gender*

```
                    Number of obs =      243    R-squared     =  0.2221
                    Root MSE      = 5.96592    Adj R-squared =  0.2123

       Source |  Partial SS    df       MS              F     Prob > F
--------------+----------------------------------------------------------
        Model |  2428.67237     3   809.557456         22.75   0.0000
              |
       belong |   1406.2366     1    1406.2366         39.51   0.0000
       gender |  408.520097     1   408.520097         11.48   0.0008
 belong*gender|  3.78016612     1   3.78016612          0.11   0.7448
              |
     Residual |  8506.54574   239   35.5922416
--------------+----------------------------------------------------------
        Total |  10935.2181   242   45.1868517
```

In this example of "two-way factorial ANOVA," the output shows significant main effects for *belong* ($P = .0000$) and *gender* ($P = .0008$), but their interaction contributes little to the model ($P = .7448$). This interaction cannot be distinguished from zero, so we might prefer to estimate a simpler model without the interaction term (results not shown):

. **anova** *drink belong gender*

To include any interaction term with **anova**, specify the variable names joined by *. Unless the number of observations with each combination of x values is the same (a condition called "balanced data"), it can be hard to interpret the main effects in a model that also includes interactions. This does not mean that the main effects in such models are unimportant, however. Regression analysis might help to make sense of complicated ANOVA results, as illustrated in the following section.

Analysis of Covariance (ANCOVA)

Analysis of Covariance (ANCOVA) extends N-way ANOVA to encompass a mix of categorical and continuous x variables. This is accomplished through the **anova** command if we specify which variables are continuous. For example, when we include *gpa* (college grade point average) among the independent variables, we find that it, too, is related to drinking behavior.

. **anova** *drink belong gender gpa*, continuous(*gpa*)

```
                    Number of obs =      218    R-squared     =  0.2970
                    Root MSE      = 5.68939    Adj R-squared =  0.2872

       Source |  Partial SS    df       MS              F     Prob > F
--------------+----------------------------------------------------------
        Model |  2927.03087     3   975.676958         30.14   0.0000
              |
       belong |  1489.31999     1   1489.31999         46.01   0.0000
       gender |  405.137843     1   405.137843         12.52   0.0005
          gpa |   407.0089      1    407.0089          12.57   0.0005
              |
     Residual |  6926.99206   214   32.3691218
--------------+----------------------------------------------------------
        Total |  9854.02294   217   45.4102439
```

From this analysis we know that a significant relation exists between *drink* and *gpa* when we control for *belong* and *gender*. Beyond their *F* tests for statistical significance, however, ANOVA or ANCOVA ordinarily do not provide much descriptive information about how variables are related. Regression, with its explicit model and parameter estimates, does a better descriptive job. Because ANOVA and ANCOVA amount to special cases of regression, we could restate these analyses in regression form. Stata does so automatically if we add the **regress** option to **anova**. For instance, we might want to see regression output in order to understand results from the following ANCOVA.

```
. anova drink belong gender belong*gender gpa, continuous(gpa)
    regress
```

Source	SS	df	MS
Model	2933.45823	4	733.364558
Residual	6920.5647	213	32.4909141
Total	9854.02294	217	45.4102439

Number of obs = 218
F(4, 213) = 22.57
Prob > F = 0.0000
R-squared = 0.2977
Adj R-squared = 0.2845
Root MSE = 5.7001

drink		Coef.	Std. Err.	t	P>|t|	[95% Conf. Interval]
_cons		27.47676	2.439962	11.26	0.000	22.6672 32.28633
belong	1	6.925384	1.286774	5.38	0.000	4.388942 9.461826
	2	(dropped)				
gender	1	-2.629057	.8917152	-2.95	0.004	-4.386774 -.8713407
	2	(dropped)				
gpa		-3.054633	.8593498	-3.55	0.000	-4.748552 -1.360713
belong*gender	1 1	-.8656158	1.946211	-0.44	0.657	-4.701916 2.970685
	1 2	(dropped)				
	2 1	(dropped)				
	2 2	(dropped)				

With the **regress** option, we get the **anova** output formatted as a regression table. The top part gives the same overall *F* test and R^2 as a standard ANOVA table. The bottom part describes the following regression:

> We construct a separate dummy variable {0,1} representing each category of each *x* variable, except for the highest categories, which are dropped. Interaction terms (if specified in the command's variable list) are constructed from the products of every possible combination of these dummy variables. Regress *y* on all these dummy variables and interactions, and also on any continuous variables specified in the command line.

The previous example therefore corresponds to a regression of *drink* on four *x* variables:

1. a dummy coded 1 = fraternity/sorority member, 0 otherwise (highest category of *belong*, nonmember, gets dropped);
2. a dummy coded 1 = female, 0 otherwise (highest category of *gender*, male, gets dropped);
3. the continuous variable *gpa*;
4. an interaction term coded 1 = sorority female, 0 otherwise.

Interpret the individual dummy variables' regression coefficients as effects on predicted or mean *y*. For example, the coefficient on the first category of *gender* (female) equals –2.629057. This informs us that the mean drinking scale levels for females are about 2.63 points lower than those of males with the same grade point average and membership status. And we know that among students of the same gender and membership status, mean drinking scale values decline by 3.054633 with each one-point increase in grades. Note also that we have confidence intervals and individual *t* tests for each coefficient; there is much more information in the **anova, regress** output than in the ANOVA table alone.

Predicted Values and Error-Bar Charts

After **anova**, the follow-up command **predict** calculates predicted values, residuals, or standard errors and diagnostic statistics. One application for such statistics is in drawing graphical representations of the model's predictions, in the form of error-bar charts. For a simple illustration, we return to the one-way ANOVA of *drink* by *year*:

```
. anova drink year
```

```
                   Number of obs =      243    R-squared     =  0.0609
                   Root MSE      = 6.55489    Adj R-squared =  0.0491

    Source |  Partial SS    df       MS              F     Prob > F
-----------+----------------------------------------------------
     Model |  666.200518     3   222.066839           5.17    0.0018
           |
      year |  666.200518     3   222.066839           5.17    0.0018
           |
  Residual |  10269.0176   239   42.9666008
-----------+----------------------------------------------------
     Total |  10935.2181   242   45.1868517
```

To calculate predicted means from the recent **anova**, type **predict** followed by a new variable name:

```
. predict drinkmean
(option xb assumed; fitted values)
. label variable drinkmean "Mean drinking scale"
```

With the **stdp** option, **predict** calculates standard errors of the predicted means:

```
. predict SEdrink, stdp
```

Using these new variables, we apply the **serrbar** command to create an error-bar chart. The **scale(2)** option tells **serrbar** to draw error bars of plus and minus two standard errors, from

$$drinkmean - 2 \times SEdrink$$

to

$$drinkmean + 2 \times SEdrink.$$

In a **serrbar** command, the first-listed variable should be the means or *y* variable; the second-listed, the standard error or standard deviation (depending on which you want to show); and the third-listed variable defines the *x* axis. Figure 5.3 shows the result of the following command:

```
. serrbar drinkmean SEdrink year, scale(2) connect(l) border sort
    ylabel(15,16 to 23) xlabel(1,2,3,4)
```

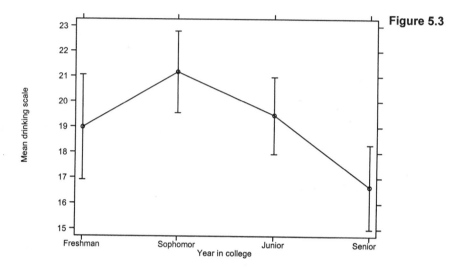

Figure 5.3

For a two-way factorial ANOVA, error-bar charts help us to visualize main and interaction effects. Below, we perform ANOVA and draw an error-bar chart to examine the relationship between students' aggressive behavior (*aggress*), gender, and year in college. Both the main effects of *gender* and *year*, and their interaction, are statistically significant.

```
. anova aggress gender year gender*year
```

		Number of obs =	243	R-squared	=	0.2503
		Root MSE	= 1.45652	Adj R-squared =		0.2280

Source	Partial SS	df	MS	F	Prob > F
Model	166.482503	7	23.7832147	11.21	0.0000
gender	94.3505972	1	94.3505972	44.47	0.0000
year	19.0404045	3	6.34680149	2.99	0.0317
gender*year	24.1029759	3	8.03432529	3.79	0.0111
Residual	498.538073	235	2.12143861		
Total	665.020576	242	2.74801891		

```
. predict aggmean
(option xb assumed; fitted values)

. label variable aggmean "Mean aggressive behavior scale"

. predict SEagg, stdp
```

```
. serrbar aggmean SEagg year, scale(2) by(gender) connect(l) sort
     ylabel(0,1,2,3) xlabel(1,2,3,4) border
     title (Aggressive behavior vs. year in college, by gender)
```

Figure 5.4

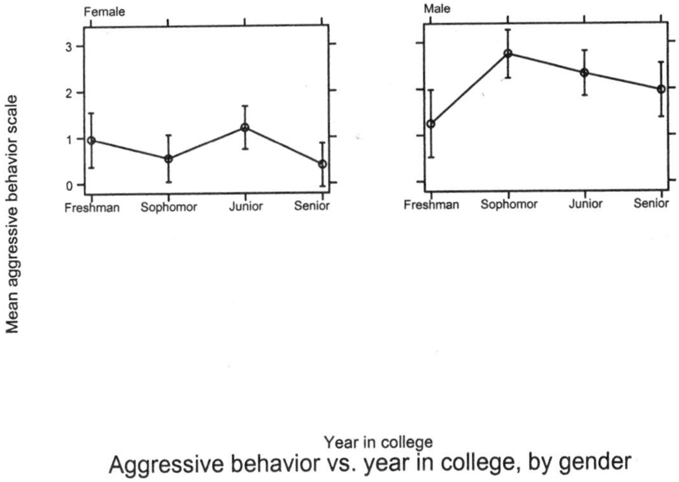

Aggressive behavior vs. year in college, by gender

The resulting error-bar chart (Figure 5.4) shows female means on the aggressive-behavior scale fluctuating at comparatively low levels during the four years of college. Male means are higher throughout, with a sophomore-year peak that resembles the pattern seen earlier for drinking (Figures 5.2 and 5.3).

predict works the same way with regression analysis (**regress**) as it does with **anova** because the two share a common mathematical framework. A list of some other **predict** options appears in Chapter 6, and further examples using these options are given in Chapter 7. The options include residuals that can be used to check assumptions regarding error distributions, and also a suite of diagnostic statistics (such as leverage, Cook's D, and *DFBETAS*) that measure the influence of individual observations on model results. The Durbin–Watson test (**dwstat**), described in Chapter 13, can also be used after **anova** to test for first-order autocorrelation. Conditional effect plotting (Chapter 7) provides a graphical approach that can aid interpretation of more complicated regression, ANOVA, or ANCOVA models.

6

Linear Regression Analysis

Stata offers an exceptionally broad range of regression procedures, from elementary to advanced. A partial list of the available commands can be seen by typing **help regress**. This chapter introduces **regress** and related commands that perform simple and multiple ordinary least squares (OLS) regression. A follow-up command, **predict**, calculates predicted values, residuals, and diagnostic statistics such as leverage or Cook's *D*. Another follow-up command, **test**, performs tests of user-specified hypotheses. **regress** can accomplish other analyses including weighted least squares and two-stage least squares. Regression with dummy variables, interaction effects, polynomial terms, and stepwise variable selection are covered briefly in this chapter, along with a first look at residual analysis.

Example Commands

. **regress y x**
Performs ordinary least squares (OLS) regression of variable *y* on one predictor, *x*.

. **regress y x if ethnic == 3 & income > 50**
Regresses *y* on *x* using only that subset of the data for which variable *ethnic* equals 3 and *income* is greater than 50.

. **predict yhat**
Generates a new variable (here arbitrarily named *yhat*) equal to the predicted values from the most recent regression.

. **predict e, resid**
Generates a new variable (arbitrarily named *e*) equal to the residuals from the most recent regression.

. **graph y yhat x, connect(.s) symbol(Ti)**
Draws a scatterplot with regression line using the variables *y*, *yhat*, and *x*.

. **graph e yhat, twoway box yline(0)**
Draws a residual versus predicted plot using the variables *e* and *yhat*. An alternative, more automatic way to draw such plots employs the **rvfplot** (residual-versus-fitted) command, discussed in Chapter 7.

. **regress y x1 x2 x3**
Performs multiple regression of *y* on three predictor variables, *x1*, *x2*, and *x3*.

. `regress y x1 x2 x3, robust`

Calculates robust (Huber/White) estimates of standard errors. See the *User's Guide* for details. The `robust` option works with some other model fitting commands as well.

. `regress y x1 x2 x3, beta`

Performs multiple regression and shows standardized regression coefficients (beta weights) in the output table.

. `correlate x1 x2 x3 y`

Displays a matrix of Pearson correlations, using only observations with no missing values on all of the variables specified. Adding the option `covariance` produces a variance–covariance matrix instead of correlations.

. `pwcorr x1 x2 x3 y, sig`

Displays a matrix of Pearson correlations, using pairwise deletion of missing values and showing probabilities from t tests of $H_0{:}\rho = 0$ on each correlation.

. `graph x1 x2 x3 y, matrix half`

Draws a scatterplot matrix. Because their variable lists are the same, this example yields a scatterplot matrix having the same organization as the correlation matrix produced by the preceding `pwcorr` command. Listing the dependent (y) variable last creates a matrix in which the bottom row forms a series of y-versus-x plots.

. `test x1 x2`

Performs an F test of the null hypothesis that coefficients on $x1$ and $x2$ both equal zero in the most recent regression model.

. `sw regress y x1 x2 x3, pr(.05)`

Performs stepwise regression using backward elimination until all remaining predictors are significant at the .05 level. All listed predictors are entered on the first iteration. Thereafter, each iteration drops one predictor with the highest P value, until all predictors remaining have probabilities below the "probability to retain," `pr(.05)`. Options permit forward or hierarchical selection. Stepwise variants exist for many other model-fitting commands as well; type `help sw` for a list.

. `regress y x1 x2 x3 [aweight = w]`

Performs weighted least squares (WLS) regression of y on $x1$, $x2$, and $x3$. Variable w holds the analytical weights, which work as if we had multiplied each variable and the constant by the square root of w, and then performed an ordinary regression. Analytical weights are often employed to correct for heteroskedasticity when the y and x variables are means, rates, or proportions, and w is the number of individuals making up each aggregate observation (e.g., city or school) in the data. If the y and x variables are individual-level, and the weights indicate numbers of replicated observations, then use frequency weights `[fweight = w]` instead. See `help svy` if the weights reflect design factors such as disproportionate sampling.

. `regress y1 y2 x (x z)`
. `regress y2 y1 z (x z)`

Estimates the reciprocal effects of $y1$ and $y2$, using instrumental variables x and z. The first parts of these commands specify the structural equations:

$$y1 = \alpha_0 + \alpha_1 y2 + \alpha_2 x + \epsilon_1$$
$$y2 = \beta_0 + \beta_1 y2 + \beta_2 w + \epsilon_2$$

The parentheses in the commands enclose variables that are exogenous to all of the structural equations. **regress** accomplishes two-stage least squares (2SLS) in this example.

. **svyreg y x1 x2 x3**

Regresses *y* on predictors *x1, x2,* and *x3,* with appropriate adjustments for a complex survey sampling design. We assume that a **svyset** command has previously been used to set up the data, by specifying the strata, clusters, and sampling probabilities. **help svy** lists the many procedures available for working with complex survey data. **help regress** outlines the syntax of this particular command; follow references to the *User's Guide* and *Reference Manuals* for details.

. **xtreg y x1 x2 x3 x4, re**

Estimates a panel (cross-sectional time series) model with random effects by generalized least squares (GLS). An observation in panel data consists of information about unit *i* at time *t*, and there are multiple observations (times) for each unit. Before using **xtreg**, the variable identifying the units was specified by an **iis** ("*i* is") command, and the variable identifying time by **tis** ("*t* is"). Once the data have been saved, these definitions are retained for future analysis by **xtreg** and other **xt** procedures. **help xt** lists available panel estimation procedures. **help xtreg** gives the syntax of this command and references to the printed documentation. If your data include many observations for each unit, a time-series approach could be more appropriate. Stata's time series procedures (introduced in Chapter 13) provide further tools for analyzing panel data.

The Regression Table

File *states.dta* contains educational data on the U.S. states and District of Columbia, including these variables:

. **describe state csat expense percent income high college region**

```
                storage  display   value
variable name   type     format    label      variable label
-------------------------------------------------------------------------
state           str20    %20s                 State
csat            int      %9.0g                Mean composite SAT score
expense         int      %9.0g                Per pupil expenditures prim&sec
percent         byte     %9.0g                % HS graduates taking SAT
income          long     %10.0g               Median household income
high            float    %9.0g                % over 25 w/HS diploma
college         float    %9.0g                % over 25 w/bachelor's degree +
region          byte     %9.0g     region     Geographical region
```

Political leaders occasionally use mean Scholastic Aptitude Test (SAT) scores to make pointed comparisons between the educational systems of different U.S. states. For example, some have raised the question of whether SAT scores are higher in states that spend more money on education. We might try to address this question by regressing mean composite SAT scores (*csat*) on per-pupil expenditures (*expense*). The appropriate Stata command has the form **regress y x**, where *y* is the predicted or dependent variable, and *x* the predictor or independent variable.

```
. regress csat expense

      Source |       SS       df       MS              Number of obs =
-------------+------------------------------           F(  1,    49) =     13.61
       Model |  48708.3001     1   48708.3001          Prob > F      =    0.0006
    Residual |  175306.21     49   3577.67775          R-squared     =    0.2174
-------------+------------------------------           Adj R-squared =    0.2015
       Total |  224014.51     50   4480.2902           Root MSE      =    59.814


        csat |      Coef.   Std. Err.      t    P>|t|     [95% Conf. Interval]
-------------+----------------------------------------------------------------
     expense |  -.0222756   .0060371    -3.69   0.001    -.0344077   -.0101436
       _cons |   1060.732   32.7009     32.44   0.000     995.0175    1126.447
```

This regression tells an unexpected story: the more money a state spends on education, the lower its students' mean SAT scores. Any causal interpretation is premature at this point, but the regression table does convey information about the linear statistical relation between *csat* and *expense*. At upper right it gives an overall *F* test, based on the sums of squares at the upper left. This *F* test evaluates the null hypothesis that coefficients on all *x* variables in the model (here there is only one *x* variable, *expense*) equal zero. The *F* statistic, 13.61 with 1 and 49 degrees of freedom, leads easily to rejection of this null hypothesis ($P = .0006$). Prob > F means "the probability of a greater *F* " statistic if we drew samples randomly from a population in which the null hypothesis is true.

At upper right, we also see the coefficient of determination, $R^2 = .2174$. Per-pupil expenditures explain about 22% of the variance in states' mean composite SAT scores. Adjusted R^2, $R^2_a = .2015$, takes into account the complexity of the model relative to the complexity of the data. This adjusted statistic is often more informative for research.

The lower half of the regression table gives the estimated model itself. We find coefficients (slope and *y*-intercept) in the first column, here yielding the prediction equation

   predicted *csat* = 1060.732 − .0222756*expense*

The second column lists estimated standard errors of the coefficients. These are used to calculate *t* tests (columns 3–4) and confidence intervals (columns 5–6) for each regression coefficient. The *t* statistics (coefficients divided by their standard errors) test null hypotheses that the corresponding population coefficients equal zero. At the $\alpha = .05$ significance level, we could reject this null hypothesis regarding both the coefficient on *expense* ($P = .001$) and the *y*-intercept (".000", really meaning $P < .0005$). Stata's modeling commands print 95% confidence intervals routinely, but we can request other levels by specifying the **level( )** option, as shown in the following:

```
. regress csat expense, level(99)
```

Because these data do not represent a random sample from some larger population of U.S. states, hypothesis tests and confidence intervals lack their usual meanings. They are discussed in this chapter anyway for purposes of illustration.

The term _cons stands for the regression constant, usually set at one. Stata automatically includes a constant unless we tell it not to. The **nocons** option causes Stata to suppress the constant, performing regression through the origin. For example,

```
. regress y x,   nocons
```

or,

. **regress y x1 x2 x3, nocons**

In certain advanced applications, you might need to specify your own constant. If the "independent variables" include a user-supplied constant (named *c*, for example), employ the **hascons** option instead of **nocons** :

. **regress y c x, hascons**

Using **nocons** in this situation would result in a misleading *F* test and R^2. Consult the *Reference Manual* or **help regress** for more about **hascons** .

Multiple Regression

Multiple regression allows us to estimate how *expense* predicts *csat*, while adjusting for a number of other possible predictor variables. We can incorporate other predictors of *csat* simply by listing these variables in the command

. **regress csat expense percent income high college**

```
      Source |       SS       df       MS              Number of obs =      51
-------------+------------------------------           F(  5,    45) =   42.23
       Model |  184663.309     5   36932.6617          Prob > F      = 0.0000
    Residual |  39351.2012    45   874.471137          R-squared     = 0.8243
-------------+------------------------------           Adj R-squared = 0.8048
       Total |   224014.51    50    4480.2902          Root MSE      = 29.571
```

```
-----------------------------------------------------------------------------
        csat |      Coef.   Std. Err.      t    P>|t|     [95% Conf. Interval]
-------------+---------------------------------------------------------------
     expense |   .0033528   .0044709     0.75   0.457    -.005652     .0123576
     percent |  -2.618177   .2538491   -10.31   0.000   -3.129455    -2.106898
      income |   .0001056   .0011661     0.09   0.928    -.002243     .0024542
        high |   1.630841    .992247     1.64   0.107    -.367647     3.629329
     college |   2.030894   1.660118     1.22   0.228   -1.312756     5.374544
       _cons |   851.5649   59.29228    14.36   0.000    732.1441     970.9857
-----------------------------------------------------------------------------
```

This gives us the multiple regression equation

$$\text{predicted } csat = 851.56 + .00335 expense - 2.618 percent + .0001 income + 1.63 high + 2.03 college$$

Controlling for four other variables weakens the coefficient on *expense* from −.0223 to .00335, which is no longer statistically distinguishable from zero. The unexpected negative relation between *expense* and *csat* found in our earlier simple regression evidently can be explained by other predictors.

Only the coefficient on *percent* (percentage of high school graduates taking the SAT) attains significance at the .05 level. We could interpret this "fourth-order partial regression coefficient" (so called because its calculation adjusts for four other predictors) as follows:

$b_2 = -2.618$: Predicted mean SAT scores decline by 2.618 points, with each one-point increase in the percentage of high school graduates taking the SAT—if *expense*, *income*, *high*, and *college* do not change.

Taken together, the five x variables in this model explain about 80% of the variance in states' mean composite SAT scores ($R_a^2 = .8048$). In contrast, our earlier simple regression with *expense* as the only predictor explained about 20% of the variance in *csat*.

To obtain standardized regression coefficients (beta weights) with any regression, add the **beta** option. Standardized coefficients are what we would see in a regression where all the variables had been transformed into standard scores (means 0, standard deviations 1).

```
. regress csat expense percent income high college, beta
```

```
      Source |       SS       df       MS              Number of obs =      51
-------------+------------------------------           F(  5,    45) =   42.23
       Model |  184663.309     5   36932.6617           Prob > F      =  0.0000
    Residual |  39351.2012    45   874.471137           R-squared     =  0.8243
-------------+------------------------------           Adj R-squared =  0.8048
       Total |   224014.51    50    4480.2902           Root MSE      =  29.571
```

```
        csat |      Coef.   Std. Err.       t    P>|t|                     Beta
-------------+----------------------------------------------         ------------
     expense |   .0033528   .0044709     0.75   0.457                    .070185
     percent |  -2.618177   .2538491   -10.31   0.000                  -1.024538
      income |   .0001056   .0011661     0.09   0.928                   .0101321
        high |   1.630841    .992247     1.64   0.107                   .1361672
     college |   2.030894   1.660118     1.22   0.228                   .1263952
       _cons |   851.5649   59.29228    14.36   0.000                          .
```

The standardized regression equation is

$$\text{predicted } csat^* = \quad .07 expense^* - 1.0245 percent^* + .01 income^* + \\ .136 high^* + .126 college^*$$

where *csat**, *expense**, etc. denote these variables in standard-score form. We might interpret the standardized coefficient on *percent*, for example, as follows:

$b_2^* = -1.0245$: Predicted mean SAT scores decline by 1.0245 standard deviations, with each one-standard-deviation increase in the percentage of high school graduates taking the SAT—if *expense*, *income*, *high*, and *college* do not change.

The F and t tests, R^2, and other aspects of the regression remain the same.

Predicted Values and Residuals

After any regression, the **predict** command can obtain predicted values, residuals, and other case statistics. Suppose we have just done a regression of composite SAT scores on their strongest single predictor:

```
. regress csat percent
```

Now, to create a new variable called *yhat* containing predicted *y* values from this regression type,

```
. predict yhat
. label variable yhat "Predicted mean SAT score"
```

Through the **resid** option, we can also create another new variable containing the residuals, here named *e*:

```
. predict e, resid
. label variable e "Residual"
```

We might instead have obtained the same predicted *y* and residuals through two **generate** commands:

```
. generate yhat0 = _b[_cons] + _b[percent]*percent
. generate e0 = csat - yhat0
```

Stata stores coefficients and other details from the recent regression. Thus _b[*varname*] holds the coefficient on independent variable *varname*. _b[_cons] holds the coefficient on _cons (usually, the *y*-intercept). These stored values are useful in programming and some advanced applications, but for most purposes, **predict** saves us the trouble of generating *yhat0* and *e0* "by hand" in this fashion.

Residuals contain information about where the model fits poorly, and so are important for diagnostic or troubleshooting analysis. Such analysis might begin just by sorting and examining the residuals. Negative residuals occur when our model overpredicts the observed values. That is, in these states the mean SAT scores are lower than we would expect, based on what percentage of students took the test. To list the states with the five lowest residuals, type

```
. sort e
. list state percent csat yhat e in 1/5
```

	state	percent	csat	yhat	e
1.	South Carolina	58	832	894.3333	-62.3333
2.	West Virginia	17	926	986.0953	-60.09526
3.	North Carolina	57	844	896.5714	-52.5714
4.	Texas	44	874	925.6666	-51.66666
5.	Nevada	25	919	968.1905	-49.19049

The four lowest residuals belong to southern states, suggesting that we might be able to improve our model, or better understand variation in mean SAT scores, by somehow taking region into account.

Positive residuals occur when actual *y* values are higher than predicted. Because the data already have been sorted by *e*, to list the five highest residuals we add the qualifier **in -5/1** (meaning 5th-from-last through last observations; **in 47/1** would accomplish the same thing).

```
. list state percent csat yhat e in -5/1
```

	state	percent	csat	yhat	e
47.	Massachusetts	79	896	847.3333	48.66673
48.	Connecticut	81	897	842.8571	54.14292
49.	North Dakota	6	1073	1010.714	62.28567
50.	New Hampshire	75	921	856.2856	64.71434
51.	Iowa	5	1093	1012.952	80.04758

predict also derives other statistics from the most recently-estimated model. Some **predict** options that can be used after **anova** or **regress** are the following:

. predict *new*	Predicted values of *y*. **predict** *new*, **xb** means the same thing (referring to **Xb**, the vector of predicted *y* values).

`. predict new, cooksd`	Cook's D influence measures.
`. predict new, covratio`	*COVRATIO* influence measures; effect of each observation on the variance–covariance matrix of estimates.
`. predict DFx1, dfbeta(x1)`	*DFBETA*s measuring each observation's influence on the coefficient of predictor $x1$.
`. predict new, dfits`	*DFITS* influence measures.
`. predict new, hat`	Diagonal elements of hat matrix (leverage).
`. predict new, resid`	Residuals.
`. predict new, rstandard`	Standardized residuals.
`. predict new, rstudent`	Studentized (jackknifed) residuals.
`. predict new, stdf`	Standard errors of predicted individual y, sometimes called the standard errors of forecast or the standard errors of prediction.
`. predict new, stdp`	Standard errors of predicted mean y.
`. predict new, stdr`	Standard errors of residuals.
`. predict new, welsch`	Welsch's distance influence measures.

Further options obtain predicted probabilities and expected values; type **help regress** for a list. All **predict** options create case statistics, which are new variables (like predicted values and residuals) that have a value for each observation in the sample.

When using **predict**, substitute a new variable name of your choosing for *new* in the commands shown above. For example, to obtain Cook's D influence measures, type

`. predict D, cooksd`

Or you can find hat matrix diagonals by typing

`. predict h, hat`

The names of variables created by **predict** (such as *yhat, e, D, h*) are arbitrary and are invented by the user. As with other elements of Stata commands, we could abbreviate the options to the minimum number of letters it takes to identify them uniquely. For example,

`. predict e, resid`

could be shortened to

`. pre e, re`

Basic Graphs for Regression

This section introduces some elementary graphs you can use to represent a regression model or examine its fit. Chapter 7 describes more specialized graphs that aid post-regression diagnostic work.

In simple regression, predicted values lie on the line defined by the regression equation. By plotting and connecting predicted values, we can make that line visible (Figure 6.1).

```
. graph csat yhat percent, connect(.s) symbol(Oi) ylabel xlabel
```

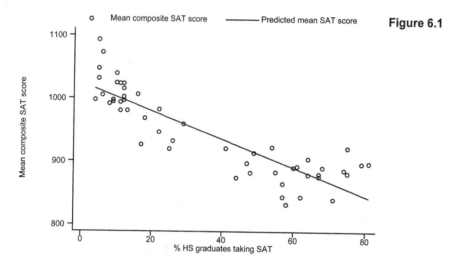

Figure 6.1

connect(.s) tells **graph** not to connect the first-named *y* variable (*csat*), but to connect the second-named (*yhat*) smoothly. **symbol(Oi)** specifies that *csat* should be plotted with large circles, and *yhat* invisibly—so we see only the connecting line. **connect(ss)** , for instance, would have connected both the *csat* and *y* values (producing a messy graph). **symbol(Td)** would have plotted *csat* as large triangles and *yhat* as small diamonds.

Residual-versus-predicted-values plots provide useful diagnostic tools. Figure 6.2 shows an example that includes marginal boxplots and a horizontal line at the mean residual, 0.

```
. graph e yhat, twoway box yline(0) ylabel xlabel
```

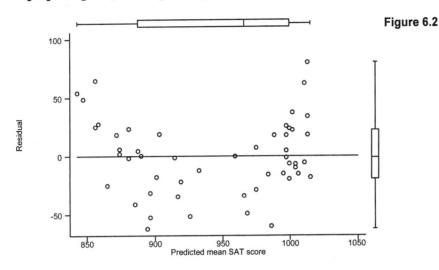

Figure 6.2

Figure 6.2 reveals that our present model overlooks an obvious pattern in the data. The residuals or prediction errors appear to be mostly positive at first (due to too-high predictions), then mostly negative, followed by mostly positive residuals again. Later sections will seek a model that better fits these data.

predict can generate two kinds of standard errors for the predicted y values, which have two different applications. These applications are sometimes distinguished by the names "confidence intervals" and "prediction intervals":

1. A "confidence interval" in this context expresses our uncertainty in estimating the conditional mean of y at a given x value (or a given combination of x values, in multiple regression). Standard errors for this purpose are obtained through

    ```
    . predict SE, stdp
    ```

    Select an appropriate t value. With 49 degrees of freedom, for 95% confidence we should use $t = 2.01$, found by looking up the t distribution or simply by asking Stata:

    ```
    . display invttail(49,.05/2)
    2.0095752
    ```

    Then the lower confidence limit is approximately

    ```
    . generate low1 = yhat - 2.01*SE
    ```

    and the upper confidence limit is

    ```
    . generate high1 = yhat + 2.01*SE
    ```

To graph these confidence limits as bands in a simple regression (Figure 6.3), issue a command such as the following. Confidence bands in simple regression have an hourglass shape, narrowest at the mean of x.

```
.  graph csat yhat low1 high1 percent, symbol(Oiii)
        connect(.ss[-]s[-]) ylabel xlabel
```

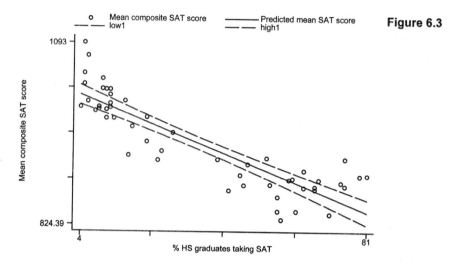

Figure 6.3

2. A "prediction interval" expresses our uncertainty in estimating the unknown value of *y* for an individual observation with known *x* value(s). Standard errors for this purpose are obtained by typing

```
. predict SEyhat, stdf
```

To find and graph these prediction bands (Figure 6.4), type

```
. generate low2 = yhat - 2.01*SEyhat
. generate high2 = yhat + 2.01*SEyhat
. graph csat yhat low2 high2 percent, symbol(Oiii)
        connect(.ss[.]s[.]) ylabel xlabel
```

Predicting the *y* values of individual observations (Figure 6.4) inherently involves greater uncertainty, and hence wider bands, than does predicting the conditional mean of *y* values (Figure 6.3). Prediction intervals, like confidence intervals, are narrowest at the mean of *x*.

As with other confidence intervals and hypothesis tests in OLS regression, the standard errors and bands just described depend on the assumption of independent and identically distributed errors. Figure 6.2 has cast doubt on this assumption, so the results in Figures 6.3 and 6.4 could be misleading.

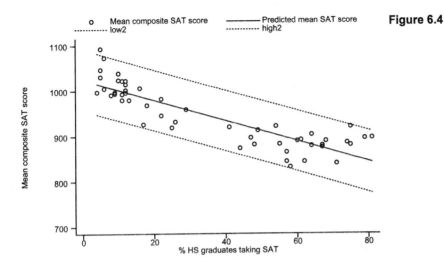

Figure 6.4

Correlations

correlate obtains Pearson product-moment correlations between variables.

. correlate *csat expense percent income high college*

(obs=51)

```
             |    csat   expense   percent    income      high  college
-------------+------------------------------------------------------------
        csat |   1.0000
     expense |  -0.4663    1.0000
     percent |  -0.8758    0.6509    1.0000
      income |  -0.4713    0.6784    0.6733    1.0000
        high |   0.0858    0.3133    0.1413    0.5099    1.0000
     college |  -0.3729    0.6400    0.6091    0.7234    0.5319    1.0000
```

    correlate uses only a subset of the data that has no missing values on any of the variables listed (with these particular variables, that does not matter because no observations have missing values). In this respect, the **correlate** command resembles **regress**, and given the same variable list, they will use the same subset of the data. Analysts not employing regression or other multi-variable techniques, however, might prefer to find correlations based upon all of the observations available for each variable pair. The command **pwcorr** (pairwise correlation) accomplishes this, and can also furnish *t*-test probabilities for the null hypotheses that each individual correlation equals zero.

. **pwcorr** *csat expense percent income high college*, sig

	csat	expense	percent	income	high	college
csat	1.0000					
expense	-0.4663	1.0000				
	0.0006					
percent	-0.8758	0.6509	1.0000			
	0.0000	0.0000				
income	-0.4713	0.6784	0.6733	1.0000		
	0.0005	0.0000	0.0000			
high	0.0858	0.3133	0.1413	0.5099	1.0000	
	0.5495	0.0252	0.3226	0.0001		
college	-0.3729	0.6400	0.6091	0.7234	0.5319	1.0000
	0.0070	0.0000	0.0000	0.0000	0.0001	

It is worth recalling here that if we drew many random samples from a population in which all variables really had 0 correlations, about 5% of the sample correlations would nonetheless test "statistically significant" at the .05 level. Analysts who review many individual hypothesis tests, such as those in a **pwcorr** matrix, to identify the handful that are significant at the .05 level, therefore run a much higher than .05 risk of making a Type I error. This problem is called the "multiple comparison fallacy." **pwcorr** offers two methods, Bonferroni and Šidák, for adjusting significance levels to take multiple comparisons into account. Of these, the Šidák method is more precise.

. **pwcorr** *csat expense percent income high college*, sidak sig

	csat	expense	percent	income	high	college
csat	1.0000					
expense	-0.4663	1.0000				
	0.0084					
percent	-0.8758	0.6509	1.0000			
	0.0000	0.0000				
income	-0.4713	0.6784	0.6733	1.0000		
	0.0072	0.0000	0.0000			
high	0.0858	0.3133	0.1413	0.5099	1.0000	
	1.0000	0.3180	0.9971	0.0020		
college	-0.3729	0.6400	0.6091	0.7234	0.5319	1.0000
	0.1004	0.0000	0.0000	0.0000	0.0009	

Comparing the test probabilities in the table above with those of the previous **pwcorr** provides some idea of how much adjustment occurs. In general, the more variables we correlate, the more the adjusted probabilities will exceed their unadjusted counterparts. See the *Reference Manual*'s discussion of **oneway** for the formulas involved.

correlate itself offers several important options. Adding the **covariance** option produces a matrix of variances and covariances instead of correlations:

. **correlate** *w x y z*, **covariance**

Typing the following after a regression analysis displays the matrix of correlations between estimated coefficients, sometimes used to diagnose multicollinearity (see Chapter 7):

. **correlate, _coef**

The following command will display the estimated coefficients' variance–covariance matrix, from which standard errors are derived:

. **correlate, _coef covariance**

Pearson correlation coefficients measure how well an OLS regression line fits the data. They consequently share the assumptions and weaknesses of OLS, and like OLS, should generally not be interpreted without first reviewing the corresponding scatterplots. A scatterplot matrix provides a quick way to do this, using the same organization as the correlation matrix. Figure 6.5 shows a scatterplot matrix corresponding to the **pwcorr** matrix given earlier. Only the lower-triangular half of the matrix is drawn. Plus signs are used as plotting symbols in Figure 6.5, and text size temporarily set to 140% of normal, to make the tiny graphs more readable.

. **set textsize 140**

. **graph** *csat expense percent income high college*, **matrix half label**
     symbol(p)

. **set textsize 100**

Figure 6.5

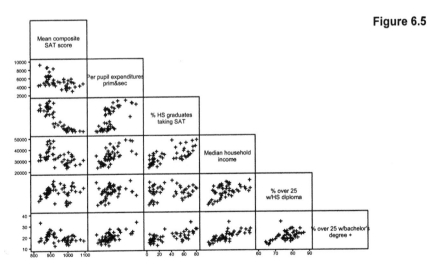

To obtain a scatterplot matrix corresponding to a **correlate** correlation matrix, from which all observations having missing values have been dropped, we would need to qualify the command. If all of the variables had some missing values, we could type a command such as

```
. graph csat expense percent income high college if csat !=.
    & expense != . & income != . & high != . & college !=. ,
    matrix half label symbol(.)
```

To reduce the likelihood of confusion and mistakes, it could make sense to create a new dataset keeping only those observations that have no missing values:

```
. keep if csat !=. & expense != . & income != . & high != .
    & college !=.
. save nmvstate
```

In this example, we immediately saved the reduced dataset with a new name, so as to avoid inadvertently writing over and losing the information in the old, more complete dataset. An alternative way to eliminate missing values uses **drop** instead of **keep**:

```
. drop if csat ==. | expense == . | income == . | high == .
    | college ==.
. save nmvstate
```

In addition to Pearson correlations, Stata also calculates several rank-based correlations. These can be employed to measure associations between ordinal variables, or as an outlier-resistant alternative to Pearson correlation for measurement variables. To obtain the Spearman rank correlation between *csat* and *expense*, equivalent to the Pearson correlation if these variables were transformed into ranks, type

```
. spearman csat expense

Number of obs =       51
Spearman's rho =    -0.4282

Test of Ho: csat and expense are independent
    Prob > |t| =      0.0017
```

Kendall's τ_a (tau-a) and τ_b (tau-b) rank correlations can be found easily for these data, although with larger datasets their calculation becomes slow:

```
. ktau csat expense

Number of obs =       51
Kendall's tau-a =   -0.2925
Kendall's tau-b =   -0.2932
Kendall's score =   -373
    SE of score =    123.095    (corrected for ties)

Test of Ho: csat and expense are independent
    Prob > |z| =      0.0025    (continuity corrected)
```

For comparison, here is the Pearson correlation with its (unadjusted) *P*-value:

```
. pwcorr csat expense, sig

             |     csat  expense
-------------+------------------
        csat |   1.0000
             |
             |
     expense |  -0.4663   1.0000
             |   0.0006
             |
```

In this example, both **spearman** (–.4282) and **pwcorr** (–.4663) yield higher correlations than **ktau** (–.2925 or –.2932). All three agree that the null hypothesis of no association can be rejected.

Hypothesis Tests

With every regression, Stata displays two kinds of hypothesis tests. Like most common hypothesis tests, they begin from the assumption that observations in the sample at hand were drawn randomly and independently from an infinitely large population. Two types of tests appear in **regress** output tables.

1. Overall F test: The F statistic at the upper right in the regression table evaluates the null hypothesis that in the population, coefficients on all the model's x variables equal zero.

2. Individual t tests: The third and fourth columns of the regression table contain t tests for each individual regression coefficient. These evaluate the null hypotheses that in the population, the coefficient on each particular x variable equals zero.

The t test probabilities are two-sided; for one-sided tests, divide these P-values in half.

In addition to these standard F and t tests, Stata can perform F tests of user-specified hypotheses. The **test** command refers back to the most recent model-fitting command such as **anova** or **regress**. For example, individual t tests from the following regression report that neither the percent of adults with at least high school diplomas (*high*) nor the percent with college degrees (*college*) has a statistically significant individual effect on composite SAT scores.

```
. regress csat expense percent income high college
```

Conceptually, however, both predictors measure the level of education attained by a state's population, and for some purposes we might want to test the null hypothesis that *both* have zero effect. To do this, we begin by repeating the multiple regression **quietly**, because we do not need to see its full output again. Then use the **test** command:

```
. quietly regress csat expense percent income high college
. test high college

 ( 1)   high = 0.0
 ( 2)   college = 0.0

       F(  2,    45) =     3.32
            Prob > F =    0.0451
```

Unlike the individual null hypotheses, the joint hypothesis that coefficients on *high* and *college* both equal zero can reasonably be rejected ($P = .0451$). Such tests on subsets of coefficients are useful when we have several conceptually related predictors or when individual coefficient estimates appear unreliable due to multicollinearity (Chapter 7).

test could duplicate the overall F test:

```
. test expense percent income high college
```

test could also duplicate the individual-coefficient tests:

```
. test expense
```

```
. test percent
. test income
```

And so forth. Applications of **test** more useful in advanced work include

1. Test whether a coefficient equals a specified constant. For example, to test the null hypothesis that the coefficient on *income* equals 1 ($H_0:\beta_3 = 1$), instead of the usual null hypothesis that it equals 0 ($H_0:\beta_3 = 0$), type

    ```
    . test income = 1
    ```

2. Test whether two coefficients are equal. For example, the following command evaluates the null hypothesis $H_0:\beta_4 = \beta_5$.

    ```
    . test high = college
    ```

3. Finally, **test** understands some algebraic expressions. We could request something like the following, which would test $H_0:\beta_3 = (\beta_4 + \beta_5)/100$:

    ```
    . test income = (high + college)/100
    ```

Consult **help test** for more information and examples.

Dummy Variables

Categorical variables can become predictors in a regression, if we first re-express their categories as one or more {0,1} dichotomies called "dummy variables." For example, we have reason to suspect that regional differences exist in states' mean SAT scores. The **tabulate** command will generate one dummy variable for each category of the tabulated variable if we add a **gen( )** option. Below, we create four dummy variables from the four-category variable *region*. The dummies are named *reg1, reg2, reg3* and *reg4*. *reg1* equals 1 for Western states and 0 for others; *reg2* equals 1 for Northeastern states and 0 for others, and so forth.

```
. tabulate region, gen(reg)

Geographica |
   l region |      Freq.        Percent          Cum.
------------+-----------------------------------------
       West |         13          26.00          26.00
    N. East |          9          18.00          44.00
      South |         16          32.00          76.00
    Midwest |         12          24.00         100.00
------------+-----------------------------------
      Total |         50         100.00
```

```
. describe reg1-reg4

              storage  display   value
variable name   type   format    label      variable label
-----------------------------------------------------------------------
reg1            byte    %8.0g                region==West
reg2            byte    %8.0g                region==N. East
reg3            byte    %8.0g                region==South
reg4            byte    %8.0g                region==Midwest
```

```
. tabulate reg1
```

region==Wes t	Freq.	Percent	Cum.
0	37	74.00	74.00
1	13	26.00	100.00
Total	50	100.00	

```
. tabulate reg2
```

region==N. East	Freq.	Percent	Cum.
0	41	82.00	82.00
1	9	18.00	100.00
Total	50	100.00	

Regressing *csat* on one dummy variable, *reg2* (Northeast), is equivalent to performing a two-sample *t* test of whether mean *csat* is the same across categories of *reg2*. That is, is the mean *csat* the same in the Northeast as in other U.S. states?

```
. regress csat reg2
```

Source	SS	df	MS		Number of obs =	50
					F(1, 48) =	9.50
Model	35191.4017	1	35191.4017		Prob > F =	0.0034
Residual	177769.978	48	3703.54121		R-squared =	0.1652
					Adj R-squared =	0.1479
Total	212961.38	49	4346.15061		Root MSE =	60.857

csat	Coef.	Std. Err.	t	P>|t|	[95% Conf. Interval]	
reg2	-69.0542	22.40167	-3.08	0.003	-114.0958	-24.01262
_cons	958.6098	9.504224	100.86	0.000	939.5002	977.7193

The dummy variable coefficient's *t* statistic ($t = -3.08$, $P = .003$) indicates a significant difference. According to this regression, mean SAT scores are 69.0542 points lower (because $b = -69.0542$) among Northeastern states. We get exactly the same result ($t = 3.08$, $P = .003$) from a simple *t* test, which also shows the means as 889.5556 (Northeast) and 958.6098 (other states), a difference of 69.0542.

```
. ttest csat, by(reg2)
```

Two-sample t test with equal variances

Group	Obs	Mean	Std. Err.	Std. Dev.	[95% Conf. Interval]	
0	41	958.6098	10.36563	66.37239	937.66	979.5595
1	9	889.5556	4.652094	13.95628	878.8278	900.2833
combined	50	946.18	9.323251	65.92534	927.4442	964.9158
diff		69.0542	22.40167		24.01262	114.0958

Degrees of freedom: 48

```
                    Ho: mean(0) - mean(1) = diff = 0
```

```
Ha: diff < 0              Ha: diff ~= 0              Ha: diff > 0
  t =    3.0825             t =    3.0825             t =    3.0825
P < t =   0.9983         P > |t| =   0.0034         P > t =   0.0017
```

This conclusion proves spurious, however, once we control for the percentage of students taking the test. We do so by a multiple regression of *csat* on both *reg2* and *percent*.

. regress csat reg2 percent

```
    Source |       SS       df       MS              Number of obs =      50
-----------+------------------------------           F(  2,    47) =  107.18
     Model | 174664.983      2   87332.4916          Prob > F      =  0.0000
  Residual | 38296.3969     47   814.816955          R-squared     =  0.8202
-----------+------------------------------           Adj R-squared =  0.8125
     Total | 212961.38      49   4346.15061          Root MSE      =  28.545

------------------------------------------------------------------------------
      csat |      Coef.   Std. Err.      t    P>|t|     [95% Conf. Interval]
-----------+------------------------------------------------------------------
      reg2 |   57.52437   14.28326     4.03   0.000     28.79016    86.25858
   percent |  -2.793009   .2134796   -13.08   0.000    -3.222475   -2.363544
     _cons |   1033.749   7.270285   142.19   0.000     1019.123    1048.374
------------------------------------------------------------------------------
```

The Northeastern region variable *reg2* now has a statistically significant *positive* coefficient ($b = 57.52437$, $P < .0005$). The earlier negative relationship was misleading. Although mean SAT scores among Northeastern states really are lower, they are lower *because higher percentages of students take this test in the Northeast.* A smaller, more "elite" group of students, often less than 20% of high school seniors, take the SAT in many of the non-Northeast states. In all Northeastern states, however, large majorities (64% to 81%) do so. Once we adjust for differences in the percentages taking the test, SAT scores actually tend to be higher in the Northeast.

To understand dummy variable regression results, it often helps to write out the regression equation, substituting zeroes and ones. For Northeastern states, the equation is approximately

$$\text{predicted } csat = 1033.7 + 57.5reg2 - 2.8percent$$
$$= 1033.7 + 57.5 \times 1 - 2.8percent$$
$$= 1091.2 - 2.8percent$$

For other states, the predicted *csat* is 57.5 points lower at any given level of *percent*:

$$\text{predicted } csat = 1033.7 + 57.5 \times 0 - 2.8percent$$
$$= 1033.7 - 2.8percent$$

Dummy variables in models such as this are termed "intercept dummy variables," because they describe a shift in the *y*-intercept or constant.

From a categorical variable with *k* categories we can define *k* dummy variables, but one of these will be redundant. Once we know a state's values on the West, Northeast, and Midwest dummy variables, for example, we can already guess its value on the South variable. For this reason, no more than $k - 1$ of the dummy variables—three, in the case of *region*—can be included in a regression. If we try to include all the possible dummies, Stata will automatically drop one because multicollinearity otherwise makes the calculation impossible.

. regress csat reg1 reg2 reg3 reg4 percent

```
      Source |       SS           df       MS                 Number of obs =       50
-------------+------------------------------                  F(  4,    45) =    64.61
       Model | 181378.099          4  45344.5247              Prob > F      =   0.0000
    Residual | 31583.2811         45  701.850691              R-squared     =   0.8517
-------------+------------------------------                  Adj R-squared =   0.8385
       Total | 212961.38          49  4346.15061              Root MSE      =   26.492


------------------------------------------------------------------------------
        csat |      Coef.   Std. Err.      t    P>|t|     [95% Conf. Interval]
-------------+----------------------------------------------------------------
        reg1 |  -23.77315   11.12578    -2.14   0.038    -46.18162   -1.364676
        reg2 |   25.79985   16.96365     1.52   0.135    -8.366693    59.96639
        reg3 |  -33.29951   10.85443    -3.07   0.004    -55.16146   -11.43757
        reg4 |  (dropped)
     percent |  -2.546058    .2140196   -11.90   0.000    -2.977116   -2.115001
       _cons |   1047.638    8.273625   126.62   0.000     1030.974    1064.302
------------------------------------------------------------------------------
```

The model's fit—including R^2, F tests, predictions, and residuals—remains essentially the same regardless of which dummy variable we (or Stata) choose to omit. Interpretation of the coefficients, however, occurs with reference to that omitted category. In this example, the Midwest dummy variable (*reg4*) was omitted. The regression coefficients on *reg1*, *reg2*, and *reg3* tell us that, at any given level of *percent*, the predicted mean SAT scores are approximately as follows:

23.8 points lower in the West (*reg1* = 1) than in the Midwest;

25.8 points higher in the Northeast (*reg2* = 1) than in the Midwest; and

33.3 points lower in the South (*reg3* = 1) than in the Midwest.

The West and South both differ significantly from the Midwest in this respect, but the Northeast does not.

An alternative command, **areg**, estimates the same model without going through dummy variable creation. Instead, it "absorbs" the effect of a *k*-category variable such as *region*. The model's fit, F test on the absorbed variable, and other key aspects of the results are the same as those we could obtain through explicit dummy variables. Note that **areg** does not provide estimates of the coefficients on individual dummy variables, however.

```
. areg csat percent, absorb(region)
```

```
                                                             Number of obs =       50
                                                             F(  1,    45) =   141.52
                                                             Prob > F      =   0.0000
                                                             R-squared     =   0.8517
                                                             Adj R-squared =   0.8385
                                                             Root MSE      =   26.492


------------------------------------------------------------------------------
        csat |      Coef.   Std. Err.      t    P>|t|     [95% Conf. Interval]
-------------+----------------------------------------------------------------
     percent |  -2.546058    .2140196   -11.90   0.000    -2.977116   -2.115001
       _cons |   1035.445    8.38689    123.46   0.000     1018.553    1052.337
-------------+----------------------------------------------------------------
      region |         F(3, 45) =         9.465   0.000            (4 categories)
------------------------------------------------------------------------------
```

Although its output is less informative than regression with explicit dummy variables, **areg** does have two advantages. It speeds up exploratory work, providing quick feedback about whether a dummy variable approach is worthwhile. Secondly, when the variable of interest has many values, creating dummies for each of them could lead to too many variables or too large a model for our particular Stata configuration. **areg** thus works around the usual limitations on dataset and matrix size.

Explicit dummy variables have other advantages, however, including ways to model interaction effects. Interaction terms called "slope dummy variables" can be formed by multiplying a dummy times a measurement variable. For example, to model an interaction between Northeast/other region and *percent*, we create a slope dummy variable called *reg2perc*.

```
. generate reg2perc = reg2 * percent
(1 missing value generated)
```

The new variable, *reg2perc*, equals *percent* for Northeastern states and zero for all other states. We can include this interaction term among the regression predictors:

```
. regress csat reg2 percent reg2perc
```

Source	SS	df	MS		Number of obs =	50
					F(3, 46) =	82.27
Model	179506.19	3	59835.3968		Prob > F =	0.0000
Residual	33455.1897	46	727.286733		R-squared =	0.8429
					Adj R-squared =	0.8327
Total	212961.38	49	4346.15061		Root MSE =	26.968

csat	Coef.	Std. Err.	t	P>|t|	[95% Conf. Interval]	
reg2	-241.3574	116.6278	-2.07	0.044	-476.117	-6.597821
percent	-2.858829	.2032947	-14.06	0.000	-3.26804	-2.449618
reg2perc	4.179666	1.620009	2.58	0.013	.9187559	7.440576
_cons	1035.519	6.902898	150.01	0.000	1021.624	1049.414

The interaction is statistically significant ($t = 2.58$, $P = .013$). Because this analysis includes both intercept (*reg2*) and slope (*reg2perc*) dummy variables, it is worthwhile to write out the equations. The regression equation for Northeastern states is approximately

$$\text{predicted } csat = 1035.5 - 241.4reg2 - 2.9percent + 4.2reg2perc$$
$$= 1035.5 - 241.4 \times 1 - 2.9percent + 4.2 \times 1 \times percent$$
$$= 794.1 + 1.3percent$$

For other states it is

$$\text{predicted } csat = 1035.5 - 241.4 \times 0 - 2.9percent + 4.2 \times 0 \times percent$$
$$= 1035.5 - 2.9percent$$

An interaction means that the effect of one variable changes, depending on the values of some other variable. From this regression, it appears that *percent* has a relatively weak and positive effect among Northeastern states, whereas its effect is stronger and negative among the rest. To visualize this model, we might draw a scatterplot with separate regression lines for the Northeast and other states, and represent those states with different plotting symbols. A quick but crude plot could be obtained by typing

```
. predict yhat1
```
```
. graph csat yhat1 percent
```

A nicer version, shown in Figure 6.6, takes more steps. We start by generating predicted values separately from the Northeast and "elsewhere" regression equations (naming them *yhatNE* and *yhatelse*). Next we generate separate variables holding *csat* values, naming these *csatNE* and *csatelse*. Finally, we put the elements together for Figure 6.6.

```
. generate yhatNE = 794.1 + 1.3*percent if reg2 == 1
(42 missing values generated)
. generate yhatelse = 1035.5 - 2.9*percent if reg2 == 0
(10 missing values generated)
. generate csatNE = csat if reg2 == 1
(42 missing values generated)
. generate csatelse = csat if reg2 == 0
(10 missing values generated)
. label variable csatNE "Northeastern states"
. label variable csatelse "other U.S. states"
. graph csatNE csatelse yhatNE yhatelse percent, connect(..ss[-])
    symbol(Spii) ylabel xlabel 11(Mean SAT score) border
```

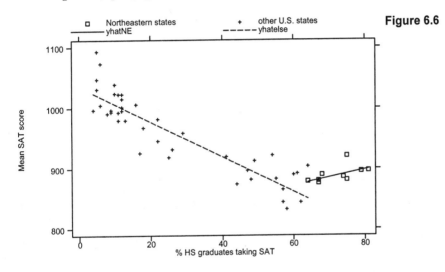

Figure 6.6

Figure 6.6 shows the striking difference, captured by our interaction effect, between Northeastern and other states. This raises the question of what other regional differences exist. Figure 6.7 explores this question by drawing a *csat–percent* scatterplot with different symbols for each region. In this plot, the Midwestern states, with one exception (Indiana), seem to have their own steeply negative regional pattern at the left side of the graph. Southern states are the most heterogeneous group.

```
. generate csatW = csat if reg1 == 1
(38 missing values generated)
. label variable csatW "Western states"
. generate csatS = csat if reg3 == 1
(35 missing values generated)
. label variable csatS "Southern states"
```

```
. generate csatMW = csat if reg4 == 1
(39 missing values generated)
. label variable csatMW "Midwestern states"
. graph csatW csatNE csatS csatMW percent, ylabel xlabel
    symbol(pSOT) 11(Mean SAT score) border
```

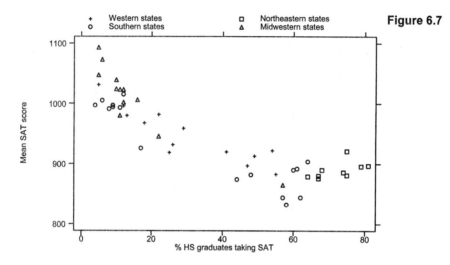

Figure 6.7

Stepwise Regression

With the regional dummy variable terms added to our data, we now have many possible predictors of *csat*. This results in an overly complicated model, with several coefficients statistically indistinguishable from zero.

```
. regress csat expense percent income college high reg1 reg2
    reg2perc reg3
```

Source	SS	df	MS
Model	195420.517	9	21713.3908
Residual	17540.863	40	438.521576
Total	212961.38	49	4346.15061

Number of obs =	50
F(9, 40) =	49.51
Prob > F =	0.0000
R-squared =	0.9176
Adj R-squared =	0.8991
Root MSE =	20.941

csat	Coef.	Std. Err.	t	P>|t|	[95% Conf. Interval]
expense	-.0022508	.0041333	-0.54	0.589	-.0106045 .006103
percent	-2.93786	.2302596	-12.76	0.000	-3.403232 -2.472488
income	-.0004919	.0010255	-0.48	0.634	-.0025645 .0015806
college	3.900087	1.719409	2.27	0.029	.4250318 7.375142
high	2.175542	1.171767	1.86	0.071	-.192688 4.543771
reg1	-33.78456	9.302983	-3.63	0.001	-52.58659 -14.98253
reg2	-143.5149	101.1244	-1.42	0.164	-347.8949 60.86509
reg2perc	2.506616	1.404483	1.78	0.082	-.3319506 5.345183
reg3	-8.799205	12.54658	-0.70	0.487	-34.15679 16.55838
_cons	839.2209	76.35942	10.99	0.000	684.8927 993.549

We might now try to simplify this model, dropping first that predictor with the highest *t* probability (*income*, *P* = .634), then re-estimating the model and deciding whether to drop something further. Through this process of backward elimination, we seek a more parsimonious model; one that is simpler but fits almost equally well. Ideally, this strategy is pursued with attention both to the statistical results and to the substantive or theoretical implications of keeping or discarding certain variables.

For analysts in a hurry, stepwise methods provide ways to automate the process of model selection. They work either by subtracting predictors from a complicated model, or by adding predictors to a simpler one according to some pre-set statistical criteria. Stepwise methods cannot consider the substantive or theoretical implications of their choices, nor can they do much troubleshooting to evaluate possible weaknesses in the models produced at each step. Despite their drawbacks, stepwise methods meet certain practical needs and have been widely used.

For automatic backward elimination, we issue a **sw regress** command that includes all of our possible predictor variables, and a maximum *P* value required to retain them. Setting the *P*-to-retain criteria as **pr(.05)** ensures that only predictors having coefficients that are significantly different from zero at the .05 level or less will be kept in the model.

```
. sw regress csat expense percent income college high reg1 reg2
     reg2perc reg3, pr(.05)

                              begin with full model
p = 0.6341 >= 0.0500    removing income
p = 0.5273 >= 0.0500    removing reg3
p = 0.4215 >= 0.0500    removing expense
p = 0.2107 >= 0.0500    removing reg2

      Source |       SS       df       MS              Number of obs =      50
-------------+------------------------------           F(  5,    44) =   91.01
       Model | 194185.761        5  38837.1521         Prob > F      =  0.0000
    Residual | 18775.6194       44  426.718624         R-squared     =  0.9118
-------------+------------------------------           Adj R-squared =  0.9018
       Total | 212961.38        49  4346.15061         Root MSE      =  20.657


        csat |      Coef.   Std. Err.       t    P>|t|     [95% Conf. Interval]
-------------+----------------------------------------------------------------
        reg1 |  -30.59218   8.479395    -3.61   0.001    -47.68128   -13.50309
     percent |  -3.119155   .1804553   -17.28   0.000    -3.482839   -2.755471
    reg2perc |   .5833272   .1545969     3.77   0.000     .2717577    .8948967
     college |   3.995495   1.359331     2.94   0.005     1.255944    6.735046
        high |   2.231294   .8178968     2.73   0.009     .5829313    3.879657
       _cons |    806.672   49.98744    16.14   0.000     705.9289    907.4151
```

sw regress dropped first *income*, then *reg3*, *expense*, and finally *reg2* before settling on the final model. Although it has four fewer coefficients, this final model has almost the same R^2 (.9118 versus .9176) and a higher R^2_a (.9018 versus .8991) compared with the earlier version.

If, instead of a *P*-to-retain, **pr(.05)**, we specify a *P*-to-enter value such as **pe(.05)**, then **sw regress** performs forward inclusion (starting with an "empty" or constant-only model) instead of backward elimination. Other stepwise options include hierarchical selection and locking certain predictors into the model. For example, the following command specifies that the first term (*x1*) should be locked into the model and not subject to possible removal:

```
. sw regress y x1 x2 x3, pr(.05) lockterm1
```

Typing the following command calls for forward inclusion of any predictors found significant at the .10 level, but with variables *x4*, *x5*, and *x6* treated as one unit—either entered or left out together:

```
. sw regress y x1 x2 x3 (x4 x5 x6), pe(.10)
```

The following command invokes hierarchical backward elimination with a $P = .20$ criterion:

```
. sw regress y x1 x2 x3 (x4 x5 x6) x7, pr(.20) hier
```

The **hier** option specifies that the terms are ordered: Consider dropping the last term (*x7*) first, and stop if it is not dropped. If *x7* is dropped, next consider the second-to-last term *(x4 x5 x6)*, and so forth.

Many other Stata commands besides **regress** also have stepwise variants that work in a similar manner. Available stepwise procedures include the following:

`sw clogit`	Conditional (fixed-effects) logistic regression
`sw cloglog`	Maximum likelihood complementary log-log estimation
`sw cnreg`	Censored normal regression
`sw cox`	Cox proportional hazard model regression
`sw ereg`	Exponential regression
`sw gamma`	Generalized log-gamma regression
`sw glm`	Generalized linear models
`sw gompertz`	Gompertz regression
`sw hetprob`	Heteroskedastic probit estimation
`sw llogistic`	Log-logistic regression
`sw lnormal`	Lognormal regression
`sw logistic`	Logistic regression (odds)
`sw logit`	Logistic regression (coefficients)
`sw nbreg`	Negative binomial regression
`sw ologit`	Ordered logistic regression
`sw oprobit`	Ordered probit regression
`sw poisson`	Poisson regression
`sw probit`	Probit regression
`sw qreg`	Quantile regression
`sw regress`	OLS regression
`sw scobit`	Skewed logit estimation
`sw tobit`	Tobit regression
`sw weibull`	Weibull regression

Type **help sw** for details about the stepwise options and logic.

Polynomial Regression

Earlier in this chapter, Figures 6.1 and 6.2 revealed an apparently curvilinear relation between mean composite SAT scores (*csat*) and the percentage of high school seniors taking the test (*percent*). Figure 6.6 illustrated one way to model the upturn in SAT scores at high *percent* values: as a phenomenon peculiar to the Northeastern states. That interaction model fit reasonably well ($R^2_a = .8327$). But Figure 6.8, a residuals versus predicted values plot for the interaction model (Figure 6.6), still exhibits signs of trouble.

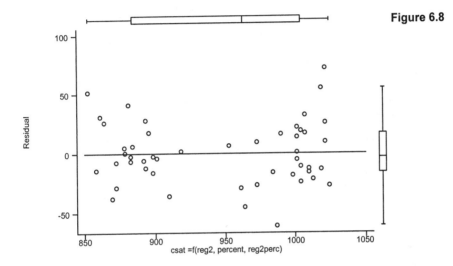

Figure 6.8

Chapter 8 presents a variety of techniques for curvilinear and nonlinear regression. "Curvilinear regression" here refers to OLS (for example, **regress**) employing nonlinear transformations of the original *y* or *x* variables. Although curvilinear regression fits a curved model with respect to the original data, this model remains linear in the transformed variables. (Nonlinear regression, also discussed in Chapter 8, applies non-OLS methods to fit models that cannot be linearized through transformation.)

One simple type of curvilinear regression, called polynomial regression, often succeeds in fitting U or inverted-U shaped curves. It includes as predictors both an independent variable and its square (and possibly higher powers if necessary). Because the *csat–percent* relation appears somewhat U-shaped, we generate a new variable equal to *percent* squared, then include *percent* and *percent*^{2} as predictors of *csat*. Figure 6.9 graphs the resulting curve.

```
. generate percent2 = percent^2
```

```
. regress csat percent percent2
```

```
      Source |       SS          df       MS                    Number of obs =        51
-------------+--------------------------------                  F(  2,     48) =    153.48
       Model |  193721.829        2   96860.9146                Prob > F       =    0.0000
    Residual |  30292.6806       48   631.097513                R-squared      =    0.8648
-------------+--------------------------------                  Adj R-squared  =    0.8591
       Total |   224014.51       50   4480.2902                 Root MSE       =    25.122
```

```
        csat |      Coef.   Std. Err.      t    P>|t|     [95% Conf. Interval]
-------------+----------------------------------------------------------------
     percent |  -6.111993   .6715406    -9.10   0.000    -7.462216   -4.76177
    percent2 |   .0495819   .0084179     5.89   0.000     .0326566   .0665072
       _cons |   1065.921   9.285379   114.80   0.000     1047.252   1084.591
```

```
. predict yhat2
(option xb assumed; fitted values)
```

```
. graph csat yhat2 percent, connect(.s) symbol(Oi) ylabel xlabel
```

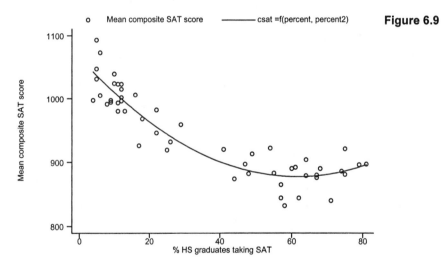

Figure 6.9

The polynomial model in Figure 6.9 matches the data slightly better than our interaction model in Figure 6.6 (R^2_a = .8591 versus .8327). Because the curvilinear pattern is now less striking in a residual versus predicted values plot (Figure 6.10), the usual assumption of independent, identically distributed errors also appears more plausible relative to this polynomial model.

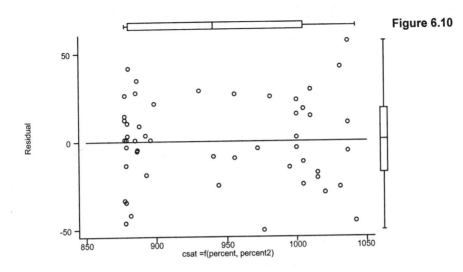

Figure 6.10

In Figures 6.6 and 6.9, we have two alternative models for the observed upturn in SAT scores at high levels of student participation. Statistical evidence seems to lean towards the polynomial model at this point. For serious research, however, we ought to choose between similar-fitting alternative models on substantive as well as statistical grounds. Which model seems more useful, or makes more sense? Which, if either, regression model suggests or corresponds to a good real-world explanation for the upturn in test scores at high levels of student participation?

Although it can closely fit sample data, polynomial regression also has important statistical weaknesses. The different powers of *x* might be highly correlated with each other, giving rise to multicollinearity. Furthermore, polynomial regression tends to track observations that have unusually large positive or negative *x* values, so a few data points can exert disproportionate influence on the results. For both reasons, polynomial regression results can sometimes be sample-specific, fitting one dataset well but generalizing poorly to other data. Chapter 7 takes a second look at this example, using tools that check for potential problems.

7

Regression Diagnostics

Do the data give us any reason to distrust our regression results? Can we find better ways to specify the model, or to estimate its parameters? Careful diagnostic work, checking for potential problems and evaluating the plausibility of key assumptions, forms a crucial step in modern data analysis. We fit an initial model, but then look closely at our results for signs of trouble or ways in which the model needs improvement. Many of the general methods introduced in earlier chapters, such as scatterplots, boxplots, normality tests, or just sorting and listing the data, prove useful for troubleshooting. Stata also provides a toolkit of specialized diagnostic techniques designed for this purpose.

Autocorrelation, a complication that often affects regression with time series data, is not covered in this chapter. Chapter 13, Time Series Analysis, introduces Stata's library of time series procedures including Durbin–Watson tests, autocorrelation graphs, lag operators, and time-series regression techniques.

Example Commands

The commands illustrated in this section all assume that you have just fit a model using either **anova** or **regress** . The commands' results refer back to that model. These follow-up commands are of three basic types:

1. **predict** options that generate new variables containing case statistics such as predicted values, residuals, standard errors, and influence statistics. Chapter 6 noted some key options; type **help regress** for a complete listing.

2. Diagnostic tests for statistical problems such as autocorrelation, heteroskedasticity, specification errors, or variance inflation (multicollinearity). Type **help regdiag** for a list.

3. Diagnostic plots such as added-variable or leverage plots, residual-versus-fitted plots, residual-versus-predictor plots, and component-versus-residual plots. Again, typing **help regdiag** obtains a full listing of regression and ANOVA diagnostic plots. General graphs for diagnosing distribution shape and normality were covered in Chapter 2; type **help diagplots** for a list of those.

predict Options

. `predict new, cooksd`

Generates a new variable equal to Cook's distance D, summarizing how much each observation influences the fitted model.

. `predict new, covratio`

Generates a new variable equal to Belsley, Kuh, and Welsch's *COVRATIO* statistic. *COVRATIO* measures the ith case's influence upon the variance–covariance matrix of the estimated coefficients.

. `predict DFx1, dfbeta(x1)`

Generates *DFBETA* case statistics measuring how much each observation affects the coefficient on predictor $x1$. The `dfbeta` command accomplishes the same thing more conveniently, and in this example will automatically name the resulting statistics *DFx1*:

. `dfbeta x1`

To create a complete set of *DFBETA*s for all predictors in the model, simply type the command `dfbeta` without arguments.

. `predict new, dfits`

Generates *DFITS* case statistics, summarizing the influence of each observation on the fitted model (similar in purpose to Cook's D and Welsch's W).

Diagnostic Tests

. `dwstat`

Calculates the Durbin–Watson test for first-order autocorrelation. Chapter 13 gives examples of this and other time series procedures.

. `hettest`

Performs Cook and Weisberg's test for heteroskedasticity. If we have reason to suspect that heteroskedasticity is a function of a particular predictor $x1$, we could focus on that predictor by typing `hettest x1`.

. `ovtest, rhs`

Performs the Ramsey regression specification error test (*RESET*) for omitted variables. The option `rhs` calls for using powers of the right-hand-side variables, instead of powers of predicted y (default).

. `vif`

Calculates variance inflation factors to check for multicollinearity.

Diagnostic Plots

. `acprplot x1, connect(m) bands(7)`

Constructs an augmented component-plus-residual plot (also known as an augmented partial residual plot), often better than `cprplot` in screening for nonlinearities. The options `connect(m) bands(7)` call for connecting with line segments the cross-medians of seven vertical bands. Alternatively, we might ask for a lowess-smoothed curve with bandwidth 0.5 by specifying the options `connect(k) bwidth(.5)`.

. **avplot** *x1*

   Constructs an added-variable plot (also called a partial-regression or leverage plot) showing the relation between *y* and *x1*, both adjusted for other *x* variables. Such plots help to notice outliers and influence points.

. **avplots**

   Draws and combines in one image all the added-variable plots from the recent **anova** or **regress** .

. **cprplot** *x1*

   Constructs a component-plus-residual plot (also known as a partial-residual plot) showing the adjusted relation between *y* and predictor *x1*. Such plots help detect nonlinearities in the data.

. **lvr2plot**

   Constructs a leverage-versus-squared-residual plot (also known as an L–R plot).

. **rvfplot**

   Graphs the residuals versus the fitted (predicted) values of *y*.

. **rvpplot** *x1*

   Graphs the residuals against values of predictor *x1*.

SAT Score Regression, Revisited

Diagnostic techniques have been described as tools for "regression criticism," because they help us examine our regression models for possible flaws and for ways that the models could be improved. In this spirit, we return now to the state Scholastic Aptitude Test regressions of Chapter 6. A relatively simple model explains about 92% of the variance in mean state SAT scores. It contains three predictors: *percent* (percent of high school graduates taking the test), *percent2* (*percent* squared), and *high* (percent of adults with a high school diploma).

. **generate** *percent2* = *percent^2*

. **regress** *csat percent percent2 high*

Source	SS	df	MS			
Model	207225.103	3	69075.0343			
Residual	16789.4069	47	357.221424			
Total	224014.51	50	4480.2902			

Number of obs =	51
F(3, 47) =	193.37
Prob > F =	0.0000
R-squared =	0.9251
Adj R-squared =	0.9203
Root MSE =	18.90

csat	Coef.	Std. Err.	t	P>|t|	[95% Conf. Interval]	
percent	-6.520312	.5095805	-12.80	0.000	-7.545455	-5.495168
percent2	.0536555	.0063678	8.43	0.000	.0408452	.0664658
high	2.986509	.4857502	6.15	0.000	2.009305	3.963712
_cons	844.8207	36.63387	23.06	0.000	771.1228	918.5185

The regression equation is

predicted *csat* = 844.82 − 6.52*percent* + .05*percent2* + 2.99*high*

The scatterplot matrix in Figure 7.1 depicts interrelations among these four variables. As noted in Chapter 6, the squared term *percent2* allows our regression model to fit the visibly curvilinear relationship between *csat* and *percent*.

```
. set textsize 140
. graph percent percent2 high csat, matrix half label symbol(p)
. set textsize 100
```

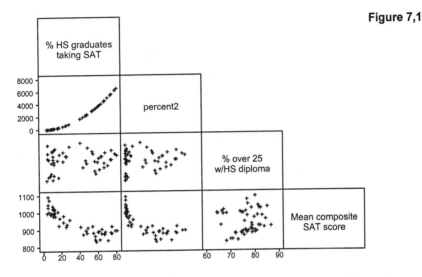

Figure 7,1

Several post-regression hypothesis tests perform checks on the model specification. The omitted-variables test **ovtest** essentially regresses *y* on the *x* variables, and also the second, third, and fourth powers of predicted *y* (after standardizing $\hat{y}$ to have mean 0 and variance 1). It then performs an *F* test of the null hypothesis that all three coefficients on those powers of $\hat{y}$ equal zero. If we reject this null hypothesis, further polynomial terms would improve the model. With the *csat* regression, we need not reject the null hypothesis.

```
. ovtest

Ramsey RESET test using powers of the fitted values of csat
        Ho:  model has no omitted variables
                 F(3, 44) =        1.48
                 Prob > F =        0.2319
```

A heteroskedasticity test, **hettest**, tests the assumption of constant error variance by examining whether squared standardized residuals are linearly related to $\hat{y}$ (see Cook and Weisberg 1994 for discussion and example). Results from the *csat* regression suggest that in this instance we should reject the null hypothesis of constant variance.

```
. hettest

Cook-Weisberg test for heteroskedasticity using fitted values of csat
        Ho: Constant variance
            chi2(1)     =        4.86
            Prob > chi2 =        0.0274
```

"Significant" heteroskedasticity implies that our standard errors and hypothesis tests might be invalid. Figure 7.2, in the next section, shows why this result occurs.

Diagnostic Plots

Chapter 6 demonstrated how **predict** can create new variables holding residual and predicted values after a **regress** command. To obtain these values from our regression of *csat* on *percent*, *percent2*, and *high*, we type the two commands:

```
. predict yhat3
. predict e3, resid
```

The new variables named *e3* (residuals) and *yhat3* (predicted values) could be displayed in a residual-versus-predicted graph by typing **graph e3 yhat**. The **rvfplot** (residual-versus-fitted) command obtains such graphs in a single step. The version in Figure 7.2 includes optional one-way scatterplots and boxplots in the margins, a horizontal line at 0 (the residual mean), neatly labeled axes, and a border drawn around the plot.

```
. rvfplot, oneway twoway box yline(0) ylabel xlabel border
```

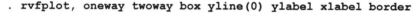

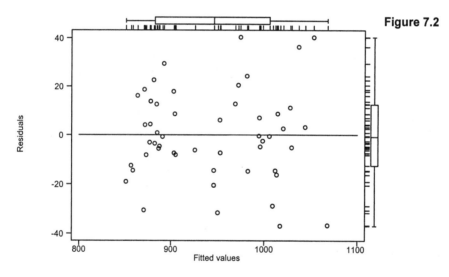

Figure 7.2

Figure 7.2 shows residuals symmetrically distributed around 0 (symmetry is consistent with the normal-errors assumption), and with no evidence of outliers or curvilinearity. The dispersion of the residuals is greater with above-average predicted values of *y*, however, which is why **hettest** earlier rejected the constant-variance hypothesis.

Residual-versus-fitted plots provide a one-graph overview of the regression residuals. For more detailed study, we can plot residuals against each predictor variable separately through a series of "residual-versus-predictor" commands. To graph the residuals against predictor *high* (not shown), type

```
. rvpplot high
```

The one-variable graphs described in Chapter 3 can also be employed for residual analysis. For example, we could use boxplots to check the residuals for outliers or skew, or quantile-normal plots to evaluate the assumption of normal errors.

Added-variable plots are valuable diagnostic tools, known by different names including partial-regression leverage plots, adjusted partial residual plots, or adjusted variable plots. They depict the relation between *y* and one *x* variable, adjusting for the effects of other *x* variables. If we regressed *y* on *x2* and *x3*, and likewise regressed *x1* on *x2* and *x3*, then took the residuals from each regression and graphed these residuals in a scatterplot, we would obtain an added-variable plot for the relation between *y* and *x1*, adjusted for *x2* and *x3*. An **avplot** command performs the necessary calculations automatically. We can draw the adjusted-variable plot for predictor *high*, for example, just by typing

```
. avplot high
```

Speeding the process further, we could type **avplots** to obtain a complete set of tiny added-variable plots with each of the predictor variables in the preceding regression. Figure 7.3 shows the results from the regression of *csat* on *percent*, *percent2*, and *high*. The lines drawn in added-variable plots have slopes equal to the corresponding partial regression coefficients. For example, the slope of the line at lower left in Figure 7.3 equals 2.99, which is the coefficient on *high*.

```
. set textsize 170
. avplots, ylabel xlabel border
. set textsize 100
```

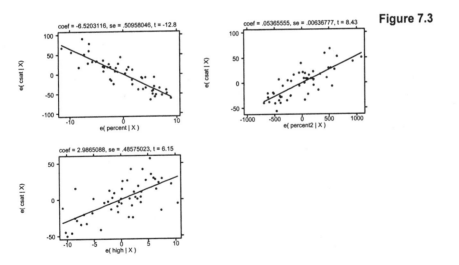

Figure 7.3

Added-variable plots help to uncover observations exerting a disproportionate influence on the regression model. In simple regression with one *x* variable, ordinary scatterplots suffice for this purpose. In multiple regression, however, the signs of influence become more subtle. An observation with an unusual combination of values on several *x* variables might have high leverage, or potential to influence the regression, even though none of its individual *x* values is unusual by itself. High-leverage observations show up in added-variable plots as points horizontally distant from the rest of the data. We see no such problems in Figure 7.3, however.

With added-variable plots, as with any scatterplot, we can use observation identifiers as the plotting symbols. Suppose that we did notice an outlier in an added-variable plot, and wanted to learn which observation this was. In the states' SAT scores example, we might do so by using state names as the plotting symbols. For readability, these names are plotted at 120% of normal size in Figure 7.4.

```
. avplot high, symbol([state]i) psize(120) ylabel xlabel
```

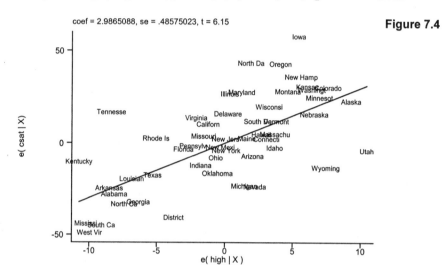

Component-plus-residual plots, produced by commands of the form **cprplot x1**, take a different approach to graphing multiple regression. The component-plus residual plot for variable *x1* graphs each observation's residual plus its component predicted from *x1*,

$$e_i + b_1 x1_i$$

against values of *x1*. Such plots might help diagnose nonlinearities and suggest alternative functional forms. An augmented component-plus-residual plot (Mallows 1986) works somewhat better, although both types often seem inconclusive. Figure 7.5 shows an augmented component-plus-residual plot from the regression of *csat* on *percent*, *percent2*, and *high*.

```
. acprplot high, connect(m) bands(5) ylabel xlabel border
```

Figure 7.5

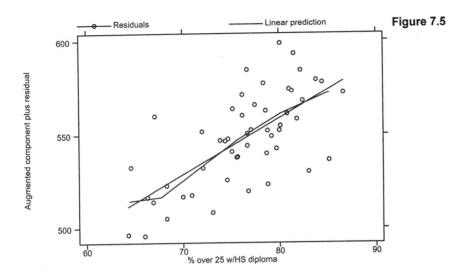

The straight line in Figure 7.5 corresponds to the regression model. The segmented curve connects cross-medians (**connect(m)**) within five vertical bands (**bands(5)**). If these cross-medians showed some curved pattern, unlike the linear regression model, we would have reason to doubt the model's adequacy. In Figure 7.5, however, the component-plus-residuals medians closely follow the regression model. This plot reinforces the conclusion we drew earlier from Figure 7.2: the present regression model adequately accounts for all nonlinearity visible in the raw data (Figure 7.1), so that none remains apparent in its residuals.

As its name implies, a leverage-versus-squared-residuals plot graphs leverage (hat matrix diagonals) against the residuals squared. Figure 7.6 shows such a plot for the *csat* regression. To identify individual outliers, we make values of *state*, at 120% of their default size, the plotting symbols.

```
. lvr2plot, ylabel xlabel border symbol([state]) psize(120)
```

Figure 7.6

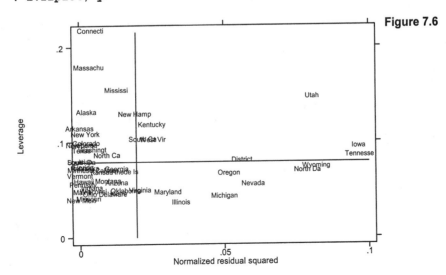

Lines in a leverage-versus-squared-residuals plot mark the means of leverage (horizontal line) and squared residuals (vertical line). Leverage tells us how much potential for influencing the regression an observation has, based on its particular combination of *x* values. Extreme *x* values or unusual combinations give an observation high leverage. A large squared residual indicates an observation with *y* value much different from that predicted by the regression model. Connecticut, Massachusetts, and Mississippi have the greatest potential leverage, but the model fits them relatively well. (This is not necessarily good. Sometimes, although not here, high-leverage observations exert so much influence that they control the regression, and it *must* fit them well.) Iowa and Tennessee are poorly fit, but have less potential influence. Utah stands out as one observation that is both ill fit and potentially influential. We can read its values by listing just this state. Because *state* is a string variable, we enclose the value "Utah" in double quotes.

```
. list csat yhat3 percent high e3 if state == "Utah"

          csat      yhat3     percent      high          e3
45.       1031     1067.712         5      85.1   -36.71239
```

Only 5% of Utah students took the SAT, and 85.1% of the state's adults graduated from high school. This unusual combination of near-extreme values on both *x* variables is the source of the state's leverage, and leads our model to predict mean SAT scores 36.7 points higher than what Utah students actually achieved. To see exactly how much difference this one observation makes, we could repeat the regression using Stata's "not equal to" qualifier **!=** to set Utah aside.

```
. regress csat percent percent2 high if state != "Utah"
```

Source	SS	df	MS			
Model	201097.423	3	67032.4744			
Residual	15214.0968	46	330.741235			
Total	216311.52	49	4414.52082			

	Number of obs	=	50
	F(3, 46)	=	202.67
	Prob > F	=	0.0000
	R-squared	=	0.9297
	Adj R-squared	=	0.9251
	Root MSE	=	18.186

csat	Coef.	Std. Err.	t	P>|t|	[95% Conf.	Interval]
percent	-6.778706	.5044217	-13.44	0.000	-7.794054	-5.763357
percent2	.0563562	.0062509	9.02	0.000	.0437738	.0689387
high	3.281765	.4865854	6.74	0.000	2.302319	4.26121
_cons	827.1159	36.17138	22.87	0.000	754.3067	899.9252

In the *n* = 50 (instead of *n* = 51) regression, all three coefficients have strengthened a bit since we deleted an ill-fit observation. The general conclusions remain unchanged, however.

Chambers et al. (1983) and Cook and Weisberg (1994) provide more detailed examples and explanations of diagnostic plots and other graphical methods for data analysis.

Diagnostic Case Statistics

After using **regress** or **anova**, we can obtain a variety of diagnostic statistics through the **predict** command (see Chapter 6 or type **help regress**). The variables created by `predict` are case statistics, meaning that they have values for each observation in the data. Diagnostic work usually begins by calculating the predicted values and residuals.

There is some overlap in purpose among other `predict` statistics. Many attempt to measure how much each observation influences regression results. "Influencing regression results," however, could refer to several different things—effects on the y-intercept, on a particular slope coefficient, on all the slope coefficients, or on the estimated standard errors, for example. Consequently, we have a variety of alternative case statistics designed to measure influence.

Standardized and studentized residuals (`rstandard` and `rstudent`) help to identify outliers among the residuals—observations that particularly contradict the regression model. Studentized residuals have the most straightforward interpretation. They correspond to the t statistic we would obtain by including in the regression a dummy predictor coded 1 for that observation and 0 for all others. Thus, they test whether a particular observation significantly shifts the y-intercept.

Hat matrix diagonals (`hat`) measure leverage, meaning the potential to influence regression coefficients. Observations possess high leverage when their x values (or their combination of x values) are unusual.

Several other statistics measure actual influence on coefficients. *DFBETA*s indicate by how many standard errors the coefficient on $x1$ would change if observation i were dropped from the regression. These can be obtained for a single predictor, $x1$, in either of two ways: through the `predict` option `dfbeta(x1)` or through the command `dfbeta` .

Cook's D (`cooksd`), Welsch's distance (`welsch`), and *DFITS* (`dfits`), unlike *DFBETA*, all summarize how much observation i influences the regression model as a whole—or equivalently, how much observation i influences the set of predicted values. *COVRATIO* measures the influence of the ith observation on the estimated standard errors. Below we generate a full set of diagnostic statistics including *DFBETA*s for all three predictors. Note that `predict` supplies variable labels automatically for the variables it creates, but `dfbeta` does not. We begin by repeating our original regression to ensure that these post-regression diagnostics refer to the proper ($n = 51$) model.

```
. quietly regress csat percent percent2 high
. predict standard, rstandard
. predict student, rstudent
. predict h, hat
. predict D, cooksd
. predict DFITS, dfits
. predict W, welsch
. predict COVRATIO, covratio
. dfbeta

         DFpercent:  DFbeta(percent)
        DFpercent2:  DFbeta(percent2)
            DFhigh:  DFbeta(high)
```

```
. describe standard - DFhigh
```

variable name	storage type	display format	value label	variable label
standard	float	%9.0g		Standardized residuals
student	float	%9.0g		Studentized residuals
h	float	%9.0g		Leverage
D	float	%9.0g		Cook's D
DFITS	float	%9.0g		Dfits
W	float	%9.0g		Welsch distance
COVRATIO	float	%9.0g		Covratio
DFpercent	float	%9.0g		
DFpercent2	float	%9.0g		
DFhigh	float	%9.0g		

```
. summarize standard - DFhigh
```

Variable	Obs	Mean	Std. Dev.	Min	Max
standard	51	-.0031359	1.010579	-2.099976	2.233379
student	51	-.00162	1.032723	-2.182423	2.336977
h	51	.0784314	.0373011	.0336437	.2151227
D	51	.0219941	.0364003	.0000135	.1860992
DFITS	51	-.0107348	.3064762	-.896658	.7444486
W	51	-.089723	2.278704	-6.854601	5.52468
COVRATIO	51	1.092452	.1316834	.7607449	1.360136
DFpercent	51	.000938	.1498813	-.5067295	.5269799
DFpercent2	51	-.0010659	.1370372	-.440771	.4253958
DFhigh	51	-.0012204	.1747835	-.6316988	.3414851

summarize shows us the minimum and maximum values of each statistic, so we can quickly check whether any are large enough to cause concern. For example, special tables could be used to determine whether the observation with the largest absolute studentized residual (*student*) constitutes a significant outlier. Alternatively, we could apply the Bonferroni inequality and t distribution table: max| *student* | is significant at level α if | t | is significant at α/n. In this example, we have max| *student* | = 2.337 (Iowa) and $n = 51$. For Iowa to be a significant outlier (cause a significant shift in intercept) at $\alpha = .05$, $t = 2.337$ must be significant at .05 / 51:

```
. display .05/51
.00098039
```

Stata's **ttail( )** function can approximate the probability of | t | > 2.337, given $df = n - K - 1 = 51 - 3 - 1 = 47$:

```
. display 2*ttail(47, 2.337)
.02375138
```

The obtained P-value ($P = .0238$) is not below $\alpha/n = .00098$, so Iowa is not a significant outlier at $\alpha = .05$.

Studentized residuals measure the ith observation's influence on the y-intercept. Cook's D, *DFITS*, and Welsch's distance all measure the ith observation's influence on all coefficients in the model (or, equivalently, on all n predicted y values). To list the 5 most influential observations as measured by Cook's D, type

```
. sort D
. list state yhat3 D DFITS W in -5/1
```

	state	yhat3	D	DFITS	W
47.	North Dakota	1036.696	.0705921	.5493086	4.020527
48.	Wyoming	1017.005	.0789454	-.5820746	-4.270465
49.	Tennessee	974.6981	.111718	.6992343	5.162398
50.	Iowa	1052.78	.1265392	.7444486	5.52468
51.	Utah	1067.712	.1860992	-.896658	-6.854601

The **in -5/1** qualifier tells Stata to list only the fifth-from-last (–5) through last (lowercase letter "l") observations. Figure 7.7 shows one way to display influence graphically: Symbols in a residual-versus-predicted plot are given sizes ("importance weights") proportional to values of Cook's *D*. Five influential observations stand out, with large positive or negative residuals and high predicted *csat* values.

```
. graph e3 yhat3 [iweight = D], ylabel xlabel yline(0)
```

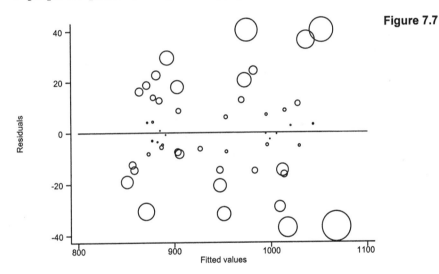

Figure 7.7

Although they have different statistical rationales, Cook's *D*, Welsch's distance, and *DFITS* are closely related. In practice they tend to flag the same observations as influential. Figure 7.8 shows their similarity in the example at hand.

```
. set textsize 120
. graph D W DFITS, matrix label half symbol(d)
. set textsize 100
```

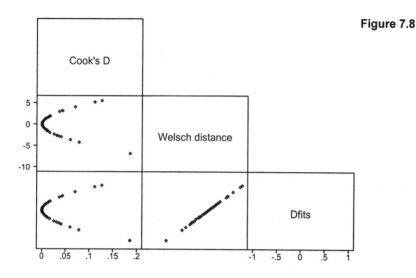

Figure 7.8

*DFBETA*s indicate how much each observation influences each regression coefficient. Typing **dfbeta** after a regression automatically generates *DFBETA*s for each predictor. In this example, they received the names *DFpercent* (*DFBETA* for predictor *percent*), *DFpercent2*, and *DFhigh*. Figure 7.9 shows their distributions as boxplots.

```
. graph DFpercent DFpercent2 DFhigh, box ylabel(-.6,-.4 to .6)
     symbol([state]) psize(120)
```

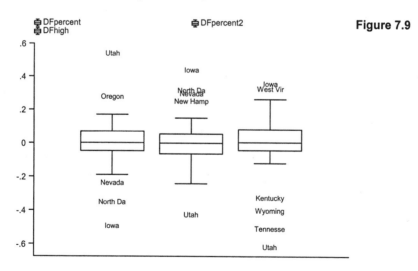

Figure 7.9

From left to right, Figure 7.9 plots the distributions of *DFBETA*s for *percent*, *percent2*, and *high*. (We could more easily distinguish them in color.) The extreme values in each plot belong to Iowa and Utah, which also have the two highest Cook's *D* values. For example, Utah's *DFhigh* = −.63. This tells us that Utah causes the coefficient on *high* to be .63 standard errors lower than it would be if Utah were set aside. Similarly, *DFpercent* = .53 indicates that with

Utah present, the coefficient on *percent* is .53 standard errors higher (because the *percent* regression coefficient is negative, "higher" means closer to 0) than it otherwise would be. Thus, Utah weakens the apparent effects of both *high* and *percent*.

The most direct way to learn how particular observations affect a regression is to repeat the regression with those observations set aside. For example, we could set aside all states that move any coefficient by half a standard error (that is, have absolute *DFBETA*s greater than .5):

```
. regress csat percent percent2 high if abs(DFpercent) < .5 &
    abs(DFpercent2) < .5 & abs(DFhigh) < .5
```

Source	SS	df	MS		
Model	175366.782	3	58455.5939		
Residual	11937.1351	44	271.298525		
Total	187303.917	47	3985.18972		

Number of obs = 48
F(3,   44) = 215.47
Prob > F   = 0.0000
R-squared   = 0.9363
Adj R-squared = 0.9319
Root MSE   = 16.471

csat	Coef.	Std. Err.	t	P>|t|	[95% Conf.	Interval]
percent	-6.510868	.4700719	-13.85	0.000	-7.458235	-5.5635
percent2	.0538131	.005779	9.31	0.000	.0421664	.0654599
high	3.35664	.4577103	7.33	0.000	2.434186	4.279095
_cons	815.0279	33.93199	24.02	0.000	746.6424	883.4133

Careful inspection will reveal the details in which this regression table (based on $n = 48$) differs from its $n = 51$ or $n = 50$ counterparts seen earlier. Our central conclusion—that mean state SAT scores are well-predicted by the percent of adults with high school diplomas and, curvilinearly, by the percent of students taking the test—remains unchanged, however.

Although diagnostic statistics draw attention to influential observations, they do not answer the question of whether we should set those observations aside. That requires a substantive decision based on careful evaluation of the data and research context. In this example, we have no substantive reason to discard any states, and even the most influential of them do not fundamentally change our conclusions.

Using any fixed definition of what constitutes an "outlier," we are liable to see more of them in larger-size samples. For this reason, sample-size-adjusted cutoffs are sometimes recommended for identifying unusual observations. After fitting a regression model with K coefficients (including the constant) based on n observations, we might look more closely at those observations for which any of the following are true:

> leverage $h > 2K/n$
>
> Cook's $D > 4/n$
>
> $DFITS > 2\sqrt{K/n}$
>
> Welsch's $W > 3\sqrt{K}$
>
> $DFBETA > 2/\sqrt{n}$
>
> $|COVRATIO - 1| \geq 3K/n$

The reasoning behind these cutoffs, and the diagnostic statistics more generally, can be found in Cook and Weisberg (1982, 1994); Belsley, Kuh, and Welsch (1980); or Fox (1991). For an introduction, see Rawlings (1988).

Multicollinearity

If perfect multicollinearity (linear relation) exists among the predictors, regression equations become unsolvable. Stata handles this by warning the user and then automatically dropping one of the offending predictors. High but not perfect multicollinearity causes more subtle problems. When we add a new x variable that is strongly related to x variables already in the model, symptoms of possible trouble include the following:

1. Substantially higher standard errors, with correspondingly lower t statistics.
2. Unexpected changes in coefficient magnitudes or signs.
3. Nonsignificant coefficients despite a high R^2.

Multiple regression attempts to estimate the independent effects of each x variable. There is little information for doing so, however, if one or more of the x variables does not have much independent variation. The symptoms listed above warn that coefficient estimates have become unreliable, and might shift drastically with small changes in the sample or model. Further troubleshooting is needed to determine whether multicollinearity really is at fault and, if so, what should be done about it.

Multicollinearity cannot necessarily be detected, or ruled out, by examining a matrix of correlations between variables. A better assessment comes from regressing each x on all of the other x variables. Then we calculate $1 - R^2$ from this regression to see what fraction of the first x variable's variance is independent of the other x variables. For example, about 97% of *high*'s variance is independent of *percent* and *percent2*:

```
. quietly regress high percent percent2
. display 1 - e(r2)
.96942331
```

After regression, e(r2) holds the value of R^2. Similar commands reveal that only 4% of *percent*'s variance is independent:

```
. quietly regress percent high percent2
. display 1 - e(r2)
.04010307
```

This finding about *percent* and *percent2* is not surprising. In polynomial regression or regression with interaction terms, some x variables are calculated directly from other x variables. Although strictly speaking their relation is nonlinear, it often is close enough to linear to raise problems of multicollinearity.

The post-regression command **vif**, for variance inflation factor, performs similar calculations automatically. This provides a quick and straightforward check for multicollinearity:

```
. vif

    Variable |      VIF      1/VIF
-------------+----------------------
     percent |    24.94    0.040103
    percent2 |    24.78    0.040354
        high |     1.03    0.969423
-------------+----------------------
    Mean VIF |    16.92
```

The 1/VIF column at right in a **vif** table gives values equal to $1 - R^2$ from the regression of each x on the other x variables, as can be seen by comparing the values for *high* (.969423) or *percent* (.040103) with our earlier **display** calculations. That is, 1/VIF (or $1 - R^2$) tells us what proportion of an *x* variable's variance is independent of all the other *x* variables. A low proportion, such as the .04 (4% independent variation) of *percent* and *percent2*, indicates potential trouble. Some analysts set a minimum level, called *tolerance*, for the 1/VIF value, and automatically exclude predictors that fall below their tolerance criterion.

The VIF column at center in a **vif** table reflects the degree to which other coefficients' variances (and standard errors) are increased due to the inclusion of that predictor. We see that *high* has virtually no impact on other variances, but *percent* and *percent2* impact the variances substantially. VIF values provide guidance but not direct measurements of the increase in coefficient variances. The following commands show the impact directly by displaying standard error estimates for the coefficient on *percent*, when *percent2* is and is not included in the model:

```
. quietly regress csat percent percent2 high
. display _se[percent]
.50958046
. quietly regress csat percent high
. display _se[percent]
.16162193
```

With *percent2* included in the model, the standard error for *percent* is three times higher:

$$.50958046 / .16162193 = 3.1529166$$

This corresponds to a tenfold increase in the coefficient's variance.

How much variance inflation is too much? Chatterjee, Hadi, and Price (2000) suggest the following as guidelines for the presence of multicollinearity:

1. The largest VIF is greater than 10; or
2. the mean VIF is larger than 1.

With our largest VIFs close to 25, and the mean almost 17, the *csat* regression clearly meets both criteria. How troublesome the problem is, and what, if anything, should be done about it, are the next questions to consider.

Because *percent* and *percent2* are closely related, we cannot estimate their separate effects with nearly as much precision as we could the effect of either predictor alone. That is why the standard error for the coefficient on *percent* increases threefold when we compare the regression of *csat* on *percent* and *high* to a polynomial regression of *csat* on *percent*, *percent2*, and *high*. Despite this loss of precision, however, we can still distinguish all the coefficients from zero. Moreover, the polynomial regression obtains a better prediction model. For these reasons, the multicollinearity in this regression does not necessarily pose a great problem, or require a solution. We could simply live with it as one feature of an otherwise acceptable model.

When solutions are needed, a simple trick called "centering" often succeeds in reducing multicollinearity in polynomial or interaction-effect models. Centering involves subtracting the mean from *x* variable values before generating polynomial or product terms. Subtracting the mean creates a new variable centered on zero and much less correlated with its own squared values. The resulting regression fits the same as an uncentered version. By reducing multicollinearity, centering often (*but not always*) yields more precise coefficient estimates with lower standard errors. The commands below generate a centered version of *percent* named *Cpercent*, and then obtain squared values of *Cpercent* named *Cpercent2*.

```
. summarize percent

    Variable |      Obs        Mean    Std. Dev.        Min        Max
-------------+--------------------------------------------------------
     percent |       51    35.76471    26.19281          4         81

. generate Cpercent = percent - r(mean)
. generate Cpercent2 = Cpercent ^2
. correlate Cpercent Cpercent2 percent percent2 high csat

(obs=51)

             | Cpercent Cperce~2  percent percent2     high     csat
-------------+------------------------------------------------------
    Cpercent |   1.0000
   Cpercent2 |   0.3791   1.0000
     percent |   1.0000   0.3791   1.0000
    percent2 |   0.9794   0.5582   0.9794   1.0000
        high |   0.1413  -0.0417   0.1413   0.1176   1.0000
        csat |  -0.8758  -0.0428  -0.8758  -0.7946   0.0858   1.0000
```

Whereas *percent* and *percent2* have a near-perfect correlation with each other ($r = .9794$), the centered versions *Cpercent* and *Cpercent2* are just moderately correlated ($r = .3791$). Otherwise, correlations involving *percent* and *Cpercent* are identical because centering is a linear transformation. Correlations involving *Cpercent2* are different from those with *percent2*, however. Figure 7.10 shows scatterplots that help to visualize these correlations, and the transformation's effects.

```
. set textsize 160
. graph Cpercent Cpercent2 percent percent2 high csat, matrix half
      symbol(p)
. set textsize 100
```

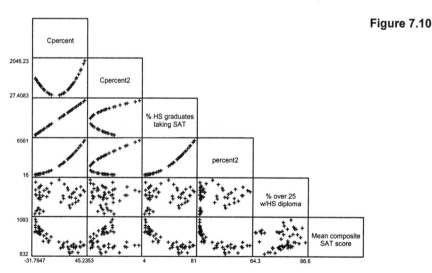

Figure 7.10

The R^2, overall F test, predictions, and many other aspects of a model should be unchanged after centering. Differences will be most noticeable in the centered variable's coefficient and standard error:

```
. regress csat Cpercent Cpercent2 high
```

```
      Source |       SS       df       MS              Number of obs =      51
-------------+------------------------------           F(  3,    47) =  193.37
       Model | 207225.103      3  69075.0343           Prob > F      =  0.0000
    Residual |  16789.407     47  357.221426           R-squared     =  0.9251
-------------+------------------------------           Adj R-squared =  0.9203
       Total |  224014.51     50  4480.2902            Root MSE      =   18.90


        csat |      Coef.   Std. Err.       t    P>|t|     [95% Conf. Interval]
-------------+----------------------------------------------------------------
    Cpercent |  -2.682362   .1119085    -23.97   0.000    -2.907493   -2.457231
   Cpercent2 |   .0536555   .0063678      8.43   0.000     .0408452    .0664659
        high |   2.986509   .4857502      6.15   0.000     2.009305    3.963712
       _cons |   680.2552   37.82329     17.99   0.000     604.1646    756.3458
```

In this example, the standard error of the coefficient on *Cpercent* is actually lower (.1119085 compared with .16162193) when *Cpercent2* is included in the model. The t statistic is correspondingly larger. Thus, it appears that centering did improve that coefficient estimate's precision. The VIF table now gives less cause for concern: each of the three predictors has more than 80% independent variation, compared with 4% for *percent* and *percent2* in the uncentered regression.

```
. vif

    Variable |       VIF       1/VIF
-------------+----------------------
    Cpercent |      1.20    0.831528
   Cpercent2 |      1.18    0.846991
        high |      1.03    0.969423
-------------+----------------------
    Mean VIF |      1.14
```

Another diagnostic table sometimes consulted to check for multicollinearity is the matrix of correlations between estimated coefficients (*not* variables). This matrix can be displayed after **regress**, **anova**, or other model-fitting procedures by typing

```
. correlate, _coef

             | Cpercent Cperce~2     high     _cons
-------------+------------------------------------------
    Cpercent |   1.0000
   Cpercent2 |  -0.3893   1.0000
        high |  -0.1700   0.1040   1.0000
       _cons |   0.2105  -0.2151  -0.9912   1.0000
```

High correlations between pairs of coefficients indicate possible collinearity problems.

By adding the option **covariance**, we can see the coefficients' variance–covariance matrix, from which standard errors are derived:

```
. correlate, _coef covariance

             |  Cpercent Cperce~2      high     _cons
-------------+----------------------------------------
    Cpercent |   .012524
   Cpercent2 |  -.000277   .000041
        high |  -.009239   .000322   .235953
       _cons |   .891126  -.051817  -18.2105    1430.6
```

8

Fitting Curves

Basic regression and correlation methods assume linear relationships. Linear models provide reasonable and simple approximations for many real phenomena, over a limited range of values. But analysts also encounter phenomena where linear approximations are too simple; these call for nonlinear alternatives. This chapter describes three broad approaches to modeling nonlinear or curvilinear relationships:

1. Nonparametric methods, including band regression and lowess smoothing.
2. Linear regression with transformed variables ("curvilinear regression"), including Box–Cox methods.
3. Nonlinear regression.

Nonparametric regression serves as an exploratory tool because it can summarize data patterns visually without requiring the analyst to specify a particular model in advance. Transformed variables extend the usefulness of linear parametric methods, such as OLS regression (**regress**), to encompass curvilinear relationships as well. Nonlinear regression, on the other hand, requires a different class of methods that can estimate parameters of intrinsically nonlinear models.

Example Commands

. **boxcox y x1 x2 x3, model(lhs)**

Finds maximum-likelihood estimates of the parameter λ (lambda) for a Box–Cox transformation of y, assuming that $y^{(\lambda)}$ is a linear function of $x1$, $x2$, and $x3$ plus Gaussian constant-variance errors. The **model(lhs)** option restricts transformation to the left-hand-side variable y. Other options could transform right-hand-side (x) variables by the same or different parameters, and control further details of the model. Type **help boxcox** for the syntax and a complete list of options. The *Reference Manual* gives technical details.

. **graph y x, bands(10) connect(m)**

Produces a y versus x scatterplot with line segments connecting the cross-medians (median x, median y points) within 10 equal-width vertical bands. This is one form of "band regression." Typing **connect(s)** instead of **connect(m)** would result in the cross-medians being connected by a smooth cubic spline curve instead of by line segments.

. **graph** *x1 x2 x3 y*, **bands(4) connect(s) matrix**

Produces a scatterplot matrix, with band regression curves (based on four bands) in each plot.

. **ksm** *y x*, **lowess bwidth(.4) gen(***newvar***)**

Draws a lowess-smoothed curve on a scatterplot of *y* versus *x*. Lowess calculations use a bandwidth of .4 (40% of the data). Smoothed values are kept and named *newvar*.

. **nl exp2** *y x*

Uses iterative nonlinear least squares to fit an exponential growth model of the form

$$\text{predicted } y = b_1 b_2^{\,x}$$

The term **exp2** refers to a separate program that specifies the model itself. You can write a program to define your own model, or use one of the common models (including exponential, logistic, and Gompertz) supplied with Stata. After **nl**, use **predict** to generate predicted values or residuals.

. **nl log4** *y x*, **init(B0=5, B1=25, B2=.1, B3=50)**

Fits a logistic growth model (**log4**) of the form

$$\text{predicted } y = b_0 + b_1 /(1 + \exp(-b_2(x - b_3)))$$

Sets initial parameter values for the iterative estimation process at $b_0 = 5$, $b_1 = 25$, $b_2 = .1$, and $b_3 = 50$.

. **regress** *lny x1 sqrtx2 invx3*

Performs curvilinear regression using the variables *lny*, *x1*, *sqrtx2*, and *invx3*. These variables were previously generated by nonlinear transformations of the raw variables *y*, *x2*, and *x3* through commands such as the following:

. **generate** *lny* = **ln(***y***)**

. **generate** *sqrtx2* = **sqrt(***x2***)**

. **generate** *invx3* = **1/***x3*

When, as in this example, the *y* variable was transformed, the predicted values generated by **predict** *yhat*, or residuals generated by **predict e, resid**, will be likewise in transformed units. For graphing or other purposes, we might want to return predicted values or residuals to raw-data units, using inverse transformations such as

. **replace** *yhat* = **exp(***yhat***)**

Band Regression

Nonparametric regression methods generally do not yield an explicit regression equation. They are primarily graphic tools for displaying the relationship, possibly nonlinear, between *y* and *x*. Stata can draw a simple kind of nonparametric regression, band regression, onto any scatterplot or scatterplot matrix. For illustration, consider these sobering Cold War data (*missile.dta*) from MacKenzie (1990). The observations are 48 types of long-range nuclear missiles, deployed by the U.S. and Soviet Union during their arms race, 1958 to 1990:

```
Contains data from C:\data\missile.dta
  obs:           48                      Missiles (MacKenzie 1990)
  vars:           6                      7 Jul 2001 10:32
  size:       1,392 (99.2% of memory free) ----------------------------------
---------------
                storage  display    value
variable name     type   format     label    variable label
---------------------------------------------------------------------
missile          str15   %15s                 Missile
country          byte    %8.0g      soviet    US or Soviet missile?
year             int     %8.0g                Year of first deployment
type             byte    %8.0g      type      ICBM or submarine-launched?
range            int     %8.0g                Range in nautical miles
CEP              float   %9.0g                Circular Error Probable (miles)
---------------------------------------------------------------------
Sorted by:  country  year
```

Variables in *missile.dta* include an accuracy measure called the "Circular Error Probable" (*CEP*). *CEP* represents the radius of a bulls eye within which 50% of the missile's warheads should land. Year by year, scientists on both sides worked to improve accuracy (Figure 8.1).

`. graph CEP year, bands(8) connect(m) ylabel xlabel`

Figure 8.1

Figure 8.1 shows *CEP* declining (accuracy increasing) over time. The options **bands(8) connect(m)** instruct **graph** to divide the scatterplot into 8 equal-width vertical bands and draw line segments connecting the points (median *x*, median *y*) within each band. This curve traces how the median of *CEP* changes with *year*.

Nonparametric regression does not require the analyst to specify a relationship's functional form in advance. Instead, it allows us to explore the data with an "open mind." This process often uncovers interesting results, such as when we view trends in U.S. and Soviet missile accuracy separately (Figure 8.2). The **by(country)** option in the following command produces a separate plot for each country, each with its own band-regression curve. **bsize(160)** causes the by-group labels (here, country names) to appear at 160% of their normal size for better readability.

```
. graph CEP year, connect(m) bands(8) ylabel xlabel by(country)
   bsize(160)
```

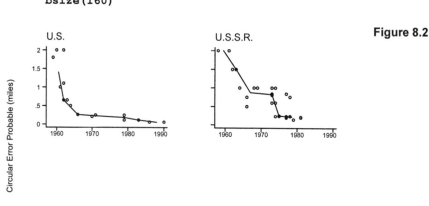

Figure 8.2

Circular Error Probable (miles)

Year of first deployment
Graphs by US or Soviet missile?

The shapes of the two curves in Figure 8.2 differ substantially. U.S. missiles became much more accurate in the 1960s, permitting a shift to smaller warheads. Three or more small warheads would fit on the same size of missile that formerly carried one large warhead. The accuracy of Soviet missiles improved more slowly, apparently stalling during the late 1960s to early 1970s, and remained a decade or so behind their American counterparts. To make up for this accuracy disadvantage, Soviet strategy emphasized larger rockets carrying high-yield warheads. Nonparametric regression can assist with a qualitative description of this sort or serve as a preliminary to fitting parametric models such as those described later.

We can add band regression curves to any scatterplot or scatterplot matrix by specifying the **connect(m) bands( )** options. Band regression's simplicity makes it a convenient exploratory tool, but it possesses one notable disadvantage—the bands have the same width across the range of x values, although some of these bands contain few or no observations. With normally distributed variables, for example, data density decreases toward the extremes. Consequently, the left and right endpoints of the band regression curve (which tend to dominate its appearance) often reflect just a few data points. The next section describes a more sophisticated, computation-intensive approach.

Lowess Smoothing

The **ksm** command accomplishes a form of nonparametric regression called lowess smoothing (for locally weighted scatterplot smoothing). The following example draws a lowess-smoothed graph of *CEP* against *year* for U.S. missiles only (*country* == 0). **ksm** understands most of the usual **graph** options.

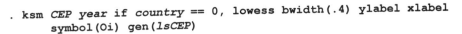

```
. ksm CEP year if country == 0, lowess bwidth(.4) ylabel xlabel
     symbol(Oi) gen(lsCEP)
```

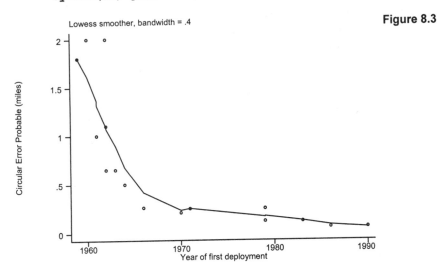

Figure 8.3

Like Figure 8.2, Figure 8.3 shows U.S. missile accuracy improving rapidly during the 1960s and progressing at a more gradual rate in the 1970s and 1980s. Lowess-smoothed values of *CEP* are generated here with the name *lsCEP*. The **bwidth(.4)** option specifies the lowess bandwidth: the fraction of the sample used in smoothing each point. The default is **bwidth(.8)**. The closer bandwidth is to 1, the greater the degree of smoothing.

Lowess predicted (smoothed) *y* values for *n* observations result from *n* weighted regressions. Let *k* represent the half-bandwidth, truncated to an integer. For each y_i, a smooth-ed value y_i^s is obtained by weighted regression involving only those observations within the interval from $i = \max(1, i - k)$ through $i = \min(i + k, n)$. The *j*th observation within this interval receives weight w_j according to a tricube function:

$$w_j = (1 - |u_j|^3)^3$$

where

$$u_j = (x_i - x_j) / \Delta$$

Δ stands for the distance between x_i and its furthest neighbor within the interval. Weights equal 1 for $x_i = x_j$, but fall off to zero at the interval's boundaries. See Chambers et al. (1983) or Cleveland (1993) for more discussion and examples of lowess methods.

ksm is a flexible command. In addition to the usual **graph** options, it offers these special options:

line	For running-line least squares smoothing; the default is running mean.
weight	For Cleveland's tricube weighting function; the default is unweighted.
lowess	Equivalent to specifying both options **line weight**.
bwidth( )	Specifies the bandwidth. Centered subsets of approximately bwidth × *n* observations are used for smoothing, except towards the endpoints where smaller, uncentered bands are used. The default is **bwidth(.8)**.

`logit`	Transforms smoothed values to logits.
`adjust`	Adjusts the mean of smoothed values to equal the mean of the original *y* variable; like `logit`, `adjust` is useful with dichotomous *y*.
`gen( )`	Creates *newvar* containing smoothed values of *y*.
`nograph`	Suppresses displaying the graph.

Because it requires *n* weighted regressions, lowess smoothing proceeds slowly with large samples.

In addition to smoothing *y*–*x* scatterplots, **ksm** can be used for exploratory time series smoothing. The file *ice.dta* contains results from the Greenland Ice Sheet 2 (GISP2) project described in Mayewski, Holdsworth, and colleagues (1993) and Mayewski, Meeker, and colleagues (1993). Researchers extracted and chemically analyzed an ice core representing more than 100,000 years of climate history. *ice.dta* includes a small fraction of this information: measured non-sea salt sulfate concentration and an index of "Polar Circulation Intensity" since AD 1500.

```
Contains data from C:\data\ice.dta
  obs:            271                     Greenland ice (Mayewski 1995)
  vars:             3                     7 Jul 2001 10:30
  size:         5,962 (99.2% of memory free)
-------------------------------------------------------------------------------
              storage  display   value
variable name  type    format    label     variable label
-------------------------------------------------------------------------------
year           int     %ty                 Year
sulfate        double  %10.0g              SO4 ion concentration, ppb
PCI            double  %6.0g               Polar Circulation Intensity
-------------------------------------------------------------------------------
Sorted by:  year
```

To retain more detail from this 271-point time series, we smooth with a relatively narrow bandwidth, only 5% of the sample:

`. ksm sulfate year, lowess bwidth(.05) gen(smooth) nograph`

Options in the previous command suppressed the default **ksm** graph but generated smoothed values as a new variable (*smooth*) so that we can construct a better-looking graph (Figure 8.4) directly:

`. graph sulfate smooth year, connect(ll) symbol(ii) ylabel`
`        xlabel(1500,1600 to 1900,1985) l1(Sulfate concentration)`

Non-sea salt sulfate (SO_4) reached the Greenland ice after being injected into the atmosphere, chiefly by volcanoes or the burning of fossil fuels such as coal and oil. Both the smoothed and raw curves in Figure 8.4 convey information. The smoothed curve shows oscillations around a slightly rising mean from 1500 through the early 1800s. After 1900, fossil fuels drive the smoothed curve upward, with temporary setbacks after 1929 (the Great Depression) and the early 1970s (combined effects of the U.S. Clean Air Act, 1970; the Arab oil embargo, 1973; and subsequent oil price hikes). Most of the sharp peaks of the raw data have been identified with known volcanic eruptions such as Iceland's Hekla (1970) or Alaska's Katmai (1912).

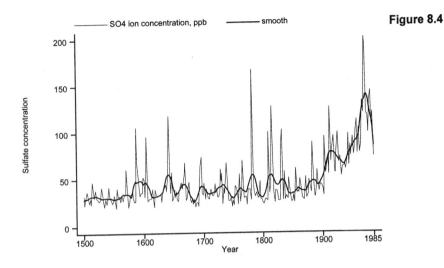

Figure 8.4

After smoothing time series data, it is often useful to study the smooth and rough (residual) series separately. Figure 8.5 presents both in a graph enhanced from Stata originals using a third-party graphics editing program, Metafile Companion (Companion Software).

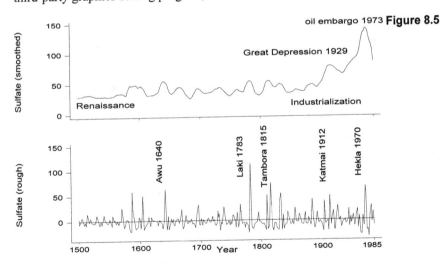

Figure 8.5

Regression with Transformed Variables — 1

By subjecting one or more variables to nonlinear transformation, and then including the transformed variable(s) in a linear regression, we implicitly fit a curvilinear model to the underlying data. Chapters 6 and 7 gave one example of this approach, polynomial regression, which incorporates second (and perhaps higher) powers of at least one x variable among the predictors. Logarithms also are used routinely in many fields. Other common transformations include those of the ladder of powers and Box–Cox transformations, introduced in Chapter 4.

   Dataset *tornado.dta* provides a simple illustration involving U.S. tornados from 1916 to 1986 (from the Council on Environmental Quality, 1988).

```
Contains data from C:\data\tornado.dta
  obs:            71                         U.S. tornados 1916-1986
                                             (Council on Env. Quality 1988)
  vars:            4                         7 Jul 2001 13:02
  size:          994 (99.2% of memory free)
--------------------------------------------------------------------------------
              storage  display   value
variable name  type    format    label     variable label
--------------------------------------------------------------------------------
year           int     %8.0g               Year
tornado        int     %8.0g               Number of tornados
lives          int     %8.0g               Number of lives lost
avlost         float   %9.0g               Average lives lost/tornado
--------------------------------------------------------------------------------
Sorted by:  year
```

   The number of fatalities decreased over this period, while the number of recognized tornados increased, because of improvements in warnings and our ability to detect more tornados, even those that do little damage. Consequently, the average lives lost per tornado (*avlost*) declined with time, but a linear regression (Figure 8.6) does not well describe this trend. The scatter descends more steeply than the regression line at first, then levels off in the mid-1950s. The regression line actually predicts negative numbers of deaths in later years. Furthermore, average tornado deaths exhibit more variation in early years than later—evidence of heteroskedasticity.

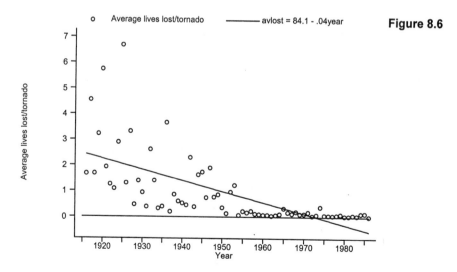

   The relation becomes linear, and heteroskedasticity vanishes if we work instead with logarithms of the average number of lives lost (Figure 8.7):

. **generate** *loglost* = **ln(***avlost***)**

```
. label variable loglost "ln(avlost)"

. regress loglost year
```

Source	SS	df	MS
Model	115.895325	1	115.895325
Residual	43.8807356	69	.63595269
Total	159.77606	70	2.28251515

```
Number of obs =      71
F(  1,    69) =  182.24
Prob > F      =  0.0000
R-squared     =  0.7254
Adj R-squared =  0.7214
Root MSE      =  .79747
```

loglost	Coef.	Std. Err.	t	P>|t|	[95% Conf. Interval]
year	-.0623418	.004618	-13.50	0.000	-.0715545 -.053129
_cons	120.5645	9.010312	13.38	0.000	102.5894 138.5395

```
. predict yhat2
(option xb assumed; fitted values)

. label variable yhat2 "ln(avlost) = 120.56 - .06year"

. label variable loglost "ln(avlost)"

. graph loglost yhat2 year, connect(.s) symbol(Si)
      ylabel(-4,-3 to 2) xlabel(1920,1930 to 1980)
      xtick(1915,1925 to 1985)
      ll(Natural log of average lives lost)
```

The resulting model is approximately

predicted $\ln(avlost) = 120.56 - .06year$

Because we regressed logarithms of lives lost on *year*, the model's predicted values are also measured in logarithmic units. Return these predicted values to their natural units (lives lost) by inverse transformation, in this case exponentiating (*e* to power) *yhat2*:

```
. replace yhat2 = exp(yhat2)
(71 real changes made)
```

Graphing these inverse-transformed predicted values reveals the curvilinear regression model, which we obtained by linear regression with a transformed *y* variable (Figure 8.8). Contrast Figures 8.7 and 8.8 with Figure 8.6 to see how transformation made the analysis both simpler and more realistic.

```
. graph avlost yhat2 year, connect(.s) symbol(Oi) ylabel(0,1 to 7)
      yline(0) xlabel(1920,1930 to 1980) xtick(1915,1925 to 1985)
```

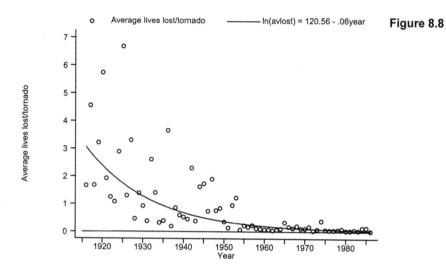

Figure 8.8

The **boxcox** command employs maximum-likelihood methods to fit curvilinear models involving Box–Cox transformations (introduced in Chapter 4). Fitting a model with Box–Cox transformation of the dependent variable (**model(lhs)** specifies left-hand side) to the tornado data, we obtain results quite similar to the model of Figures 8.7 and 8.8. The **nolog** option in the following command does not affect the model, but suppresses display of log likelihood after each iteration of the fitting process.

```
. boxcox avlost year, model(lhs) nolog
```

```
                                      Number of obs   =          71
                                      LR chi2(1)      =       92.28
Log likelihood = -7.7185533           Prob > chi2     =       0.000

------------------------------------------------------------------------
     avlost |     Coef.   Std. Err.      z    P>|z|    [95% Conf. Interval]
------------+-----------------------------------------------------------
     /theta |  -.0560959   .0646726   -0.87   0.386    -.1828519      .07066
------------------------------------------------------------------------

Estimates of scale-variant parameters
----------------------------
            |     Coef.
------------+---------------
Notrans     |
       year |   -.0661891
      _cons |   127.9713
------------+---------------
     /sigma |    .8301177
----------------------------

    Test        Restricted     LR statistic      P-Value
    H0:        log likelihood      chi2        Prob > chi2
----------------------------------------------------------
theta = -1     -84.928791        154.42          0.000
theta =  0      -8.0941678         0.75          0.386
theta =  1    -101.50385         187.57          0.000
----------------------------------------------------------
```

The **boxcox** output shows theta = −.056 as the optimal Box–Cox parameter for transforming *avlost*, in order to linearize its relationship with *year*. Therefore, the left-hand-side transformation is

$$alvlost^{(-.056)} = (alvlost^{-.056} - 1)/-.056$$

Box–Cox transformation by a parameter close to zero, such as −.056, produces results similar to the natural-logarithm transformation we applied earlier to this variable "by hand." It is therefore not surprising that the **boxcox** regression model also resembles the earlier model drawn in Figures 8.7 and 8.8:

$$\text{predicted } alvlost^{(-.056)} = 127.97 - .07 year$$

The **boxcox** procedure assumes normal, independent, and identically distributed errors. It does not select transformations with the aim of normalizing residuals, however.

boxcox can estimate several types of models, including multiple regressions in which some or all of the right-hand-side variables are transformed by a parameter different from the *y*-variable transformation. It cannot apply different transformations to each separate right-hand-side predictor. To do that, we return to a "by hand" curvilinear-regression approach, as illustrated in the next section.

Regression with Transformed Variables — 2

For a multiple-regression example, we will use data on living conditions in 109 countries found in dataset *nations.dta* (from World Bank 1987; World Resources Institute 1993).

```
Contains data from C:\data\nations.dta
  obs:           109                  Data on 109 nations, ca. 1985
  vars:           15                  7 Jul 2001 10:33
  size:         4,033 (98.9% of memory free)
-------------------------------------------------------------------------
              storage  display   value
variable name  type    format    label    variable label
-------------------------------------------------------------------------
country        str8    %9s                 Country
pop            float   %9.0g               1985 population in millions
birth          byte    %8.0g               Crude birth rate/1000 people
death          byte    %8.0g               Crude death rate/1000 people
chldmort       byte    %8.0g               Child (1-4 yr) mortality 1985
infmort        int     %8.0g               Infant (<1 yr) mortality 1985
life           byte    %8.0g               Life expectancy at birth 1985
food           int     %8.0g               Per capita daily calories 1985
energy         int     %8.0g               Per cap energy consumed, kg oil
gnpcap         int     %8.0g               Per capita GNP 1985
gnpgro         float   %9.0g               Annual GNP growth % 65-85
urban          byte    %8.0g               % population urban 1985
school1        int     %8.0g               Primary enrollment % age-group
school2        byte    %8.0g               Secondary enroll % age-group
school3        byte    %8.0g               Higher ed. enroll % age-group
-------------------------------------------------------------------------
```

Relationships among birth rate, per capita gross national product (GNP), and child mortality are not linear, as can be seen clearly in the scatterplot matrix of Figure 8.9. The skewed *gnpcap* and *chldmort* distributions also present potential leverage and influence problems.

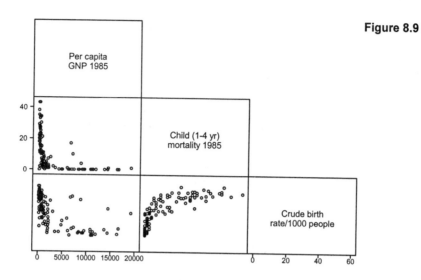

Figure 8.9

Experimenting with ladder-of-powers transformations reveals that the log of *gnpcap* and the square root of *chldmort* have distributions more symmetrical, with fewer outliers or potential leverage points, than the raw variables. More importantly, these transformations largely eliminate the nonlinearities, as can be seen in Figure 8.10.

```
. generate loggnp = log10(gnpcap)
. label variable loggnp "Log-10 of per cap GNP"
. generate srmort = sqrt(chldmort)
. label variable srmort "Square root child mortality"
. graph loggnp srmort birth, matrix half label
```

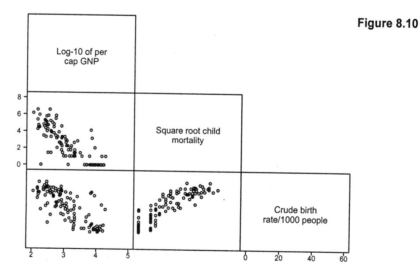

Figure 8.10

We can now apply linear regression using the transformed variables:

```
. regress birth loggnp srmort

      Source |       SS       df       MS              Number of obs =     109
-------------+------------------------------           F(  2,    106) =  198.06
       Model | 15837.9603       2   7918.98016         Prob > F      =  0.0000
    Residual | 4238.18646     106   39.9828911         R-squared     =  0.7889
-------------+------------------------------           Adj R-squared =  0.7849
       Total | 20076.1468     108   185.890248         Root MSE      =  6.3232


-----------------------------------------------------------------------------
       birth |      Coef.   Std. Err.       t    P>|t|     [95% Conf. Interval]
-------------+---------------------------------------------------------------
      loggnp | -2.353738   1.686255     -1.40    0.166    -5.696903    .9894259
       srmort |  5.577359    .533567     10.45    0.000     4.51951    6.635207
       _cons |  26.19488   6.362687      4.12    0.000     13.58024    38.80953
-----------------------------------------------------------------------------
```

Unlike the raw-data regression (not shown), this transformed-variables version finds that per capita gross national product does not significantly affect birth rate once we control for child mortality. The transformed-variables regression fits slightly better: R^2_a = .7849 instead of .6715. (We can compare R^2_a across models here only because both have the same untransformed *y* variable.) Leverage plots would confirm that transformations have much reduced the curvilinearity of the raw-data regression.

Conditional Effect Plots

Conditional effect plots trace the predicted values of *y* as a function of one *x* variable, with other *x* variables held constant at arbitrary values such as means, medians, quartiles, or extremes. Such plots help with interpreting results from transformed-variables regression.

Continuing with the previous example, we can calculate predicted birth rates as a function of *loggnp*, with *srmort* held at its mean (2.49):

```
. generate yhat1 = _b[_cons] + _b[loggnp]*loggnp + _b[srmort]*2.49
. label variable yhat1 "birth = f(gnpcap | srmort = 2.49)
```

The _b[*varname*] terms refer to the regression coefficient on *varname* from this session's most recent regression. _b[_cons] is the *y*-intercept or constant.

For a conditional effect plot, graph *yhat1* (after inverse transformation if needed, although it is not needed here) against the untransformed *x* variable (Figure 8.11):

```
. graph yhat1 gnpcap, connect(1) symbol(.) ylabel(30,31 to 35)
     xlabel(0,4000 to 20000) sort border
```

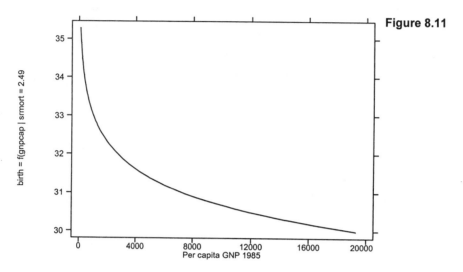

Figure 8.11

Similarly, Figure 8.12 depicts predicted birth rates as a function of *srmort*, with *loggnp* held at its mean (3.09):

```
. generate yhat2 = _b[_cons] + _b[loggnp]*3.09 + _b[srmort]*srmort
. label variable yhat2 "birth = f(chldmort | loggnp = 3.09)"
. graph yhat2 chldmort, connect(1) symbol(.) ylabel(20,25 to 55)
     xlabel(0,5 to 45) sort border
```

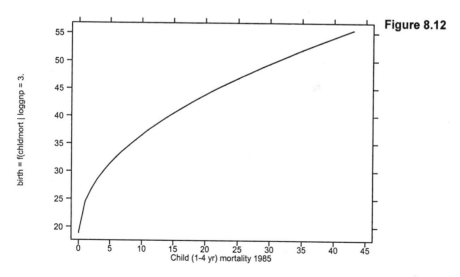

Figure 8.12

How can we compare the strength of different *x* variables' effects? Standardized regression coefficients (beta weights) are sometimes used for this purpose, but they imply a specialized definition of "strength" and can easily be misleading. A more substantively meaningful comparison might come from looking at conditional effect plots drawn with identical *y* scales, and with vertical lines marking the 10th and 90th percentiles of the *x*-variable distributions (found by **summarize, detail**). The vertical distances traveled by the predicted values curve over the range or the middle 80% of *x* values provide a visual comparison of effect magnitude. For example, we might create two separate plots using **yscale** to control vertical scales and **xline** to draw 10th and 90th-percentile lines:

```
. set textsize 180
. graph yhat1 gnpcap, connect(1) sort symbol(.) ylabel xlabel
       yscale(20,60) xline(230,10890) border saving(fig8_13a)
. graph yhat2 chldmort, connect(1) sort symbol(.) ylabel xlabel
       yscale(20,60) xline(0,27) border saving(fig8_13b)
set textsize 100
```

These commands set a large textsize (180% of normal) and explicitly saved the two graphs as files *fig8_13a* and *fig8_13b* to facilitate combining both into a single image with the **graph using** command. The final result appears in Figure 8.13.

```
. graph using fig8_13a fig8_13b
```

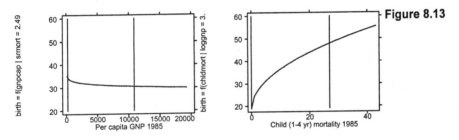

Figure 8.13

Combining several conditional effects plots into one image with common vertical scales, as done in Figure 8.13, allows quick visual comparison of the strength of different effects. Figure 8.13 makes obvious how much stronger is the effect of child mortality on birth rates—as separate plots (Figures 8.11 and 8.12) did not.

Nonlinear Regression — 1

Variable transformations allow fitting some curvilinear relationships using the familiar techniques of intrinsically linear models. Intrinsically nonlinear models, on the other hand, require a different class of fitting techniques. The **nl** command performs nonlinear regression by iterative least squares. This section introduces it using a dataset of simple examples, *nonlin.dta*:

```
Contains data from C:\data\nonlin.dta
  obs:              100                    Nonlinear model examples
                                          (artificial data)
  vars:               5                    11 Aug 2001 13:54
  size:           2,100  (99.8% of memory free)
-------------------------------------------------------------------------------
              storage  display    value
variable name   type   format     label      variable label
-------------------------------------------------------------------------------
x             byte     %9.0g                 Independent variable
y1            float    %9.0g                 y1 = 10 * 1.03^x + e
y2            float    %9.0g                 y2 = 10 * (1 - .95^x) + e
y3            float    %9.0g                 y3 = 5 + 25/(1+exp(-.1*(x-50)))
                                              + e
y4            float    %9.0g                 y4 = 5 +
                                              25*exp(-exp(-.1*(x-50))) + e
-------------------------------------------------------------------------------
Sorted by:   x
```

The *nonlin.dta* data are manufactured, with y variables defined as various nonlinear functions of x, plus random Gaussian errors. $y1$, for example, represents the exponential growth process $y1 = 10 \times 1.03^x$. Estimating these parameters from the data, **nl** obtains $y1 = 11.20 \times 1.03^x$, which is reasonably close to the true model:

. **nl exp2 y1 x**

(obs = 100)

```
Iteration 0:   residual SS =   27625.96
Iteration 1:   residual SS =   26547.42
Iteration 2:   residual SS =   26138.3
Iteration 3:   residual SS =   26138.29
```

Source	SS	df	MS
Model	667018.255	2	333509.128
Residual	26138.2933	98	266.717278
Total	693156.549	100	6931.56549

```
Number of obs =       100
F(  2,    98) =   1250.42
Prob > F      =    0.0000
R-squared     =    0.9623
Adj R-squared =    0.9615
Root MSE      =  16.33148
Res. dev.     =  840.3864
```

2-param. exp. growth curve, y1=b1*b2^x

y1	Coef.	Std. Err.	t	P>|t|	[95% Conf. Interval]
b1	11.20416	1.146682	9.77	0.000	8.928602 13.47971
b2	1.028838	.0012404	829.41	0.000	1.026376 1.031299

(SE's, P values, CI's, and correlations are asymptotic approximations)

The **predict** command obtains predicted values and residuals for a nonlinear model estimated by **nl** . Figure 8.14 graphs predicted values from the previous example, showing the close fit ($R^2 = .96$) between model and data.

```
. predict yhat1
(option yhat assumed; fitted values)

. graph y1 yhat1 x, connect(.s) symbol(Oi) ylabel xlabel
```

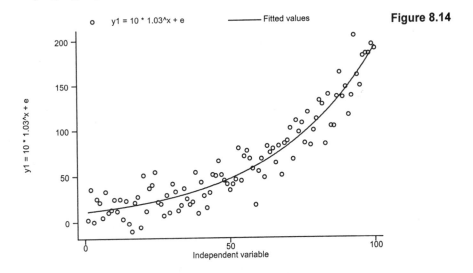

Figure 8.14

The **exp2** part of our **nl exp2 y1 x** command specified a particular exponential growth function by calling a brief program named *nlexp2.ado*. Stata includes several such programs, defining the following functions:

exp3 Exponential: $y = b_0 + b_1 b_2{}^x$

exp2 Exponential: $y = b_1 b_2{}^x$

exp2a Negative exponential: $y = b_1(1 - b_2{}^x)$

log4 Logistic with b_0 starting level and $(b_0 + b_1)$ asymptotic upper limit:
$$y = b_0 + b_1/(1 + \exp(-b_2(x - b_3)))$$

log3 Logistic with 0 starting level and b_1 asymptotic upper limit:
$$y = b_1/(1 + \exp(-b_2(x - b_3)))$$

gom4 Gompertz with b_0 starting level and $(b_0 + b_1)$ asymptotic upper limit:
$$y = b_0 + b_1 \exp(-\exp(-b_2(x - b_3)))$$

gom3 Gompertz with 0 starting level and $(b_0 + b_1)$ asymptotic upper limit:
$$y = b_1 \exp(-\exp(-b_2(x - b_3)))$$

nonlin.dta contains examples corresponding to **exp2** (*y1*), **exp2a** (*y2*), **log4** (*y3*), and **gom4** (*y4*) functions. Figure 8.15 shows curves fit by **nl** to *y2*, *y3*, and *y4*.

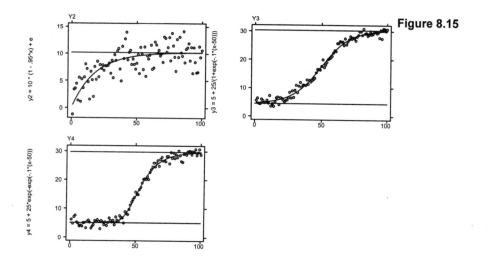

Figure 8.15

Users can write further nl*function* programs of their own. Here is the code for the
nlexp2.ado program defining an exponential growth model:

```
program define nlexp2
    version 7.0
    if "`1'"=="?" {
        global S_2 "2-param. exp. growth curve, `e(depvar)'=b1*b2^`2'"
        global S_1 "b1 b2"
        *
        * Approximate initial values by regression of log Y on X.
        *
        local exp "[`e(wtype)' `e(wexp)']"
        tempvar Y
        quietly {
            gen `Y' = log(`e(depvar)') if e(sample)
            reg `Y' `2' `exp' if e(sample)
        }
        global b1 = exp(_b[_cons])
        global b2 = exp(_b[`2'])
        exit
    }
    replace `1'=$b1*($b2)^`2'
end
```

This program finds some approximate initial values of the parameters to be estimated,
storing these as "global macros" named b1 and b2 . It then calculates an initial set of
predicted values, as a "local macro" named 1 , employing the initial parameter estimates and
the model equation:

```
replace `1' = $b1 * ($b2)^`2'
```

Subsequent iterations of **nl** will return to this line, calculating new predicted values (replacing
the contents of macro 1) as they refine the parameter estimates b1 and b2 . In Stata

programs, the notation $b1 means "the contents of global macro b1." Similarly, the notation `1' means "the contents of local macro 1."

Before attempting to write your own nonlinear function, examine nllog4.ado , nlgom4.ado , and others as examples, and consult the manual or **help nl** for explanations. Chapter 14 gives further examples of macros and other aspects of Stata programming.

Nonlinear Regression — 2

Our second example involves real data, and illustrates some steps that can help in research. Dataset *lichen.dta* contains measurements of lichen growth observed on the Norwegian arctic island of Svalbard (from Werner 1990). These slow-growing symbionts are often used to date rock monuments and other deposits, so their growth rates interest scientists in several fields.

```
Contains data from C:\data\lichen.dta
  obs:            11                   Lichen growth (Werner 1990)
  vars:            8                   7 Jul 2001 10:31
  size:          572 (99.8% of memory free)
-------------------------------------------------------------------------
              storage  display    value
variable name  type    format     label    variable label
-------------------------------------------------------------------------
locale        str31    %31s                 Locality and feature
point         str1     %9s                  Control point
date          int      %8.0g                Date
age           int      %8.0g                Age in years
rshort        float    %9.0g                Rhizocarpon short axis mm
rlong         float    %9.0g                Rhizocarpon long axis mm
pshort        int      %8.0g                P.minuscula short axis mm
plong         int      %8.0g                P.minuscula long axis mm
-------------------------------------------------------------------------
Sorted by:
```

Lichens characteristically exhibit a period of relatively fast early growth, gradually slowing, as suggested by the lowess-smoothed curve in Figure 8.16.

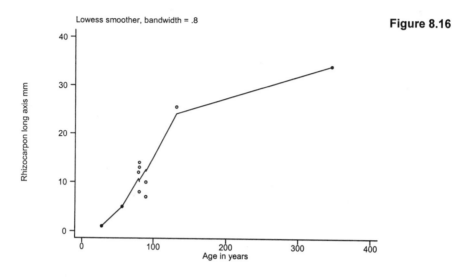

Lowess smoother, bandwidth = .8

Figure 8.16

Lichenometricians seek to summarize and compare such patterns by drawing growth curves. Their growth curves might not employ an explicit mathematical model, but we can fit one here to illustrate the process of nonlinear regression. Gompertz curves are asymmetrical S-curves, which are widely used to model biological growth:

$$y = b_1 \exp(-\exp(-b_2(x - b_3)))$$

They might provide a reasonable model for lichen growth.

If we intend to graph a nonlinear model, the data should contain a good range of closely spaced x values. Actual ages of the 11 lichen samples in *lichen.dta* range from 28 to 346 years. We can create 89 additional artificial observations, with "ages" from 0 to 352 in 4-year increments, by the following commands:

```
. range newage 0 396 100
obs was 11, now 100
```

```
. replace age = newage[_n-11] if age == .
(89 real changes made)
```

The first command created a new variable, *newage*, with 100 values ranging from 0 to 396 in 4-year increments. In so doing, we also created 89 new artificial observations, with missing values on all variables except *newage*. The **replace** command substitutes the missing artificial-case *age* values with *newage* values, starting at 0. The first 15 observations in our data now look like this:

```
. list rlong age newage in 1/15
```

	rlong	age	newage
1.	1	28	0
2.	5	56	4
3.	12	79	8
4.	14	80	12
5.	13	80	16
6.	8	80	20
7.	7	89	24
8.	10	89	28
9.	34	346	32
10.	34	346	36

```
11.        25.5        131        40
12.          .           0        44
13.          .           4        48
14.          .           8        52
15.          .          12        56
```

```
. summarize rlong age newage
```

Variable	Obs	Mean	Std. Dev.	Min	Max
rlong	11	14.86364	11.31391	1	34
age	100	170.68	104.7042	0	352
newage	100	198	116.046	0	396

We now could **drop newage**. Only the original 11 observations have nonmissing *rlong* values, so only they will enter into model estimation. Stata calculates predicted values for any observation with nonmissing *x* values, however. We can therefore obtain such predictions for both the 11 real observations and the 89 artificial ones, which will allow us to graph the regression curve accurately.

Lichen growth starts with a size close to zero, so we chose the **gom3** Gompertz function rather than **gom4** (which incorporates a nonzero takeoff level, the parameter b_0). Figure 8.16 suggests an asymptotic upper limit somewhere near 34, and we can include this as our initial estimate of the parameter b_1. Estimation of this model is accomplished by

```
. nl gom3 rlong age, init(B1=34) nolog
```

(obs = 11)

Source	SS	df	MS		
Model	3633.16112	3	1211.05371	Number of obs =	11
Residual	77.0888815	8	9.63611018	F(3, 8) =	125.68
				Prob > F =	0.0000
				R-squared =	0.9792
				Adj R-squared =	0.9714
Total	3710.25	11	337.295455	Root MSE =	3.104208
				Res. dev. =	52.63435

3-parameter Gompertz function, rlong=b1*exp(-exp(-b2*(age-b3)))

rlong	Coef.	Std. Err.	t	P>|t|	[95% Conf. Interval]	
b1	34.36637	2.267186	15.16	0.000	29.13823	39.59451
b2	.0217685	.0060806	3.58	0.007	.0077465	.0357904
b3	88.79701	5.632545	15.76	0.000	75.80834	101.7857

(SE's, P values, CI's, and correlations are asymptotic approximations)

A **nolog** option suppresses displaying a log of iterations with the output. All three parameter estimates differ significantly from 1.

We obtain predicted values using **predict**, and graph these to see the regression curve. The **yline** option draws in the lower and estimated upper limits (0 and 34.366) of this curve (Figure 8.17).

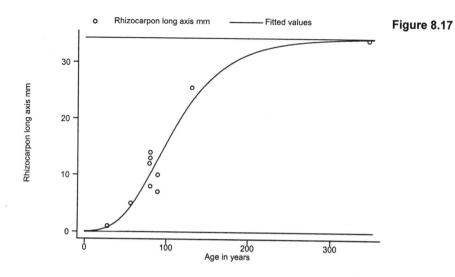

Figure 8.17

Especially when working with sparse data or a relatively complex model, nonlinear regression programs can be quite sensitive to their initial parameter estimates. The `init` option with `nl` permits researchers to suggest their own initial values if the default values supplied by an nl*function* program do not seem to work. Previous experience with similar data, or publications by other researchers, could help supply suitable initial values. Alternatively, we could estimate through trial and error by employing `generate` to calculate predicted values based on arbitrarily-chosen sets of parameter values and `graph` to compare the resulting predictions with the data.

9

Robust Regression

Stata's basic **regress** and **anova** commands perform ordinary least squares (OLS) regression. The popularity of OLS derives in part from its theoretical advantages given "ideal" data. If errors are normally, independently, and identically distributed (normal i.i.d.), then OLS is more efficient than any other unbiased estimator. The flip side of this statement often gets overlooked: if errors are not normal, or not i.i.d., then other unbiased estimators might outperform OLS. In fact, the efficiency of OLS degrades quickly in the face of heavy-tailed (outlier-prone) error distributions. Yet such distributions are common in many fields.

OLS tends to track outliers, fitting them at the expense of the rest of the sample. Over the long run, this leads to greater sample-to-sample variation or inefficiency when samples often contain outliers. Robust regression methods aim to achieve almost the efficiency of OLS with ideal data and substantially better-than-OLS efficiency in non-ideal (for example, non-normal errors) situations. "Robust regression" encompasses a variety of different techniques, each with advantages and drawbacks for dealing with problematic data. This chapter introduces two varieties of robust regression, **rreg** and **qreg**, and briefly compares their results with those of OLS (**regress**).

rreg and **qreg** resist the pull of outliers, giving them better-than-OLS efficiency in the face of nonnormal, heavy-tailed error distributions. They share the OLS assumption that errors are independent and identically distributed, however. As a result, their standard errors, tests, and confidence intervals are not trustworthy in the presence of heteroskedasticity or correlated errors. To relax the assumption of independent, identically distributed errors when using **regress** or certain other modeling commands (although not **rreg** or **qreg**), Stata offers options that estimate robust standard errors.

For clarity, this chapter focuses mostly on two-variable examples, but robust multiple regression or *n*-way ANOVA are straightforward using the same commands. Chapter 14 returns to the topic of robustness, showing how we can use Monte Carlo experiments to evaluate competing statistical techniques.

Example Commands

. `rreg y x1 x2 x3`

Performs robust regression of *y* on three predictors, using iteratively reweighted least squares with Huber and biweight functions tuned for 95% Gaussian efficiency. Given appropriately configured data, **rreg** can also obtain robust means, confidence intervals, difference of means tests, and ANOVA/ANCOVA.

. `rreg y x1 x2 x3, nolog tune(6) genwt(rweight) iterate(10)`

Performs robust regression of *y* on three predictors. The options shown above tell Stata not to print the iteration log, to use a tuning constant of 6 (which downweights outliers more steeply than the default 7), to generate a new variable (arbitrarily named *rweight*) holding the final-iteration robust weights for each observation, and to limit the maximum number of iterations to 10.

. `qreg y x1 x2 x3`

Performs quantile regression, also known as least absolute value (LAV) or minimum *L1*-norm regression, of *y* on three predictors. By default, `qreg` models the conditional .5 quantile (approximate median) of *y* as a linear function of the predictor variables, and thus provides "median regression."

. `qreg y x1 x2 x3, quantile(.25)`

Performs quantile regression modeling the conditional .25 quantile (first quartile) of *y* as a linear function of *x1*, *x2*, and *x3*.

. `bsqreg y x1 x2 x3, rep(100)`

Performs quantile regression, with standard errors estimated by bootstrap data resampling with 100 repetitions (default is `rep(20)`).

. `predict e, resid`

Calculates residual values (arbitrarily named *e*) after any `regress`, `rreg`, `qreg`, or `bsqreg` command. Similarly, `predict yhat` calculates the predicted values of *y*. Other `predict` options apply, with some restrictions.

. `regress y x1 x2 x3, robust`

Performs OLS regression of *y* on three predictors. Coefficient variances, and hence standard errors, are estimated by a robust method (Huber/White or sandwich) that does not assume identically distributed errors. With the `cluster()` option, one source of correlation among the errors can be accommodated as well. The *User's Guide* (23.11) describes the reasoning behind these methods.

Regression with Ideal Data

To clarify the issue of robustness, we will explore the small (n = 20) contrived dataset *robust1.dta*:

```
Contains data from C:\data\robust1.dta
  obs:            20                     Robust regression examples 1
                                           (artificial data)
  vars:           10                     8 Jul 2001 13:15
  size:          880 (99.9% of memory free)
-------------------------------------------------------------------------------
                storage  display   value
variable name    type    format    label      variable label
-------------------------------------------------------------------------------
x                float   %4.2f                Normal X
e1               float   %4.2f                Normal errors
y1               float   %4.2f                y1 = 10 + 2*x + e1
e2               float   %4.2f                Normal errors with 1 outlier
y2               float   %4.2f                y2 = 10 + 2*x + e2
x3               float   %4.2f                Normal X with 1 leverage obs.
e3               float   %4.2f                Normal errors with 1 extreme
```

```
y3              float   %4.2f                y3 = 10 + 2*x3 + e3
e4              float   %4.2f                Skewed errors
y4              float   %4.2f                y4 = 10 + 2*x + e4
---------------------------------------------------------------------------
Sorted by:
```

The variables x and $e1$ each contain 20 random values from independent standard normal distributions. $y1$ contains 20 values produced by the regression model:

$$y1 = 10 + 2x + e1$$

The commands that manufactured these first three variables are

```
. clear
. set obs 20
. generate x = invnorm(uniform())
. generate e1 = invnorm(uniform())
. generate y1 = 10 + 2*x + e1
```

With real data, coding mistakes and measurement errors sometimes create wildly incorrect values. To simulate this, we might shift the second observation's error from –0.89 to 19.89:

```
. generate e2 = e1
. replace e2 = 19.89 in 2
. generate y2 = 10 + 2*x + e2
```

Similar manipulations produce the other variables in *robust1.dta*.

$y1$ and x present an ideal regression problem: the expected value of $y1$ really is a linear function of x, and errors come from normal, independent, and identical distributions—because we defined them that way. OLS does a good job of estimating the true intercept (10) and slope (2), obtaining the line shown in Figure 9.1.

```
. regress y1 x
```

Source	SS	df	MS		Number of obs =	20
					F(1, 18) =	108.25
Model	134.059351	1	134.059351		Prob > F =	0.0000
Residual	22.29157	18	1.23842055		R-squared =	0.8574
					Adj R-squared =	0.8495
Total	156.350921	19	8.22899586		Root MSE =	1.1128

y1	Coef.	Std. Err.	t	P>|t|	[95% Conf. Interval]	
x	2.048057	.1968465	10.40	0.000	1.634498	2.461616
_cons	9.963161	.2499861	39.85	0.000	9.43796	10.48836

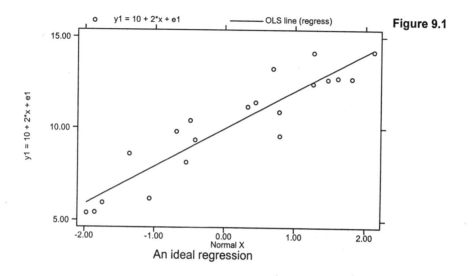

An ideal regression

An iteratively reweighted least squares (IRLS) procedure, **rreg**, obtains robust regression estimates. The first **rreg** iteration begins with OLS. Any observations so influential as to have Cook's *D* values greater than 1 are automatically set aside after this first step. Next, weights are calculated for each observation using a Huber function, which downweights observations that have larger residuals, and weighted least squares is performed. After several WLS iterations, the weight function shifts to a Tukey biweight (as suggested by Li 1985), tuned for 95% Gaussian efficiency (see *Regression with Graphics* for details). **rreg** estimates standard errors and tests hypotheses using a pseudovalues method (Street, Carrol, and Ruppert 1988) that does not assume normality.

```
. rreg y1 x

  Huber iteration 1:  maximum difference in weights = .35774407
  Huber iteration 2:  maximum difference in weights = .02181578
Biweight iteration 3:  maximum difference in weights = .14421371
Biweight iteration 4:  maximum difference in weights = .01320276
Biweight iteration 5:  maximum difference in weights = .00265408

Robust regression estimates                     Number of obs =      20
                                                F(  1,    18) =   79.96
                                                Prob > F      =  0.0000

-------------------------------------------------------------------------
    y1 |     Coef.    Std. Err.       t    P>|t|    [95% Conf. Interval]
-------+-----------------------------------------------------------------
     x |   2.047813   .2290049     8.94   0.000     1.566692    2.528935
 _cons |   9.936163   .2908259    34.17   0.000     9.325161    10.54717
-------------------------------------------------------------------------
```

This "ideal data" example includes no serious outliers, so here **rreg** is unneeded. The **rreg** intercept and slope estimates resemble those obtained by **regress** (and are not far from the true values 10 and 2), but they have slightly larger estimated standard errors. Given normal i.i.d. errors, as in this example, **rreg** theoretically possesses about 95% of the efficiency of OLS.

rreg and **regress** both belong to the family of *M*-estimators (for maximum-likelihood). An alternative order-statistic strategy called *L*-estimation fits quantiles of *y*, rather than its expectation or mean. For example, we could model how the median (.5 quantile) of *y* changes with *x*. **qreg** , an *L1*-type estimator, accomplishes such quantile regression and provides another method with good resistance to outliers:

```
. qreg y1 x

Iteration  1:  WLS sum of weighted deviations =  17.711531

Iteration  1:  sum of abs. weighted deviations =  17.130001
Iteration  2:  sum of abs. weighted deviations =  16.858602

Median regression                            Number of obs =         20
  Raw sum of deviations     46.84 (about 10.4)
  Min sum of deviations  16.8586                Pseudo R2     =     0.6401

------------------------------------------------------------------------
      y1 |     Coef.    Std. Err.       t     P>|t|    [95% Conf. Interval]
---------+--------------------------------------------------------------
       x |   2.139896    .2590447     8.26    0.000     1.595664    2.684129
   _cons |    9.65342    .3564108    27.09    0.000     8.904628   10.40221
------------------------------------------------------------------------
```

Although **qreg** obtains reasonable parameter estimates, its standard errors here exceed those of **regress** (OLS) and **rreg**. Given ideal data, **qreg** is the least efficient of these three estimators. The following sections view their performance with less ideal data.

Y Outliers

The variable *y2* is identical to *y1*, but with one outlier caused by the "wild" error of observation #2. OLS has little resistance to outliers, so this shift in observation #2 (at upper left in Figure 9.2) substantially changes the **regress** results:

```
. regress y2 x

    Source |       SS       df       MS              Number of obs =       20
-----------+------------------------------           F(  1,    18) =     0.97
     Model |  18.764271        1   18.764271          Prob > F      =   0.3378
  Residual | 348.233471       18  19.3463039          R-squared     =   0.0511
-----------+------------------------------           Adj R-squared =  -0.0016
     Total | 366.997742       19  19.3156706          Root MSE      =   4.3984

------------------------------------------------------------------------
      y2 |     Coef.    Std. Err.       t     P>|t|    [95% Conf. Interval]
---------+--------------------------------------------------------------
       x |   .7662304    .7780232     0.98    0.338    -.8683356    2.400796
   _cons |    11.1579    .9880542    11.29    0.000     9.082078   13.23373
------------------------------------------------------------------------
```

The outlier raises the OLS intercept (from 9.936 to 11.1579) and lessens the slope (from 2.048 to 0.766). R^2 has dropped from .8574 to .0511. Standard errors quadrupled, and the OLS slope (solid line in Figure 9.2) no longer significantly differs from zero.

The outlier has little impact on **rreg**, however, as shown by the dashed line in Figure 9.2. The robust coefficients barely change, and remain close to the true parameters 10 and 2. Nor do the robust standard errors increase much.

. **rreg y2 x, nolog genwt(rweight2)**

Robust regression estimates

	Number of obs =	19
	F(1, 17) =	63.01
	Prob > F =	0.0000

y2	Coef.	Std. Err.	t	P>|t|	[95% Conf. Interval]
x	1.979015	.2493146	7.94	0.000	1.453007 2.505023
_cons	10.00897	.3071265	32.59	0.000	9.360986 10.65695

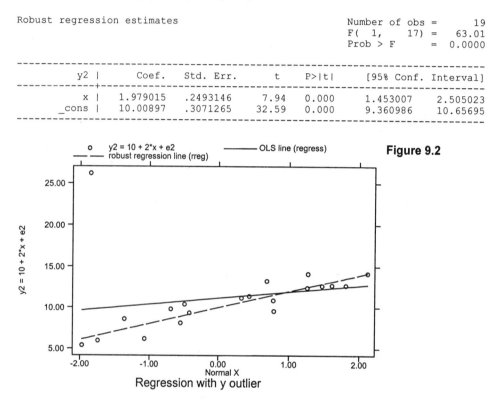

Figure 9.2

Regression with y outlier

The **nolog** option above caused Stata not to print the iteration log. The **genwt(rweight2)** option saved robust weights as a variable named *rweight2*.

. **predict resid2, resid**

. **list y2 x resid2 rweight2**

```
       y2      x     resid2    rweight2
 1.   5.37  -1.97  -.7403071  .94644465
 2.  26.19  -1.85  19.84221         .
 3.   5.93  -1.74  -.6354806  .96037073
 4.   8.58  -1.36   1.262494  .84933840
 5.   6.16  -1.07  -1.731421  .72576310
 6.   9.80  -0.69   1.156554  .87273631
 7.   8.12  -0.55  -.8005085  .93758391
 8.  10.40  -0.49    1.36075  .82606386
 9.   9.35  -0.42     .17222  .99712388
10.  11.16   0.33   .4979582  .97581674
11.  11.40   0.44   .5202664  .97360863
12.  13.26   0.69   1.885513  .68048066
13.  10.88   0.78  -.6725982  .95572833
14.   9.58   0.79  -1.992389  .64644918
```

```
15. 12.41   1.26  -.0925257   .99913568
16. 14.14   1.27   1.617685   .75887073
17. 12.66   1.47  -.2581189   .99338589
18. 12.74   1.61  -.4551811   .97957817
19. 12.70   1.81  -.8909839   .92307041
20. 14.19   2.12  -.0144787   .99997651
```

Residuals near zero produce weights near one; farther-out residuals get progressively lower weights. Observation #2 has been automatically set aside as too influential because of Cook's $D > 1$. **rreg** assigns its *rweight2* as "missing," so this observation has no effect on the final estimates. The same final estimates, although not the correct standard errors or tests, could be obtained using **regress** with analytical weights (results not shown):

. **regress y2 x [aweight = rweight2]**

Applied to the regression of *y2* on *x*, **qreg** also resists the outlier's influence and performs much better than **regress**, but not as well as **rreg**. **qreg** appears less efficient than **rreg**, and in this sample its coefficient estimates are slightly farther from the true values of 10 and 2.

. **qreg y2 x, nolog**

```
Median regression                           Number of obs =        20
  Raw sum of deviations    56.68 (about 10.88)
  Min sum of deviations 36.20036            Pseudo R2     =    0.3613

------------------------------------------------------------------------
       y2 |    Coef.   Std. Err.       t    P>|t|    [95% Conf. Interval]
----------+-------------------------------------------------------------
        x |  1.821428   .4105944     4.44   0.000    .9588014    2.684055
    _cons |    10.115   .5088526    19.88   0.000    9.045941    11.18406
------------------------------------------------------------------------
```

Monte Carlo researchers have also noticed that the standard errors calculated by **qreg** sometimes underestimate the true sample-to-sample variation, particularly with smaller samples. As an alternative, Stata provides the command **bsqreg**, which performs the same median or quantile regression as **qreg**, but employs bootstrapping (data resampling) to estimate the standard errors empirically. The option **rep( )** controls the number of bootstrap repetitions. Its default is **rep(20)**, which is enough for exploratory work. Before reaching "final" conclusions, we might take the time to draw 200 or more bootstrap samples, as done below. Both **qreg** and **bsqreg** fit identical models. Bootstrapping here obtains a slightly larger standard error estimate for the coefficient on *x* (.44 vs. .41), and a smaller standard error for the y-intercept (.48 vs. .51). Chapter 14 discusses bootstrapping further.

```
. bsqreg y2 x
```

```
(estimating base model)
(bootstrapping ....................)
```

```
Median regression, bootstrap(20) SEs          Number of obs =        20
   Raw sum of deviations     56.68 (about 10.88)
   Min sum of deviations 36.20036                Pseudo R2     =    0.3613
```

y2	Coef.	Std. Err.	t	P>|t|	[95% Conf. Interval]
x	1.821428	.4444694	4.10	0.001	.8876326 2.755224
_cons	10.115	.4793223	21.10	0.000	9.107981 11.12202

X Outliers (Leverage)

rreg, **qreg**, and **bsqreg** deal comfortably with *y*-outliers, unless the observations with unusual *y* values have unusual *x* values (leverage) too. The *y3* and *x3* variables in *robust.dta* present an extreme example of leverage. Apart from the leverage observation (#2), these variables equal *y1* and *x*.

The high leverage of observation #2, combined with its exceptional *y3* value, make it influential: **regress** and **qreg** both track this outlier, reporting that the "best-fitting" line has a negative slope (Figure 9.3).

```
. regress y3 x3
```

Source	SS	df	MS		
Model	139.306724	1	139.306724		
Residual	227.691018	18	12.649501		
Total	366.997742	19	19.3156706		

Number of obs = 20
F(1, 18) = 11.01
Prob > F = 0.0038
R-squared = 0.3796
Adj R-squared = 0.3451
Root MSE = 3.5566

y3	Coef.	Std. Err.	t	P>|t|	[95% Conf. Interval]
x3	-.6212248	.1871973	-3.32	0.004	-1.014512 -.227938
_cons	10.80931	.8063436	13.41	0.000	9.115244 12.50337

```
. qreg y3 x3, nolog
```

```
Median regression                             Number of obs =        20
   Raw sum of deviations     56.68 (about 10.88)
   Min sum of deviations 56.19466                Pseudo R2     =    0.0086
```

y3	Coef.	Std. Err.	t	P>|t|	[95% Conf. Interval]
x3	-.6222217	.347103	-1.79	0.090	-1.351458 .1070146
_cons	11.36533	1.419214	8.01	0.000	8.383676 14.34699

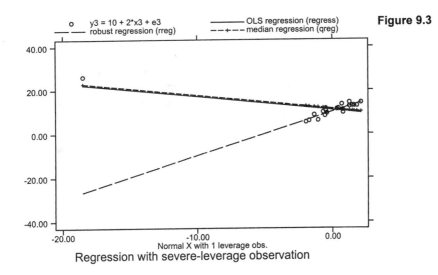

Figure 9.3

Regression with severe-leverage observation

Figure 9.3 illustrates that **regress** and **qreg** are not robust against leverage (*x*-outliers). The **rreg** program, however, not only downweights large-residual observations (which by itself gives little protection against leverage), but also automatically sets aside observations with Cook's *D* (influence) statistics greater than 1. This happens when we regress *y3* on *x3*; **rreg** ignores the one influential observation and produces a more reasonable regression line based on the remaining 19 observations:

`. rreg y3 x3, nolog`

```
Robust regression estimates                 Number of obs =       19
                                             F(  1,    17) =    63.01
                                             Prob > F      =   0.0000

-----------------------------------------------------------------------
     y3 |      Coef.   Std. Err.       t    P>|t|    [95% Conf. Interval]
--------+--------------------------------------------------------------
     x3 |   1.979015   .2493146     7.94   0.000     1.453007    2.505023
  _cons |   10.00897   .3071265    32.59   0.000     9.360986    10.65695
-----------------------------------------------------------------------
```

Setting aside high-influence observations, as done by **rreg**, provides a simple but not foolproof way to deal with leverage. More comprehensive methods, termed bounded-influence regression, also exist and could be implemented in a Stata program.

The examples in Figures 9.2 and 9.3 involve single outliers, but robust procedures can handle more. Too many severe outliers, or a cluster of similar outliers, might cause them to break down. But in such situations, which often are noticeable in diagnostic plots, the analyst must question whether fitting a linear model makes sense. It might be worthwhile to seek an explicit model for what is causing the outliers to be different.

Monte Carlo experiments (illustrated in Chapter 14) confirm that estimators like **rreg** and **qreg** generally remain unbiased, with better-than-OLS efficiency, when applied to heavy-tailed (outlier-prone) but symmetrical error distributions. The next section illustrates what can happen when errors have asymmetrical distributions.

Asymmetrical Error Distributions

The variable *e4* in *robust2.dta* has a skewed and outlier-filled distribution: *e4* equals *e1* (a standard normal variable) raised to the fourth power, and then adjusted to have 0 mean. These skewed errors, plus the linear relationship with *x*, define the variable *y4* = 10 + 2*x* + *e4*. Regardless of an error distribution's shape, OLS remains an unbiased estimator. Over the long run, its estimates should center on the true parameter values.

```
. regress y4 x

      Source |       SS       df       MS                  Number of obs =      20
-------------+------------------------------               F(  1,    18) =    6.97
       Model |  155.870383      1   155.870383             Prob > F      =  0.0166
    Residual |  402.341909     18   22.3523283             R-squared     =  0.2792
-------------+------------------------------               Adj R-squared =  0.2392
       Total |  558.212291     19   29.3795943             Root MSE      =  4.7278

------------------------------------------------------------------------------
          y4 |      Coef.   Std. Err.       t    P>|t|     [95% Conf. Interval]
-------------+----------------------------------------------------------------
           x |   2.208388   .8362862      2.64   0.017     .4514157    3.96536
       _cons |   9.975681   1.062046      9.39   0.000     7.744406   12.20696
------------------------------------------------------------------------------
```

The same is not true for most robust estimators. Unless errors are symmetrical, the median line estimated by **qreg**, or the biweight line estimated by **rreg**, do not theoretically coincide with the expected-*y* line estimated by **regress**. So long as the errors' skew reflects only a small fraction of their distribution, **rreg** might exhibit little bias. But when the entire distribution is skewed, as with *e4*, **rreg** will downweight mostly one side, resulting in noticeably biased *y*-intercept estimates.

```
. rreg y4 x, nolog

Robust regression estimates                               Number of obs =      20
                                                          F(  1,    18) = 1319.29
                                                          Prob > F      =  0.0000

------------------------------------------------------------------------------
          y4 |      Coef.   Std. Err.       t    P>|t|     [95% Conf. Interval]
-------------+----------------------------------------------------------------
           x |   1.952073   .0537435     36.32   0.000     1.839163    2.064984
       _cons |   7.476669   .0682518    109.55   0.000     7.333278    7.620061
------------------------------------------------------------------------------
```

Although the **rreg** estimated *y*-intercept in Figure 9.4 is too low, the slope remains parallel to the OLS line and the true model. In fact, being less affected by outliers, the **rreg** slope (1.95) is closer to the true slope (2) and has a much smaller standard error than **regress**. This illustrates the tradeoff of using **rreg** or similar estimators with skewed errors: we risk getting biased estimates of the *y*-intercept, but can still expect unbiased and relatively precise estimates of other regression coefficients. In many applications, such coefficients are substantively more interesting than the *y*-intercept, so this tradeoff is worthwhile. Moreover, the robust *t* and *F* tests, unlike those of OLS, do not assume normal errors.

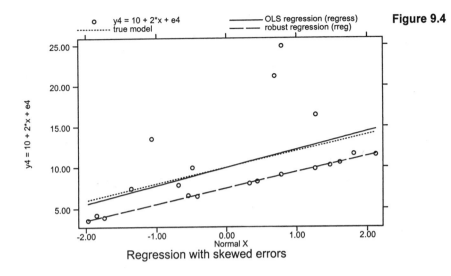

Figure 9.4

Regression with skewed errors

Robust Analysis of Variance

rreg can also perform robust analysis of variance or covariance once the analysis is recast in regression form. For illustration, consider these data (*faculty.dta*) on college faculty salaries:

```
Contains data from C:\data\faculty.dta
  obs:            226                          College faculty salaries
  vars:             6                          8 Jul 2001 15:53
  size:         2,938 (99.8% of memory free)
-------------------------------------------------------------------------
              storage  display    value
variable name   type   format     label     variable label
-------------------------------------------------------------------------
rank            byte    %8.0g      rank      Academic rank
gender          byte    %8.0g      sex       Gender (dummy variable)
female          byte    %8.0g                Gender (effect coded)
assoc           byte    %8.0g                Assoc Professor (effect coded)
full            byte    %8.0g                Full Professor (effect coded)
pay             float   %9.0g                Annual salary
-------------------------------------------------------------------------
Sorted by:
```

Faculty salaries increase with rank. In this sample, men have higher average salaries:

```
. table gender rank, contents(mean pay)
```

```
-----------------------------------------
Gender      |
(dummy      |        Academic rank
variable)   |   Assist      Assoc      Full
------------+----------------------------
     Male   |    29280   38622.22   52084.9
   Female   | 28711.04   38019.05     47190
-----------------------------------------
```

An ordinary (OLS) analysis of variance indicates that both *rank* and *gender* significantly affect salary. Their interaction is not significant:

`. anova pay rank gender rank*gender`

```
                      Number of obs =     226    R-squared     =  0.7305
                      Root MSE      = 5108.21    Adj R-squared =  0.7244

        Source |  Partial SS    df        MS            F      Prob > F
   ------------+----------------------------------------------------------
         Model |  1.5560e+10     5   3.1120e+09       119.26     0.0000
               |
          rank |  7.6124e+09     2   3.8062e+09       145.87     0.0000
        gender |   127361829     1    127361829         4.88     0.0282
   rank*gender |  87997720.1     2   43998860.1         1.69     0.1876
               |
      Residual |  5.7406e+09   220   26093824.5
   ------------+----------------------------------------------------------
         Total |  2.1300e+10   225   94668810.3
```

But salary is not normally distributed, and the senior-rank averages reflect the influence of a few highly paid outliers. Suppose we want to check these results by performing a robust analysis of variance. We need effect-coded versions of the *rank* and *gender* variables, which this dataset also contains:

`. tabulate gender female`

```
    Gender |
    (dummy | Gender (effect coded)
  variable) |      -1          1 |     Total
 -----------+----------------------+----------
      Male |     149          0 |      149
    Female |       0         77 |       77
 -----------+----------------------+----------
     Total |     149         77 |      226
```

`. tabulate rank assoc`

```
  Academic | Assoc Professor (effect coded)
      rank |      -1          0          1 |     Total
 ----------+---------------------------------+----------
    Assist |      64          0          0 |       64
     Assoc |       0          0        105 |      105
      Full |       0         57          0 |       57
 ----------+---------------------------------+----------
     Total |      64         57        105 |      226
```

`. tab rank full`

```
  Academic | Full Professor (effect coded)
      rank |      -1          0          1 |     Total
 ----------+---------------------------------+----------
    Assist |      64          0          0 |       64
     Assoc |       0        105          0 |      105
      Full |       0          0         57 |       57
 ----------+---------------------------------+----------
     Total |      64        105         57 |      226
```

If *faculty.dta* did not already have these effect-coded variables (*female, assoc,* and *full*), we could create them from *gender* and *rank* using a series of **generate** and **replace** statements. We also need two interaction terms representing female associate professors and female full professors:

```
. generate femassoc = female*assoc
. generate femfull = female*full
```

Males and assistant professors are "omitted categories" in this example. Now we can duplicate the previous ANOVA using regression:

```
. regress pay assoc full female femassoc femfull
```

Source	SS	df	MS
Model	1.5560e+10	5	3.1120e+09
Residual	5.7406e+09	220	26093824.5
Total	2.1300e+10	225	94668810.3

```
Number of obs =      226
F(  5,    220) =   119.26
Prob > F       =   0.0000
R-squared      =   0.7305
Adj R-squared  =   0.7244
Root MSE       =   5108.2
```

pay	Coef.	Std. Err.	t	P>|t|	[95% Conf. Interval]	
assoc	-663.8995	543.8499	-1.22	0.223	-1735.722	407.9229
full	10652.92	783.9227	13.59	0.000	9107.957	12197.88
female	-1011.174	457.6938	-2.21	0.028	-1913.199	-109.1483
femassoc	709.5864	543.8499	1.30	0.193	-362.2359	1781.409
femfull	-1436.277	783.9227	-1.83	0.068	-2981.236	108.6819
_cons	38984.53	457.6938	85.18	0.000	38082.51	39886.56

```
. test assoc full

 ( 1)  assoc = 0.0
 ( 2)  full = 0.0

       F(  2,    220) =   145.87
              Prob > F =    0.0000

. test female

 ( 1)  female = 0.0

       F(  1,    220) =     4.88
              Prob > F =    0.0282

. test femassoc femfull

 ( 1)  femassoc = 0.0
 ( 2)  femfull = 0.0

       F(  2,    220) =     1.69
              Prob > F =    0.1876
```

regress followed by the appropriate **test** commands obtains exactly the same R^2 and F test results that we found earlier using **anova**. Predicted values from this regression equal the mean salaries.

```
. predict predpay1
(option xb assumed; fitted values)
. label variable predpay1 "OLS predicted salary"
. table gender rank, contents(mean predpay1)
```

```
----------------------------------------
Gender   |
(dummy   |           Academic rank
variable) |  Assist      Assoc       Full
----------+-----------------------------------
   Male  |   29280    38622.22     52084.9
 Female  | 28711.04   38019.05       47190
----------------------------------------
```

Predicted values (means), R^2, and F tests would also be the same regardless of which categories we chose to omit from the regression. Our "omitted categories," males and assistant professors, are not really absent. Their information is implied by the included categories: if a faculty member is not female, he must be male, and so forth.

To perform a robust analysis of variance, apply **rreg** to this model:

```
. rreg pay assoc full female femassoc femfull, nolog
```

```
Robust regression estimates                    Number of obs =      226
                                               F( 5,   220) =   138.25
                                               Prob > F      =   0.0000
```

pay	Coef.	Std. Err.	t	P>|t|	[95% Conf. Interval]
assoc	-315.6463	458.1588	-0.69	0.492	-1218.588 587.2956
full	9765.296	660.4048	14.79	0.000	8463.767 11066.83
female	-749.4949	385.5778	-1.94	0.053	-1509.394 10.40395
femassoc	197.7833	458.1588	0.43	0.666	-705.1587 1100.725
femfull	-913.348	660.4048	-1.38	0.168	-2214.878 388.1815
_cons	38331.87	385.5778	99.41	0.000	37571.97 39091.77

```
. test assoc full

 ( 1)   assoc = 0.0
 ( 2)   full = 0.0

        F( 2,   220) =   182.67
            Prob > F =    0.0000

. test female

 ( 1)   female = 0.0

        F( 1,   220) =     3.78
            Prob > F =    0.0532

. test femassoc femfull

 ( 1)   femassoc = 0.0
 ( 2)   femfull = 0.0

        F( 2,   220) =     1.16
            Prob > F =    0.3144
```

rreg downweights several outliers, mainly highly-paid male full professors. To see the robust means, again use predicted values:

```
. predict predpay2
(option xb assumed; fitted values)

. label variable predpay2 "Robust predicted salary"

. table gender rank, contents(mean predpay2)
```

```
----------------------------------------
Gender    |
(dummy    |       Academic rank
variable) |  Assist    Assoc     Full
----------+-----------------------------
     Male | 28916.15  38567.93  49760.01
   Female | 28848.29  37464.51  46434.32
----------------------------------------
```

The male–female salary gap among assistant and full professors appears smaller if we use robust means. It does not entirely vanish, however, and the gender gap among associate professors slightly widens.

With effect coding and suitable interaction terms, **regress** can duplicate ANOVA exactly. **rreg** can do parallel analyses, testing for differences among robust means instead of ordinary means (as **regress** and **anova** do). Used in similar fashion, **qreg** opens the third possibility of testing for differences among medians. For comparison, here is a quantile regression version of the faculty pay analysis:

```
. qreg pay assoc full female femassoc femfull, nolog
```

```
Median regression                      Number of obs =       226
  Raw sum of deviations  1738010 (about 37360)
  Min sum of deviations   798870        Pseudo R2     =    0.5404

---------------------------------------------------------------------
     pay |     Coef.   Std. Err.      t    P>|t|   [95% Conf. Interval]
---------+-----------------------------------------------------------
   assoc |      -760   440.1693   -1.73   0.086   -1627.488   107.4881
    full |     10335   615.7735   16.78   0.000    9121.43    11548.57
  female | -623.3333   365.1262   -1.71   0.089   -1342.926    96.2594
femassoc | -156.6667   440.1693   -0.36   0.722   -1024.155   710.8214
 femfull | -691.6667   615.7735   -1.12   0.263   -1905.236   521.9031
   _cons |     38300   365.1262  104.90   0.000    37580.41   39019.59
---------------------------------------------------------------------
```

```
. test assoc full

 ( 1)   assoc = 0.0
 ( 2)   full = 0.0

       F(  2,   220) =  208.94
            Prob > F =    0.0000
```

```
. test female

 ( 1)   female = 0.0

       F(  1,   220) =    2.91
            Prob > F =    0.0892
```

```
. test femassoc femfull

( 1)  femassoc = 0.0
( 2)  femfull = 0.0

       F(  2,   220) =    1.60
            Prob > F =    0.2039

. predict predpay3
(option xb assumed; fitted values)

. label variable predpay3 "Median predicted salary"

. table gender rank, contents(mean predpay3)

----------------------------------
Gender    |
(dummy    |        Academic rank
variable) | Assist    Assoc     Full
----------+-----------------------
   Male   | 28500    38320    49950
   Female | 28950    36760    47320
----------------------------------
```

Predicted values from this quantile regression closely resemble the median salaries in each subgroup, as we can verify directly:

```
. table gender rank, contents(median pay)

----------------------------------
Gender    |
(dummy    |        Academic rank
variable) | Assist    Assoc     Full
----------+-----------------------
   Male   | 28500    38320    49950
   Female | 28950    36590    46530
----------------------------------
```

qreg thus allows us to fit models analogous to *n*-way ANOVA or ANCOVA, but involving .5 quantiles or approximate medians instead of the usual means. In theory, .5 quantiles and medians are the same. In practice, quantiles are approximated from actual sample data values, whereas the median is calculated by averaging the two central values, if a subgroup contains an even number of observations. The sample median and .5 quantile approximations then can be different, but in a way that does not much affect model interpretation.

Further rreg and qreg Applications

Diagnostic statistics and plots (Chapter 7) and nonlinear transformations (Chapter 8) extend the usefulness of robust procedures as they do in ordinary regression. With transformed variables, **rreg** or **qreg** fit curvilinear regression models. **rreg** can also robustly perform simpler types of analysis. To obtain a 90% confidence interval for the mean of a single variable, *y*, we could type either the usual confidence-interval command **ci** :

```
. ci y, level(90)
```

Or, we could get exactly the same mean and interval through a regression with no *x* variables:

```
. regress y, level(90)
```

Similarly, we can obtain robust mean with 90% confidence interval by typing

```
. rreg y, level(90)
```

qreg could be used in the same way, but keep in mind the previous section's note about how a .5 quantile found by **qreg** might differ from a sample median. In any of these commands, the **level( )** option specifies the desired degree of confidence. If we omit this option, Stata automatically displays a 95% confidence interval.

To compare two means, analysts typically employ a two-sample *t* test (**ttest**) or one-way analysis of variance (**oneway** or **anova**). As seen earlier, we can perform equivalent tests (yielding identical *t* and *F* statistics) with regression, for example, by regressing the measurement variable on a dummy variable (here called *group*) representing the two categories:

```
. regress y group
```

A robust version of this test results from typing the following command:

```
. rreg y group
```

qreg performs median regression by default, but it is actually a more general tool. It can fit a linear model for any quantile of *y*, not just the median (.5 quantile). For example, commands such as the following can analyze how the first quartile (.25 quantile) of *y* changes with *x*:

```
. qreg y x, quant(.25)
```

Assuming constant error variance, the slopes of the .25 and .75 quantile lines should be roughly the same. **qreg** could thus perform a check for heteroskedasticity or subtle kinds of nonlinearity.

Robust Estimates of Variance

Both **rreg** and **qreg** tend to perform better than OLS (**regress**) in the presence of outlier-prone, nonnormal errors. All three procedures share the common assumption that errors follow independent and identical distributions, however. If the distributions of errors vary across *x* values or observations, then the standard errors calculated by **regress**, **rreg**, or **qreg** probably will understate the true sample-to-sample variation, and yield unrealistically narrow confidence intervals.

regress and some other model fitting commands (although not **rreg** or **qreg**) have an option that estimates standard errors without relying on the strong and sometimes implausible assumptions of independent, identically distributed errors. This option uses an approach derived independently by Huber, White, and others that is sometimes referred to as a sandwich estimator of variance. We will demonstrate using another artificial dataset (*robust2.dta*) containing ill-behaved data.

```
Contains data from C:\data\robust2.dta
   obs:            500                       Robust regression examples 2
                                             (artificial data)
   vars:            12                       8 Jul 2001 18:20
   size:         24,500 (99.5% of memory free)
-------------------------------------------------------------------------------
               storage   display   value
variable name   type     format    label    variable label
-------------------------------------------------------------------------------
x               float    %9.0g              Standard normal x
e5              float    %9.0g              Standard normal errors
y5              float    %9.0g              y5 = 10 + 2*x + e5 (normal
                                               i.i.d. errors)
e6              float    %9.0g              Contaminated normal errors:
                                               95% N(0,1), 5%(N(0,10)
y6              float    %9.0g              y6 = 10 + 2*x + e6
                                               (Contaminated normal errors)
e7              float    %9.0g              Centered chi-square(1) errors
y7              float    %9.0g              y7 = 10 + 2*x + e7 (skewed
                                               errors)
e8              float    %9.0g              Normal errors, variance
                                               increases with x
y8              float    %9.0g              y8 = 10 + 2*x + e8
                                               (heteroskedasticity)
group           byte     %9.0g
e9              float    %9.0g              Normal errors, variance
                                               increases with x, mean &
                                               variance increase with cluster
y9              float    %9.0g              y9 = 10 + 2*x + e9
                                               (heteroskedasticity &
                                               correlated errors)
-------------------------------------------------------------------------------
Sorted by:
```

When we regress *y8* on *x*, we obtain a significant positive slope. A scatterplot shows strong heteroskedasticity, however (Figure 9.5). Variation around the regression line increases with *x*. Because errors do not appear to be identically distributed at all values of *x*, the standard errors, confidence intervals, and tests printed by **regress** are untrustworthy.   **rreg** or **qreg** would face the same problem.

```
. regress y8 x

      Source |       SS       df       MS                Number of obs =     500
-------------+------------------------------              F(  1,   498) =  133.96
       Model |  1607.35658      1   1607.35658            Prob > F      =  0.0000
    Residual |  5975.19162    498   11.9983767            R-squared     =  0.2120
-------------+------------------------------              Adj R-squared =  0.2104
       Total |   7582.5482    499   15.1954874            Root MSE      =  3.4639

-------------------------------------------------------------------------------
          y8 |      Coef.   Std. Err.       t     P>|t|     [95% Conf. Interval]
-------------+-----------------------------------------------------------------
           x |   1.819032   .1571612      11.57   0.000     1.510251    2.127813
       _cons |   10.06642    .154919      64.98   0.000     9.762047     10.3708
-------------------------------------------------------------------------------
```

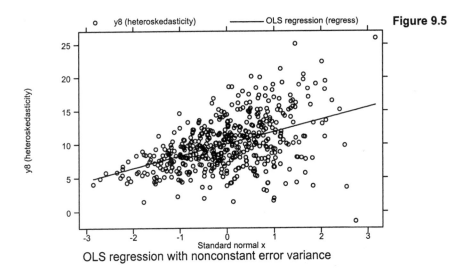

Figure 9.5

OLS regression with nonconstant error variance

More credible standard errors and confidence intervals for this OLS regression can be obtained by using the **robust** option:

`. regress y8 x, robust`

```
Regression with robust standard errors        Number of obs =      500
                                               F(  1,   498) =    83.80
                                               Prob > F      =   0.0000
                                               R-squared     =   0.2120
                                               Root MSE      =   3.4639
```

y8	Coef.	Robust Std. Err.	t	P>|t|	[95% Conf. Interval]
x	1.819032	.1987122	9.15	0.000	1.428614 2.209449
_cons	10.06642	.1561846	64.45	0.000	9.759561 10.37328

Although the fitted model remains unchanged, the robust standard error for the slope is 27% larger (.199 vs. .157) than its nonrobust counterpart. With the **robust** option, the regression output does not show the usual ANOVA sums of squares because these no longer have their customary interpretation.

The rationale underlying these robust standard-error estimates is explained in the *User's Guide* (section 23.11). Briefly, we give up on the classical goal of estimating true population parameters (β's) for a model such as

$$y_i = \beta_0 + \beta_1 x_i + \epsilon_i$$

Instead, we pursue the less ambitious goal of simply estimating the sample-to-sample variation that our b coefficients might have, if we drew many random samples and applied OLS repeatedly to calculate b values for a model such as

$$y_i = b_0 + b_1 x_i + e_i$$

We do not assume that these b estimates will converge on some "true" population parameter. Confidence intervals formed using the robust standard errors therefore lack the classical interpretation of having a certain likelihood (across repeated sampling) of containing the true value of β. Rather, the robust confidence intervals have a certain likelihood (across repeated sampling) of containing b, defined as the value upon which sample b estimates converge. Thus, we pay for relaxing the identically-distributed-errors assumption by settling for a less impressive conclusion.

A further option, `cluster()`, allows us to relax the independent-errors assumption in a limited way. If errors are correlated within subgroups or clusters of the data, as is the case with variable *y9* in *robust2.dta*, then robust standard error estimates (again, for the same OLS model) can be obtained by a command such as

`.  regress y9 x, robust cluster(group)`

The variable *group* in this example identifies the subgroups or clusters. Both the mean and variance of errors affecting the relationship between *y9* and *x* change with the value of *group*, which makes the `cluster()` option necessary.

OLS regression with robust standard errors, estimated by `regress` with the `robust` option, should not be confused with the robust regression estimated by `rreg`. Despite similar-sounding names, the two procedures are unrelated.

10

Logistic Regression

The regression and ANOVA methods described in Chapters 5 through 9 require measured y variables. Stata also offers a full range of techniques for modeling categorical, ordinal, and censored dependent variables. A list of some relevant commands follows. For more details on any of these, type **help** *command* .

binreg   Binomial regression (generalized linear models).

blogit   Logit estimation with grouped (blocked) data.

bprobit   Probit estimation with grouped (blocked) data.

clogit   Conditional fixed-effects logistic regression.

cloglog   Complementary log-log estimation.

cnreg   Censored-normal regression, assuming that y follows a Gaussian distribution but is censored at a point that might vary from observation to observation.

constraint   Defines, lists, and drops linear constraints.

dprobit   Probit regression giving changes in probabilities instead of coefficients.

glm   Generalized linear models. Includes option to model logistic, probit, or complementary log-log links. Allows response variable to be binary or proportional for grouped data.

glogit   Logit regression for grouped data.

gprobit   Probit regression for grouped data.

heckprob Probit estimation with selection.

hetprob   Heteroskedastic probit estimation.

intreg   Interval regression, where y is either point data, interval data, left-censored data, or right-censored data.

logistic Logistic regression, giving odds ratios.

logit   Logistic regression—similar to **logistic**, but giving coefficients instead of odds ratios.

mlogit   Multinomial logistic regression, with polytomous y variable.

nlogit   Nested logit estimation.

ologit   Logistic regression with ordinal y variable.

oprobit   Probit regression with ordinal y variable.

probit   Probit regression, with dichotomous y variable.

scobit   Skewed probit estimation.

svylogit   Logistic regression with complex survey data. Survey (**svy**) versions of **mlogit**, **ologit**, **oprobit**, and **probit** also exist.

tobit   Tobit regression, assuming *y* follows a Gaussian distribution but is censored at a known, fixed point (see **cnreg** for a more general version).

xtclog   Random-effects and population-averaged cloglog models. Panel (**xt**) versions of **logit**, **probit**, and population-averaged generalized linear models (**xtgee**) also exist.

After most model-fitting commands, **predict** can calculate predicted values or probabilities. **predict** also obtains appropriate diagnostic statistics, such as those described for logistic regression in Hosmer and Lemeshow (2000). Specific **predict** options depend on the type of model just estimated.

Examples of several of these commands appear in the next section. The remainder of this chapter concentrates on the important family of methods called logit or logistic regression. We review basic logit methods for dichotomous, ordinal, and polytomous dependent variables.

Example Commands

. logistic y x1 x2 x3

Performs logistic regression of {0,1} variable *y* on predictors *x1*, *x2*, and *x3*. Predictor variable effects are reported as odds ratios. A closely related command,
. logit y x1 x2 x3
performs essentially the same analysis, but reports effects as logit regression coefficients. The models fit by **logistic** and **logit** are the same, so subsequent predictions or diagnostic tests will be identical.

. lfit

Presents a Pearson chi-squared goodness-of-fit test for the estimated logistic model: observed versus expected frequencies of *y* = 1, using cells defined by the covariate (*x*-variable) patterns. When a large number of *x* patterns exist, we might want to group them according to estimated probabilities. **lfit, group(10)** would perform the test with 10 approximately equal-size groups.

. lstat

Presents classification statistics and classification table. **lstat**, **lroc**, and **lsens** (see later) are particularly useful when the point of analysis is classification. These commands all refer to the previously-estimated **logistic** model.

. lroc

Graphs the receiver operating characteristic (ROC) curve, and calculates area under the curve.

. lsens

Graphs both sensitivity and specificity versus the probability cutoff.

. predict phat

Generates a new variable (arbitrarily named *phat*) equal to predicted probabilities that *y* = 1 based on the most recent **logistic** model.

. **predict** *dX2*, **dx2**

Generates a new variable (arbitrarily named *dX2*), the diagnostic statistic measuring change in Pearson chi-squared, from the most recent **logistic** analysis.

. **mlogit** *y x1 x2 x3*, **base(3) rrr nolog**

Performs multinomial logistic regression of multiple-category variable *y* on three *x* variables. Option **base(3)** specifies *y* = 3 as the base category for comparison; **rrr** calls for relative risk ratios instead of regression coefficients; and **nolog** suppresses display of the log likelihood on each iteration.

. **predict** *P2*, **outcome(2)**

Generates a new variable (arbitrarily named *P2*) representing the predicted probability that *y* = 2, based on the most recent **mlogit** analysis.

. **glm** *success x1 x2 x3*, **family(binomial trials) eform**

Performs a logistic regression via generalized linear modeling using tabulated rather than individual-observation data. The variable *success* gives the number of times that the outcome of interest occurred, and *trials* gives the number of times it could have occurred for each combination of the predictors *x1*, *x2*, and *x3*. That is, *success/trials* would equal the proportion of times that an outcome such as "patient recovers" occurred. The **eform** option asks for results in the form of odds ratios ("exponentiated form") rather than logit coefficients.

. **cnreg** *y x1 x2 x3*, **censored(cen)**

Performs censored-normal regression of measurement variable *y* on three predictors *x1*, *x2*, and *x3*. If an observation's true *y* value is unknown due to left or right censoring, it is replaced for this regression by the nearest *y* value at which censoring occurs. The censoring variable *cen* is a {−1,0,1} indicator of whether each observation's value of *y* has been left censored, not censored, or right censored.

Space Shuttle Data

Our main example for this chapter, *shuttle.dta*, involves data covering the first 25 flights of the U.S. space shuttle. These data contain evidence that, if properly analyzed, might have persuaded NASA officials not to launch *Challenger* on its final flight (the 25th shuttle flight, designated STS 51-L). The data are drawn from the *Report of the Presidential Commission on the Space Shuttle Challenger Accident* (1986) and from Tufte (1997). Tufte's book contains an excellent discussion about data and analytical issues. His comments regarding specific shuttle flights are included as a string variable in these data.

```
Contains data from C:\data\shuttle.dta
  obs:           25                      First 25 space shuttle flights
  vars:           8                      9 Jul 2001 16:55
  size:       1,675 (99.0% of memory free)
-------------------------------------------------------------------------------
              storage  display    value
variable name  type    format     label     variable label
-------------------------------------------------------------------------------
flight         byte    %8.0g      flbl      Flight
month          byte    %8.0g                Month of launch
day            byte    %8.0g                Day of launch
year           int     %8.0g                Year of launch
distress       byte    %8.0g      dlbl      Thermal distress incidents
```

```
temp              byte    %8.0g         Joint temperature, degrees F
damage            byte    %9.0g         Damage severity index (Tufte
                                        1997)
comments          str55   %55s          Comments (Tufte 1997)
-------------------------------------------------------------------------
Sorted by:
```

`. list flight-temp`

```
        flight      month       day     year  distress     temp
  1.     STS-1          4        12     1981      none        66
  2.     STS-2         11        12     1981   1 or 2         70
  3.     STS-3          3        22     1982      none        69
  4.     STS-4          6        27     1982        .         80
  5.     STS-5         11        11     1982      none        68
  6.     STS-6          4         4     1983   1 or 2         67
  7.     STS-7          6        18     1983      none        72
  8.     STS-8          8        30     1983      none        73
  9.     STS-9         11        28     1983      none        70
 10.   STS_41-B         2         3     1984   1 or 2         57
 11.   STS_41-C         4         6     1984   3 plus         63
 12.   STS_41-D         8        30     1984   3 plus         70
 13.   STS_41-G        10         5     1984      none        78
 14.   STS_51-A        11         8     1984      none        67
 15.   STS_51-C         1        24     1985   3 plus         53
 16.   STS_51-D         4        12     1985   3 plus         67
 17.   STS_51-B         4        29     1985   3 plus         75
 18.   STS_51-G         6        17     1985   3 plus         70
 19.   STS_51-F         7        29     1985   1 or 2         81
 20.   STS_51-I         8        27     1985   1 or 2         76
 21.   STS_51-J        10         3     1985      none        79
 22.   STS_61-A        10        30     1985   3 plus         75
 23.   STS_61-B        11        26     1985   1 or 2         76
 24.   STS_61-C         1        12     1986   3 plus         58
 25.   STS_51-L         1        28     1986        .         31
```

This chapter studies three variables:

distress   The number of "thermal distress incidents," in which hot gas blow-through or charring damaged joint seals of a flight's booster rockets. Burn-through of a booster joint seal precipitated the *Challenger* disaster. Many previous flights had experienced less severe damage, so the joint seals were known to be a source of possible danger.

temp   The calculated joint temperature at launch time, in degrees Fahrenheit. Temperature depends largely on weather. Rubber O-rings sealing the booster rocket joints become less flexible when cold.

date   Date, measured in days elapsed since January 1, 1960 (an arbitrary starting point). *date* is generated from the month, day, and year of launch using the **mdy** (month-day-year to elapsed time; see **help dates**) function:

`. generate date = mdy(month, day, year)`

`. label variable date "Date (days since 1/1/60)"`

   Launch date matters because several changes over the course of the shuttle program might have made it riskier. Booster rocket walls were thinned to save weight and increase payloads, and joint seals were subjected to higher-pressure testing. Furthermore, the reusable shuttle hardware was aging. So we might ask, did the probability of booster joint damage (one or more distress incidents) increase with launch date?

distress is a labeled numeric variable:

. tabulate distress

```
Thermal     |
distress    |
incidents   |      Freq.      Percent        Cum.
------------+-----------------------------------
     none   |          9        39.13       39.13
   1 or 2   |          6        26.09       65.22
   3 plus   |          8        34.78      100.00
------------+-----------------------------------
    Total   |         23       100.00
```

Ordinarily, **tabulate** displays the labels, but the **nolabel** option reveals that the underlying numerical codes are 0 = "none", 1 = "1 or 2", and 2 = "3 plus":

. tabulate distress, nolabel

```
Thermal     |
distress    |
incidents   |      Freq.      Percent        Cum.
------------+-----------------------------------
        0   |          9        39.13       39.13
        1   |          6        26.09       65.22
        2   |          8        34.78      100.00
------------+-----------------------------------
    Total   |         23       100.00
```

We can use these codes to create a new dummy variable, *any*, coded 0 for no distress and 1 for one or more distress incidents:

. generate any = distress
(2 missing values generated)

. replace any = 1 if distress == 2
(8 real changes made)

. label variable any "Any thermal distress"

To see what this accomplished,

. tabulate distress any

```
Thermal     | Any thermal distress
distress    |
incidents   |         0            1 |      Total
------------+-----------------------+----------
     none   |         9            0 |          9
   1 or 2   |         0            6 |          6
   3 plus   |         0            8 |          8
------------+-----------------------+----------
    Total   |         9           14 |         23
```

Logistic regression models how a {0,1} dichotomy such as *any* depends on one or more *x* variables. The syntax of **logit** resembles that of **regress** and most other model-fitting commands, with the dependent variable listed first:

```
. logit any date, coef

Iteration 0:    log likelihood = -15.394543
Iteration 1:    log likelihood =  -13.01923
Iteration 2:    log likelihood = -12.991146
Iteration 3:    log likelihood = -12.991096
Logit estimates                                 Number of obs   =         23
                                                LR chi2(1)      =       4.81
                                                Prob > chi2     =     0.0283
Log likelihood = -12.991096                     Pseudo R2       =     0.1561

------------------------------------------------------------------------------
     any |      Coef.   Std. Err.       z     P>|z|     [95% Conf. Interval]
---------+--------------------------------------------------------------------
    date |   .0020907   .0010703     1.95     0.051    -6.93e-06     .0041884
   _cons |  -18.13116   9.517217    -1.91     0.057    -36.78456     .5222396
------------------------------------------------------------------------------
```

The **logit** iterative estimation procedure maximizes the logarithm of the likelihood function, shown at the output's top. At iteration 0, the log likelihood describes the fit of a model including only the constant. The last log likelihood describes the fit of the final model,

$$L = -18.13116 + .0020907date \qquad [10.1]$$

where L represents the predicted logit, or log odds, of any distress incidents:

$$L = \ln[P(any = 1) / P(any = 0)] \qquad [10.2]$$

An overall χ^2 test at the upper right evaluates the null hypothesis that all coefficients in the model, except the constant, equal zero,

$$\chi^2 = -2(\ln \mathcal{L}_i - \ln \mathcal{L}_f) \qquad [10.3]$$

where $\ln \mathcal{L}_i$ is the initial or iteration 0 (model with constant only) log likelihood, and $\ln \mathcal{L}_f$ is the final iteration's log likelihood. Here,

$$\chi^2 = -2[-15.394543 - (-12.991096)]$$
$$= 4.81$$

The probability of a greater χ^2, with 1 degree of freedom (the difference in complexity between initial and final models), is low enough (.0283) to reject the null hypothesis in this example. Consequently, *date* does have a significant effect.

Less accurate, though convenient, tests are provided by the asymptotic z (standard normal) statistics displayed with **logit** results. With one predictor variable, that predictor's z statistic and the overall χ^2 statistic test equivalent hypotheses, analogous to the usual t and F statistics in simple OLS regression. Unlike their OLS counterparts, the logit z approximation and χ^2 tests sometimes disagree (they do here). The χ^2 test has more general validity.

Like Stata's other maximum-likelihood estimation procedures, **logit** displays a pseudo R^2 with its output:

$$\text{pseudo } R^2 = 1 - \ln \mathcal{L}_f / \ln \mathcal{L}_i \qquad . \qquad [10.4]$$

For this example,

$$\text{pseudo } R^2 = 1 - (-12.991096) / (-15.394543)$$
$$= .1561$$

Although they provide a quick way to describe or compare the fit of different models for the same dependent variable, pseudo R^2 statistics lack the straightforward explained-variance interpretation of true R^2 in OLS regression.

After **logit**, the **predict** command (with no options) obtains predicted probabilities,

$$Phat = 1 / (1 + e^{-L})$$ [10.5]

Graphed against *date*, these probabilities follow an S-shaped logistic curve as seen in Figure 10.1.

```
. predict Phat
. label variable Phat "Predicted P(distress>=1)"
. graph Phat date, connect(s)
```

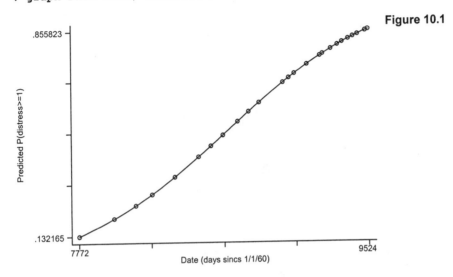

Figure 10.1

The coefficient given by **logit** (.0020907) describes *date*'s effect on the logit or log odds of any thermal distress incidents. Each additional day increased the predicted log odds of thermal distress incidents by .0020907. Equivalently, we could say that each additional day multiplied predicted odds of thermal distress by $e^{.0020907} = 1.0020929$; each 100 days therefore multiplied the odds by $(e^{.0020907})^{100} = 1.23$. ($e \approx 2.71828$, the base number for natural logarithms.) Stata can make these calculations utilizing the _b[*varname*] coefficients stored after any estimation:

```
. display exp(_b[date])
1.0020929
```

```
. display exp(_b[date])^100
1.2325359
```

Or, we could simply include an **or** (odds ratio) option on the **logit** command line. An alternative way to obtain odds ratios employs the **logistic** command described in the next section. **logistic** estimates exactly the same model as **logit**, but its default output displays a table containing odds ratios rather than coefficients.

Using Logistic Regression

Here is the same regression seen earlier, but using **logistic** instead of **logit**:

. **logistic** *any date*

```
Logit estimates                              Number of obs   =         23
                                             LR chi2(1)      =       4.81
                                             Prob > chi2     =     0.0283
Log likelihood = -12.991096                  Pseudo R2       =     0.1561

-------------------------------------------------------------------------
       any | Odds Ratio   Std. Err.      z     P>|z|    [95% Conf. Interval]
-----------+-------------------------------------------------------------
      date |   1.002093   .0010725    1.95   0.051     .9999931    1.004197
-------------------------------------------------------------------------
```

Note the identical log likelihoods and χ^2 statistics. Instead of coefficients (b), **logistic** displays odds ratios (e^b). The numbers in the "Odds Ratio" column of the **logistic** output are amounts by which the odds favoring $y = 1$ are multiplied, with each 1-unit increase in that x variable (if other x variables' values stay the same).

After fitting a model, we can obtain a classification table and related statistics by typing

. **lstat**

```
Logistic model for any

                -------- True --------
Classified |         D              ~D  |       Total
-----------+----------------------------+-----------
     +     |        12               4  |        16
     -     |         2               5  |         7
-----------+----------------------------+-----------
  Total    |        14               9  |        23

Classified + if predicted Pr(D) >= .5
True D defined as any ~= 0
-------------------------------------------------------
Sensitivity                     Pr( +| D)    85.71%
Specificity                     Pr( -|~D)    55.56%
Positive predictive value       Pr( D| +)    75.00%
Negative predictive value       Pr(~D| -)    71.43%
-------------------------------------------------------
False + rate for true ~D        Pr( +|~D)    44.44%
False - rate for true D         Pr( -| D)    14.29%
False + rate for classified +   Pr(~D| +)    25.00%
False - rate for classified -   Pr( D| -)    28.57%
-------------------------------------------------------
Correctly classified                         73.91%
-------------------------------------------------------
```

By default, **lstat** employs a probability of .5 as its cutoff (although we can change this by adding a **cutoff( )** option). Symbols in the classification table have the following meanings:

D   The event of interest did occur (that is, $y = 1$) for that observation. In this example, D indicates that thermal distress occurred.

~D   The event of interest did not occur (that is, $y = 0$) for that observation. In this example, ~D corresponds to flights having no thermal distress.

+ The model's predicted probability is greater than or equal to the cutoff point. Since we used the default cutoff, + here indicates that the model predicts a .5 or higher probability of thermal distress.

– The predicted probability is less than the cutoff. Here, – means a predicted probability of thermal distress below .5.

Thus for 12 flights, classifications are accurate in the sense that the model estimated at least a .5 probability of thermal distress, and distress did in fact occur. For 5 other flights, the model predicted less than a .5 probability, and distress did not occur. The overall "correctly classified" rate is therefore 12 + 5 = 17 out of 23, or 73.91%. The table also gives conditional probabilities such as "sensitivity" or the percentage of observations with $P \geq .5$ given that thermal distress occurred (12 out of 14 or 85.71%).

After **logistic** or **logit**, the follow-up command **predict** calculates various prediction and diagnostic statistics. Discussion of the diagnostic statistics can be found in Hosmer and Lemeshow (2000).

predict *newvar*	Predicted probability that $y = 1$
predict *newvar*, **xb**	Linear prediction (predicted log odds that $y = 1$)
predict *newvar*, **stdp**	Standard error of the linear prediction
predict *newvar*, **dbeta**	ΔB influence statistic, analogous to Cook's D
predict *newvar*, **deviance**	Deviance residual for jth x pattern, d_j
predict *newvar*, **dx2**	Change in Pearson χ^2, written as $\Delta\chi^2$ or $\Delta\chi^2_P$
predict *newvar*, **ddeviance**	Change in deviance χ^2, written as ΔD or $\Delta\chi^2_D$
predict *newvar*, **hat**	Leverage of the jth x pattern, h_j
predict *newvar*, **number**	Assigns numbers to x patterns, $j = 1,2,3 \dots J$
predict *newvar*, **resid**	Pearson residual for jth x pattern, r_j
predict *newvar*, **rstandard**	Standardized Pearson residual

Statistics obtained by the **dbeta**, **dx2**, **ddeviance**, and **hat** options do not measure the influence of individual observations, as their counterparts in ordinary regression do. Rather, these statistics measure the influence of "covariate patterns"; that is, the consequences of dropping all observations with that particular combination of x values. See Hosmer and Lemeshow (2000) for details. A later section of this chapter shows these statistics in use.

Does booster joint temperature also affect the probability of any distress incidents? We could investigate by including *temp* as a second predictor variable:

```
. logistic any date temp
```

```
Logit estimates                                    Number of obs  =          23
                                                   LR chi2(2)     =        8.09
                                                   Prob > chi2    =      0.0175
Log likelihood = -11.350748                        Pseudo R2      =      0.2627
```

```
------------------------------------------------------------------------------
         any | Odds Ratio  Std. Err.       z    P>|z|     [95% Conf. Interval]
-------------+----------------------------------------------------------------
        date |   1.00297    .0013675     2.17   0.030     1.000293    1.005653
        temp |  .8408309    .0987887    -1.48   0.140     .6678848    1.058561
------------------------------------------------------------------------------
```

The classification table indicates that including temperature as a predictor improved our correct classification rate to 78.26%.

```
. lstat
```

```
Logistic model for any
```

```
              -------- True --------
Classified |         D          ~D   |     Total
-----------+------------------------+-----------
     +     |        12           3   |       15
     -     |         2           6   |        8
-----------+------------------------+-----------
   Total   |        14           9   |       23
```

```
Classified + if predicted Pr(D) >= .5
True D defined as any ~= 0
--------------------------------------------------
Sensitivity                     Pr( +| D)   85.71%
Specificity                     Pr( -|~D)   66.67%
Positive predictive value       Pr( D| +)   80.00%
Negative predictive value       Pr(~D| -)   75.00%
--------------------------------------------------
False + rate for true ~D        Pr( +|~D)   33.33%
False - rate for true D         Pr( -| D)   14.29%
False + rate for classified +   Pr(~D| +)   20.00%
False - rate for classified -   Pr( D| -)   25.00%
--------------------------------------------------
Correctly classified                        78.26%
--------------------------------------------------
```

According to the estimated model, each 1-degree increase in joint temperature multiplies the odds of booster joint damage by .84 (in other words, each 1-degree warming reduces the odds of damage by about 16%). Although this effect seems strong enough to cause concern, the asymptotic z test says that it is not statistically significant ($z = -1.476$, $P = .140$). A more definitive test, however, employs the likelihood-ratio χ^2. The **lrtest** command compares nested models estimated by maximum likelihood. First, estimate a "full" model containing all variables of interest, as done above with the `logistic any date temp` command. Next, type the command

```
. lrtest, saving(0)
```

Now estimate a reduced model, including only a subset of the x variables from the full model. (Such reduced models are said to be "nested.") Finally, type **lrtest** again. For example (using the **quietly** prefix, because we already saw this output once),

```
. quietly logistic any date
```

```
. lrtest
```

```
Logistic:  likelihood-ratio test          chi2(1)    =      3.28
                                           Prob > chi2 =    0.0701
```

This second **lrtest** command tests the recent (presumably nested) model against the model previously saved by **lrtest, saving(0)**. It employs a general test statistic for nested maximum-likelihood models,

$$\chi^2 = -2(\ln \mathcal{L}_1 - \ln \mathcal{L}_0) \qquad [10.6]$$

where $\ln \mathcal{L}_0$ is the log likelihood for the first model (with all x variables), and $\ln \mathcal{L}_1$ is the log likelihood for the second model (with a subset of those x variables). Compare the resulting test statistic to a χ^2 distribution with degrees of freedom equal to the difference in complexity (number of x variables dropped) between models 0 and 1. Type **help lrtest** for more about this command, which works with any of Stata's maximum-likelihood estimation procedures (**logit**, **mlogit**, **stcox**, and many others). The overall χ^2 statistic routinely given by **logit** or **logistic** output (equation [10.3]) is a special case of [10.6].

The previous **lrtest** example performed this calculation:

$$\chi^{2\prime} = -2[-12.991096 - (-11.350748)]$$
$$= 3.28$$

with 1 degree of freedom, yielding $P = .0701$; the effect of *temp* is significant at $\alpha = .10$. Given the small sample and fatal consequences of a Type II error, $\alpha = .10$ seems a more prudent cutoff than the usual $\alpha = .05$.

Conditional Effect Plots

Conditional effect plots help in understanding what a logistic model implies about probabilities. The idea behind such plots is to draw a curve showing how the model's prediction of y changes as a function of one x variable, while holding all other x variables constant at chosen values such as their means, quartiles, or extremes. For example, we could find the predicted probability of any thermal distress incidents as a function of *temp*, holding *date* constant at its 25th percentile, 8569 (found by **summarize date, detail**):

```
. quietly logit any date temp
. generate L1 = _b[_cons] + _b[date]*8569 + _b[temp]*temp
. generate Phat1 = 1/(1 + exp(-L1))
. label variable Phat1 "P(distress>=1 | date=8569)"
```

L1 is the predicted logit, and *Phat1* equals the corresponding predicted probability that *distress* ≥ 1, calculated according to equation [10.5]. Similar steps find the predicted probability of any distress with *date* fixed at its 75th percentile (9341):

```
. generate L2 = _b[_cons] + _b[date]*9341 + _b[temp]*temp
. generate Phat2 = 1/(1 + exp(-L2))
. label variable Phat2 "P(distress>=1 | date=9341)"
```

We can now graph the relation between *temp* and the probability of any distress, for the two levels of *date*, as shown in Figure 10.2:

```
. graph Phat1 Phat2 temp, connect(ss) ylabel xlabel
     11(Probability of distress > = 1) border
```

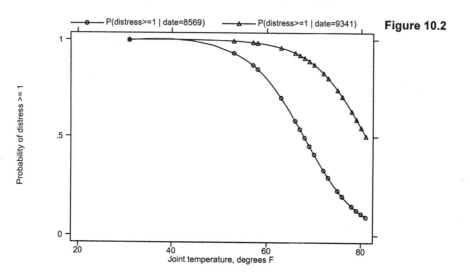

Figure 10.2

Among earlier flights (*date* = 8569, left curve), the probability of thermal distress goes from very low, at around 80° F, to near 1, below 50° F. Among later flights (*date* = 9341, right curve), however, the probability of any distress exceeds .5 even in warm weather, and climbs toward 1 on flights below 70° F. Note that *Challenger*'s launch temperature, 31° F, places it at top left in Figure 10.2. This analysis predicts almost certain booster joint damage.

Diagnostic Statistics and Plots

As mentioned earlier, the logistic regression influence and diagnostic statistics obtained by **predict** refer not to individual observations, as do the OLS regression diagnostics of Chapter 7. Rather, logistic diagnostics refer to *x* patterns. With the space shuttle data, how-ever, each *x* pattern is unique—no two flights share the same combination of *date* and *temp* (naturally, because no two were launched the same day). Before using **predict**, we quietly re-estimate the recent model, to be sure that model is what we think:

```
. quietly logistic any date temp
. predict Phat3
(option p assumed; Pr(any))
. label variable Phat3 "Predicted probability"
. predict dX2, dx2
(2 missing values generated)
. label variable dX2 "Change in Pearson chi-squared"
```

```
. predict dB, dbeta
(2 missing values generated)

. label variable dB "Influence"

. predict dD, ddeviance
(2 missing values generated)

. label variable dD "Change in deviance"
```

Hosmer and Lemeshow (2000) suggest plots that help in reading these diagnostics. To graph change in Pearson χ^2 versus probability of distress (Figure 10.3), type:

```
. graph dX2 Phat3, ylabel xlabel
```

Figure 10.3

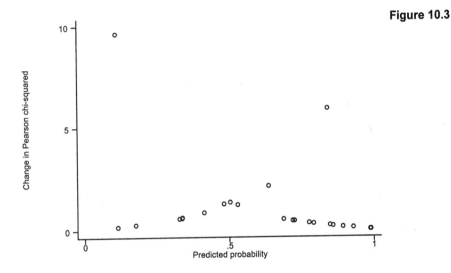

Two poorly fit *x* patterns, at upper right and left, stand out. We can identify these two flights (STS-2 and STS 51-A) by drawing the graph with flight numbers as plotting symbols. This is done by adding to the **graph** command the options **symbol([flight])** **psize(140)**, with the results shown in Figure 10.4.

```
. graph dX2 Phat3, ylabel xlabel symbol([flight]) psize(140)
```

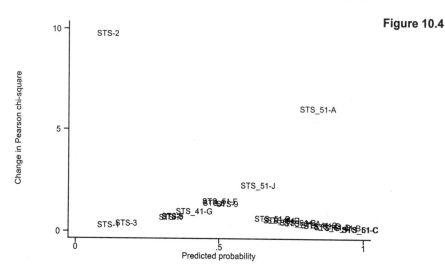

Figure 10.4

```
. list flight any date temp dX2 Phat3 if dX2 > 5
```

	flight	any	date	temp	dX2	Phat3
2.	STS-2	1	7986	70	9.630337	.1091805
4.	STS-4	.	8213	80	.	.0407113
14.	STS_51-A	0	9078	67	5.899742	.8400974
25.	STS_51-L	.	9524	31	.	.9999012

Flight STS 51-A experienced no thermal distress, despite a late launch date and cool temperature (see Figure 10.2). The model predicts a .84 probability of distress for this flight. All points along the up-to-right curve in Figure 10.4 have *any* = 0, meaning no thermal distress. Atop the up-to-left (*any* = 1) curve, flight STS-2 experienced thermal distress despite being one of the earliest flights, and launched in slightly milder weather. The model predicts only a .109 probability of distress. (Recall that Stata considers missing values as "high" numbers, which is why it listed the two missing-values flights, including *Challenger*, among those with *dX2* > 5.)

Similar findings result from plotting *dD* versus predicted probability, as seen in Figure 10.5. Again, flights STS-2 (top left) and STS 51-A (top right) stand out as poorly fit.

. **graph** *dD* **Phat3, ylabel xlabel symbol([[***flight***]]) psize(140)**

Figure 10.5

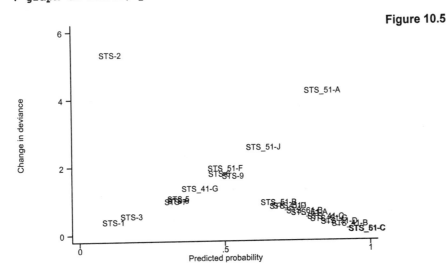

dB measures an *x* pattern's influence in logistic regression, as Cook's *D* measures an individual observation's influence in OLS. For a logistic-regression analogue to the OLS diagnostic plot in Figure 7.7, we can make the plotting symbols proportional to influence as done in Figure 10.6. Figure 10.6 reveals that the two worst-fit observations are also the most influential.

. **graph** *dD* **Phat3 [iweight = ***dB***], ylabel xlabel**

Figure 10.6

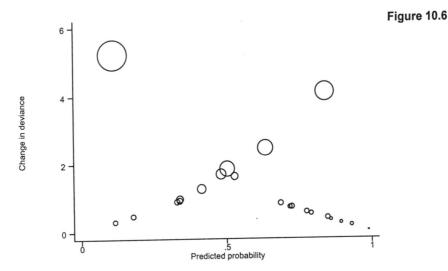

Poorly fit and influential observations deserve special attention because they both contradict the main pattern of the data and pull model estimates in their contrary direction. Of course, simply removing such outliers allows a "better fit" with the remaining data—but this is circular reasoning. A more thoughtful reaction would be to investigate what makes the outliers unusual. Why did shuttle flight STS-2, but not STS 51-A, experience booster joint damage? Seeking an answer might lead investigators to previously overlooked variables or to otherwise respecify the model.

Logistic Regression with Ordered-Category *y*

logit and **logistic** estimate only models that have two-category {0,1} *y* variables. We need other methods for models in which *y* takes on more than two categories. For example,

ologit Ordered logistic regression, where *y* is an ordinal (ordered-category) variable. The numerical values representing the categories do not matter, except that higher numbers mean "more." For example, the *y* categories might be {1 = "poor," 2 = "fair," 3 = "excellent"}.

mlogit Multinomial logistic regression, where *y* has multiple but unordered categories such as {1 = "Democrat," 2 = "Republican," 3 = "undeclared"}.

If *y* is {0,1}, **logit** (or **logistic**), **ologit**, and **mlogit** all produce essentially the same estimates.

We earlier simplified the three-category ordinal variable *distress* into a dichotomy, *any*. **logit** and **logistic** require {0,1} dependent variables. **ologit**, on the other hand, is designed for ordinal variables like *distress* that have more than two categories. The numerical codes representing these categories do not matter, so long as higher numerical values mean "more" of whatever is being measured. Recall that *distress* has categories 0 = "none," 1 = "1 or 2," and 2 = "3 plus" incidents of booster-joint distress.

Ordered logistic regression indicates that *date* and *temp* both affect *distress*, with the same signs (positive for *date*, negative for *temp*) seen in our earlier analyses:

`. ologit distress date temp, nolog`

```
Ordered logit estimates                          Number of obs   =        23
                                                 LR chi2(2)      =     12.32
                                                 Prob > chi2     =    0.0021
Log likelihood = -18.79706                       Pseudo R2       =    0.2468

------------------------------------------------------------------------------
    distress |     Coef.    Std. Err.      z     P>|z|    [95% Conf. Interval]
-------------+----------------------------------------------------------------
        date |    .003286   .0012662     2.60    0.009    .0008043    .0057677
        temp |  -.1733752   .0834473    -2.08    0.038    -.336929   -.0098215
-------------+----------------------------------------------------------------
       _cut1 |   16.42813   9.554813            (Ancillary parameters)
       _cut2 |   18.12227   9.722293
------------------------------------------------------------------------------
```

Likelihood-ratio tests are more accurate than the asymptotic *z* tests shown. First, have **lrtest** save results from the full model (including both predictors) just estimated:

`. lrtest, saving(0)`

Next, estimate a simpler model without *temp*:

`. quietly ologit distress date`
`. lrtest`

```
Ologit: likelihood-ratio test                chi2(1)    =       6.12
                                              Prob > chi2 =    0.0133
```

The likelihood-ratio test indicates that *temp*'s effect is significant. Similar steps find that *date* also has a significant effect:

```
. quietly ologit distress temp
. lrtest
```

```
Ologit: likelihood-ratio test                chi2(1)    =      10.33
                                              Prob > chi2 =    0.0013
```

The ordered-logit model estimates a score, S, as a linear function of *date* and *temp*:

$$S = .003286date - .1733752temp$$

Predicted probabilities depend on the value of S, plus a logistically distributed disturbance u, relative to the estimated cut points:

$P(distress="none") = P(S+u \leq \_cut1) = P(S+u \leq 16.42813)$

$P(distress="1 or 2") = P(\_cut1 < S+u \leq \_cut2) = P(16.42813 < S+u \leq 18.12227)$

$P(distress="3 plus") = P(\_cut2 < S+u) = P(18.12227 < S+u)$

After **ologit**, **predict** calculates predicted probabilities for each category of the dependent variable. We supply **predict** with names for these probabilities. For example: *none* could denote the probability of no distress incidents (first category of *distress*); *onetwo* the probability of 1 or 2 incidents (second category of *distress*); and *threeplus* the probability of 3 or more incidents (third and last category of *distress*):

```
. quietly ologit distress date temp
. predict none onetwo threeplus
(option p assumed; predicted probabilities)
```

This creates three new variables:

```
. describe none onetwo threeplus
```

variable name	storage type	display format	value label	variable label
none	float	%9.0g		Pr(distress==0)
onetwo	float	%9.0g		Pr(distress==1)
threeplus	float	%9.0g		Pr(distress==2)

Predicted probabilities for *Challenger*'s last flight, the 25th in these data, are unsettling:

```
. list flight none onetwo threeplus if flight == 25
```

flight	none	onetwo	threeplus
25. STS_51-L	.0000754	.0003346	.99959

Our model, based on the analysis of 23 pre-*Challenger* shuttle flights, predicts little chance ($P = .000075$) of *Challenger* experiencing no booster joint damage, a scarcely greater likelihood of one or two incidents ($P = .0003$), but virtual certainty ($P = .9996$) of three or more damage incidents.

See Long (1997) or Hosmer and Lemeshow (2000) for more about ordered logistic regression and related techniques. The *Reference Manuals* explain Stata's implementation.

Multinomial Logistic Regression

When the dependent variable's categories have no natural ordering, we resort to multinomial logistic regression, also called polytomous logistic regression. The **mlogit** command makes this straightforward. If y has only two categories, **mlogit** fits the same model as **logistic**. Otherwise, though, an **mlogit** model is more complex. This section presents an extended example interpreting **mlogit** results, using data (*NWarctic.dta*) from a survey of high school students in Alaska's Northwest Arctic borough (Hamilton and Seyfrit 1993).

```
Contains data from C:\data\NWarctic.dta
  obs:           259                    NW Arctic high school students
                                        (Hamilton & Seyfrit 1993)
  vars:            3                    9 Jul 2001 12:45
  size:        2,590 (100.0% of memory free)
-------------------------------------------------------------------------
              storage  display    value
variable name   type   format     label    variable label
-------------------------------------------------------------------------
life           byte    %8.0g      migrate   Expect to live most of life?
ties           float   %9.0g                Social ties to community scale
kotz           byte    %8.0g      kotz      Live in Kotzebue or smaller
                                              village?
-------------------------------------------------------------------------
```

Variable *life* indicates where students say they expect to live most of the rest of their lives: in the same region (Northwest Arctic), elsewhere in Alaska, or outside of Alaska:

```
. tabulate life, plot

   Expect to |
   live most |
    of life? |    Freq.
-------------+------------+----------------------------------------------
        same |       92   |****************************************
    other AK |      120   |*****************************************************
    leave AK |       47   |********************
-------------+------------+----------------------------------------------
       Total |      259
```

Kotzebue (population near 3,000) is the Northwest Arctic's regional hub and largest city. More than a third of these students live in Kotzebue. The rest live in smaller villages of 200 to 700 people. The relatively cosmopolitan Kotzebue students less often expect to stay where they are, and lean more towards leaving the state:

```
. tabulate life kotz, chi2
```

```
Expect to |  Live in Kotzebue or
live most |    smaller village?
 of life? |    village  Kotzebue |    Total
----------+----------------------+----------
     same |       75        17 |       92
 other AK |       80        40 |      120
 leave AK |       11        36 |       47
----------+----------------------+----------
    Total |      166        93 |      259

         Pearson chi2(2) =  46.2992   Pr = 0.000
```

mlogit can replicate this simple analysis (though its likelihood-ratio chi-squared need not exactly equal the Pearson chi-squared found by **tabulate**):

. **mlogit** *life kotz*, **nolog base(1) rrr**

```
Multinomial regression                     Number of obs   =       259
                                           LR chi2(2)      =     46.23
                                           Prob > chi2     =    0.0000
Log likelihood = -244.64465                Pseudo R2       =    0.0863

------------------------------------------------------------------------
       life |     RRR    Std. Err.      z    P>|z|    [95% Conf. Interval]
------------+-----------------------------------------------------------
other AK    |
       kotz | 2.205882   .7304664    2.39   0.017    1.152687    4.221369
------------+-----------------------------------------------------------
leave AK    |
       kotz | 14.4385    6.307555    6.11   0.000    6.132946    33.99188
------------------------------------------------------------------------
(Outcome life==same is the comparison group)
```

base(1) specifies that category 1 of y (*life* = "same") is the base category for comparison. The **rrr** option instructs **mlogit** to show relative risk ratios, which resemble the odds ratios given by **logistic**.

Referring back to the **tabulate** output, we can calculate that among Kotzebue students the odds favoring "leave Alaska" over "stay in the same area" are

P(leave AK) / P(same) = (36/93) / (17/93)

  = 2.1176471

Among other students the odds favoring "leave Alaska" over "same area" are

P(leave AK) / P(same) = (11/166) / (75/166)

  = .1466667

Thus, the odds favoring "leave Alaska" over "same area" are 14.4385 times higher for Kotzebue students than for others:

2.1176471 / .1466667 = 14.4385

This multiplier, a ratio of two odds, equals the relative risk ratio (14.4385) displayed by **mlogit**.

In general, the relative risk ratio for category j of y, and predictor x_k, equals the amount by which predicted odds favoring $y = j$ (compared with y = base) are multiplied, per 1-unit increase in x_k, other things being equal. In other words, the relative risk ratio rrr_{jk} is a multiplier such that, if all x variables except x_k stay the same,

$$\text{rrr}_{jk} \times \frac{P(y=j \mid x_k)}{P(y=\text{base} \mid x_k)} = \frac{P(y=j \mid x_k+1)}{P(y=\text{base} \mid x_k+1)}$$

ties is a continuous scale indicating the strength of students' social ties to family and community. We include *ties* as a second predictor:

```
. mlogit life kotz ties, nolog base(1) rrr
```

```
Multinomial regression                          Number of obs  =        259
                                                LR chi2(4)     =      91.96
                                                Prob > chi2    =     0.0000
Log likelihood = -221.77969                     Pseudo R2      =     0.1717
```

life	RRR	Std. Err.	z	P>|z|	[95% Conf. Interval]
other AK					
kotz	2.214184	.7724996	2.28	0.023	1.117483 4.387193
ties	.4802486	.0799184	-4.41	0.000	.3465911 .6654492
leave AK					
kotz	14.84604	7.146824	5.60	0.000	5.778907 38.13955
ties	.230262	.059085	-5.72	0.000	.1392531 .38075

(Outcome life==same is the comparison group)

Asymptotic *z* tests here indicate that the four relative risk ratios, describing two *x* variables' effects, all differ significantly from 1.0. If a *y* variable has *J* categories, then **mlogit** models the effects of each predictor (*x*) variable with *J* – 1 relative risk ratios or coefficients, and hence also employs *J* – 1 *z* tests—evaluating two or more separate null hypotheses for each predictor. Likelihood-ratio tests evaluate the overall effect of each predictor. First, save as "0" the results from the full model:

```
. lrtest, saving(0)
```

Then estimate a simpler model with one *x* variable omitted, and perform a likelihood-ratio test. For example, to test the effect of *ties*,

```
. quietly mlogit life kotz
```

```
. lrtest
```

```
Mlogit:  likelihood-ratio test                  chi2(2)      =      45.73
                                                Prob > chi2  =     0.0000
```

To then test the effect of *kotz*:

```
. quietly mlogit life ties
```

```
. lrtest
```

```
Mlogit:  likelihood-ratio test                  chi2(2)      =      39.05
                                                Prob > chi2  =     0.0000
```

A word of caution about such likelihood-ratio testing: If our data contained missing values, the three **mlogit** commands just shown might have analyzed three overlapping subsets of observations. The full model would use only observations with nonmissing *life*, *kotz*, and *ties* values; the *ties*-omitted model would bring back in any observations missing only *ties* values;

and the *kotz*-omitted model would bring back observations missing only *kotz* values. When this happens, Stata will say "Warning: observations differ" but still present the likelihood-ratio test. In that case, however, the test would be invalid. Analysts must either screen observations with `if` qualifiers attached to modeling commands, such as

```
. mlogit life kotz ties, nolog base(1) rrr
. lrtest, saving(0)
. quietly mlogit life kotz if ties != .
. lrtest
. quietly mlogit life ties if kotz != .
. lrtest
```

or simply drop all observations having missing values before proceeding:

```
. drop if life == . | kotz == . | ties == .
```

Dataset *NWarctic.dta* has already been screened in this fashion to drop observations with missing values.

Both *kotz* and *ties* significantly predict *life*. What else can we say from this output? To interpret specific effects, recall that *life* = "same" is the base category. The relative risk ratios tell us that:

Odds that a student expects migration to elsewhere in Alaska rather than staying in the same area are 2.21 times greater (increase about 121%) among Kotzebue students (*kotz*=1), adjusting for social ties to community.

Odds that a student expects to leave Alaska rather than stay in the same area are 14.85 times greater (increase about 1385%) among Kotzebue students (*kotz*=1), adjusting for social ties to community.

Odds that a student expects migration to elsewhere in Alaska rather than staying are multiplied by .48 (decrease about 52%) with each 1-unit (since *ties* is standardized, its units equal standard deviations) increase in social ties, controlling for Kotzebue/village residence.

Odds that a student expects to leave Alaska rather than staying are multiplied by .23 (decrease about 77%) with each 1-unit increase in social ties, controlling for Kotzebue/village residence.

`predict` can calculate predicted probabilities from `mlogit`. The `outcome(#)` option specifies for which *y* category we want probabilities. For example, to get predicted probabilities that *life* = "leave AK" (category 3),

```
. quietly mlogit life kotz ties
. predict PleaveAK, outcome(3)
(option p assumed; predicted probability)
. label variable PleaveAK "P(life = 3 | kotz, ties)"
```

Tabulating predicted probabilities for each value of the dependent variable shows how the model fits:

```
. table life, contents(mean PleaveAK) row
```

```
------------------------------
Expect to |
live most |
of life?  | mean(PleaveAK)
----------+-------------------
    same  |     .0811267
other AK  |     .1770225
leave AK  |     .3892264
          |
   Total  |     .1814672
------------------------------
```

A minority of these students (47/259 = 18%) expect to leave Alaska. The model averages only a .39 probability of leaving Alaska even for those who actually chose this response—reflecting the fact that although our predictors have significant effects, most variation in migration plans remains unexplained.

Conditional effect plots help to visualize what a model implies regarding continuous predictors. We can draw them using estimated coefficients (not risk ratios) to calculate probabilities:

```
. mlogit life kotz ties, nolog base(1)
```

```
Multinomial regression                    Number of obs   =        259
                                          LR chi2(4)      =      91.96
                                          Prob > chi2     =     0.0000
Log likelihood = -221.77969               Pseudo R2       =     0.1717
```

life	Coef.	Std. Err.	z	P>|z|	[95% Conf. Interval]
other AK					
kotz	.794884	.3488868	2.28	0.023	.1110784 1.47869
ties	-.7334513	.1664104	-4.41	0.000	-1.05961 -.407293
_cons	.206402	.1728053	1.19	0.232	-.1322902 .5450942
leave AK					
kotz	2.697733	.4813959	5.60	0.000	1.754215 3.641252
ties	-1.468537	.2565991	-5.72	0.000	-1.971462 -.9656124
_cons	-2.115025	.3758163	-5.63	0.000	-2.851611 -1.378439

(Outcome life==same is the comparison group)

The following commands calculate predicted logits, and then the probabilities needed for conditional effect plots. *L2village* represents the predicted logit of *life* = 2 (other Alaska) for village students. *L3kotz* is the predicted logit of *life* = 3 (leave Alaska) for Kotzebue students, and so forth:

```
. generate L2village = .206402 +.794884*0 -.7334513*ties
```
```
. generate L2kotz = .206402 +.794884*1 -.7334513*ties
```
```
. generate L3village = -2.115025 +2.697733*0 -1.468537*ties
```
```
. generate L3kotz = -2.115025 +2.697733*1 -1.468537*ties
```

Like other Stata modeling commands, **mlogit** saves coefficient estimates as macros. For example, [2]_b[*kotz*] refers to the coefficient on *kotz* in the model's second (*life* = 2) equation. Therefore, we could have generated the same predicted logits as follows. *L2v* will be identical to *L2village* defined earlier, *L3k* the same as *L3kotz*, and so forth:

```
.  generate L2v = [2]_b[_cons] +[2]_b[kotz]*0 +[2]_b[ties]*ties
.  generate L2k = [2]_b[_cons] +[2]_b[kotz]*1 +[2]_b[ties]*ties
.  generate L3v = [3]_b[_cons] +[3]_b[kotz]*0 + [3]_b[ties]*ties
.  generate L3k = [3]_b[_cons] +[3]_b[kotz]*1 + [3]_b[ties]*ties
```

From either set of logits, we next calculate the predicted probabilities:

```
.  generate P1village = 1/(1 +exp(L2village) +exp(L3village))
.  label variable P1village "same area"
.  generate P2village = exp(L2village)/(1+exp(L2village)+exp(L3village))
.  label variable P2village "other Alaska"
.  generate P3village = exp(L3village)/(1+exp(L2village)+exp(L3village))
.  label variable P3village "leave Alaska"
.  generate P1kotz = 1/(1 +exp(L2kotz) +exp(L3kotz))
.  label variable P1kotz "same area"
.  generate P2kotz = exp(L2kotz)/(1 +exp(L2kotz) +exp(L3kotz))
.  label variable P2kotz "other Alaska"
.  generate P3kotz = exp(L3kotz)/(1 +exp(L2kotz) +exp(L3kotz))
.  label variable P3kotz "leave Alaska"
```

Figures 10.7 and 10.8 show conditional effect plots for village and Kotzebue students separately:

```
.  graph P1village P2village P3village ties, symbol(iio)
       l1(Probability) b2("Social ties to place (village students)")
       ylabel(0,.2 to 1) xlabel(-3,-2 to 3) yline(0,1) xline(0)
       noaxis con(l[_]ll[-]) sort
```

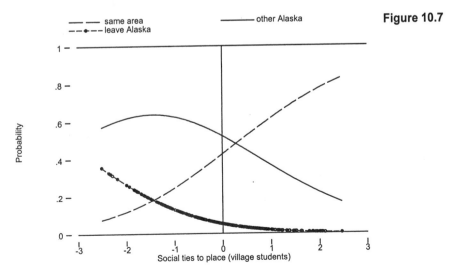

Figure 10.7

```
. graph P1kotz P2kotz P3kotz ties, symbol(iio) l1(Probability)
    b2("Social ties to place (Kotzebue students)")
    ylabel(0,.2 to 1) xlabel(-3,-2 to 3) yline(0,1) xline(0)
    noaxis con(l[_]l1[-]) sort
```

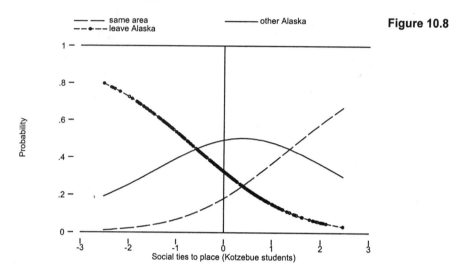

Figure 10.8

The plots indicate that among village students, social ties increase the probability of staying rather than moving elsewhere in Alaska. Relatively few village students expect to leave Alaska. In contrast, among Kotzebue students, *ties* particularly affects the probability of leaving Alaska, rather than simply moving elsewhere in the state. Only if they feel very strong social ties do Kotzebue students tend to favor staying put.

11

Survival and Event-Count Models

This chapter presents methods for analyzing event data. *Survival analysis* encompasses several related techniques that focus on times until the event of interest occurs. Although the event could be good or bad, by convention we refer to that event as a "failure." The time until failure is "survival time." Survival analysis is important in biomedical research, but it can equally well apply to other fields from engineering to social science—for example, in modeling the time until an unemployed person gets a job, or a single person gets married. Stata offers a full range of survival analysis procedures, only a few of which are illustrated in this chapter.

We also look briefly at Poisson regression and its relatives. These methods focus not on survival times but, rather, on the rates or counts of events over a specified interval of time. Event-count methods include Poisson regression and negative binomial regression. Such models can be fit either through specialized commands, or through the broader approach of generalized linear modeling (GLM).

Consult the *Reference Manuals'* **st** entries, starting with **st stset**, for more information about Stata's extensive capabilities. Selvin (1995) provides well-illustrated introductions to survival analysis and Poisson regression. I have borrowed (with permission) several of his examples. Other good introductions to survival analysis include a chapter in Rosner (1995), and the more comprehensive treatments by Hosmer and Lemeshow (1999) and Lee (1992). McCullagh and Nelder (1989) describe generalized linear models. Long (1997) has a chapter on regression models for count data (including Poisson and negative binomial), and also has some material on generalized linear models. An extensive and current treatment of generalized linear models is found in Hardin and Hilbe (2001).

Example Commands

Most of Stata's survival-analysis (**st***) commands require that the data have previously been identified as survival-time by issuing an **stset** command (see following). **stset** need only be run once, and the data subsequently saved.

. **stset** *timevar*, **failure**(*failvar*)

Identifies single-record survival-time data. Variable *timevar* indicates the time elapsed before either a particular event (called a "failure") occurred, or the period of observation ended ("censoring"). Variable *failvar* indicates whether a failure (*failvar* = 1) or censoring (*failvar* = 0) occurred at *timevar*. The dataset contains only one record per individual. The dataset must be **stset** before any further **st*** commands will work. If we subsequently **save** the dataset, however, the **stset** definitions are saved as well. **stset** creates new variables named *_st, _d, _t,* and *_t0* that encode information necessary for subsequent `st*` commands.

. `stset timevar, failure(failvar) id(patient) enter(time start)`

Identifies multiple-record survival-time data. In this example, the variable *timevar* indicates elapsed time before failure or censoring; *failvar* indicates whether failure (1) or censoring (0) occurred at this time. *patient* is an identification number. The same individual might contribute more than one record to the data, but always has the same identification number. *start* contains the time when each individual came under observation.

. `stdes`

Describes survival-time data, listing the definitions set by `stset` and other characteristics of the data.

. `stsum`

Obtains summary statistics: the total time at risk, incidence rate, number of subjects, and percentiles of survival time.

. `ctset time nfail ncensor nenter, by(ethnic sex)`

Identifies count-time data. In this example, the variable *time* is a measure of time; *nfail* is the number of failures occurring at *time*. We also specified *ncensor* (number of censored observations at *time*) and *nenter* (number entering at *time*), although these can be optional. *ethnic* and *sex* are other categorical variables defining observations in these data.

. `cttost`

Converts count-time data, previously identified by a `ctset` command, into survival-time form that can be analyzed by `st*` commands.

. `sts graph`

Graphs the Kaplan–Meier survivor function. To visually compare two or more survivor functions, such as one for each value of the categorical variable *sex*, use the `by()` option,

. `sts graph, by(sex)`

To adjust, through Cox regression, for the effects of a continuous independent variable such as *age*, use the `adjustfor()` option,

. `sts graph, by(sex) adjustfor(age)`

Note: the `by()` and `adjustfor()` options work similarly with the other `sts` commands `sts list`, `sts generate`, and `sts test`.

. `sts list`

Lists the estimated Kaplan–Meier survivor (failure) function.

. `sts test sex`

Tests the equality of the Kaplan–Meier survivor function across categories of *sex*.

. `sts generate survfunc = S`

Creates a new variable arbitrarily named *survfunc*, containing the estimated Kaplan–Meier survivor function.

. `stcox x1 x2 x3`

Estimates a Cox proportional hazard model, regressing time to failure on continuous or dummy variable predictors *x1–x3*.

. `stcox x1 x2 x3, strata(x4) basechazard(hazard) robust`

Estimates a Cox proportional hazard model, stratified by *x4*. Stores the group-specific baseline cumulative hazard function as a new variable named *hazard*. (Baseline survivor function estimates could be obtained through a `basesur(survive)` option.) Obtains robust standard error estimates. See Chapter 9 or, for a more complete explanation of robust standard errors, consult the *User's Guide* 23.11.

. `stphplot, by(sex)`

Plots −ln(−ln(survival)) versus ln(analysis time) for each level of the categorical variable *sex*, from the previous `stcox` model. Roughly parallel curves support the Cox model assumption that the hazard ratio does not change with time. Other checks on the Cox assumptions are performed by the commands `stcoxkm` (compares Cox predicted curves with Kaplan–Meier observed survival curves) and `stphtest` (performs test based on Schoenfeld residuals). See `help stcox` for syntax and options.

. `streg x1 x2, dist(weibull)`

Estimates Weibull-distribution model regression of time-to-failure on continuous or dummy variable predictors *x1* and *x2*.

. `streg x1 x2 x3 x4, dist(exponential) robust`

Estimates exponential-distribution model regression of time-to-failure on continuous or dummy predictors *x1–x4*. Obtains heteroskedasticity-robust standard error estimates. In addition to Weibull and exponential, other `dist()` specifications for `streg` include lognormal, log-logistic, Gompertz, or generalized gamma distributions. Type `help streg` for more information.

. `stcurve, survival`

After `streg` , plots the survival function from this model at mean values of all the *x* variables.

. `stcurve, cumhaz at(x3=50, x4=0)`

After `streg` , plots the cumulative hazard function from this model at mean values of *x1* and *x2*, *x3* set at 50, and *x4* set at 0.

. `poisson count x1 x2 x3, irr exposure(x4)`

Performs Poisson regression of event-count variable *count* (assumed to follow a Poisson distribution) on continuous or dummy independent variables *x1–x3*. Independent-variable effects will be reported as incidence rate ratios (`irr`). The `exposure()` option identifies a variable indicating the amount of exposure, if this is not the same for all observations. Note: A Poisson model assumes that the event probability remains constant, regardless of how many times an event occurs for each observation. If the probability does not remain constant, we should consider using `nbreg` (negative binomial regression) or `gnbreg` (generalized negative binomial regression) instead.

. `glm count x1 x2 x3, link(log) family(poisson) lnoffset(x4) eform`

Performs the same regression specified in the `poisson` example above, but as a generalized linear model (GLM). `glm` can fit Poisson, negative binomial, logit, and many other types of models, depending on what `link()` (link function) and `family()` (distribution family) options we employ.

Survival-Time Data

Survival-time data contain, at a minimum, one variable measuring how much time elapsed before a certain event occurred to each observation. The literature often terms this event of interest a "failure," regardless of its substantive meaning. When failure has not occurred to an observation by the time data collection ends, that observation is said to be "censored." The **stset** command sets up a dataset for survival-time analysis by identifying which variable measures time and (if necessary) which variable is a dummy indicating whether the observation failed or was censored. The dataset can also contain any number of other measurement or categorical variables, and individuals (for example, medical patients) can be represented by more than one observation.

To illustrate the use of **stset**, we will begin with an example from Selvin (1995:453) concerning 51 individuals diagnosed with HIV. The data initially reside in a raw-data file (*aids.raw*) that looks like this:

```
1           1          1         34
2          17          1         42
3'         37          0         47
        (rows 4–50 omitted)
51         81          0         29
```

The first column values are case numbers (1, 2, 3, ..., 51). The second column tells how many months elapsed after the diagnosis, before that person either developed symptoms of AIDS or the study ended (1, 17, 37, ...). The third column holds a 1 if the individual developed AIDS symptoms (failure), or a 0 if no symptoms had appeared by the end of the study (censoring). The last column reports the individual's age at the time of diagnosis.

We can read the raw data into memory using **infile**, then label the variables and data and save in Stata format as file *aids1.dta*:

```
. infile case time aids age using aids.raw, clear
(51 observations read)

. label variable case "Case ID number"

. label variable time "Months since HIV diagnosis"

. label variable aids "Developed AIDS symptoms"

. label variable age "Age in years"

. label data "AIDS (Selvin 1995:453)"

. compress
case was float now byte
time was float now byte
aids was float now byte
age was float now byte

. save aids1
file c:\data\aids1.dta saved
```

The next step is to identify which variable measures time and which indicates failure/censoring. Although not necessary with these single-record data, we can also note which variable holds individual case identification numbers. In an **stset** command, the first-named variable measures time. Subsequently, we identify with **failure()** the dummy representing whether an observation failed (1) or was censored (0). After using **stset**, we save the data again to preserve this information.

```
. stset time, failure(aids) id(case)

               id:  case
    failure event:  aids ~= 0 & aids ~= .
obs. time interval:  (time[_n-1], time]
 exit on or before:  failure

-----------------------------------------------------------------------
        51  total obs.
         0  exclusions
-----------------------------------------------------------------------
        51  obs. remaining, representing
        51  subjects
        25  failures in single failure-per-subject data
      3164  total analysis time at risk, at risk from t =         0
                              earliest observed entry t =         0
                                last observed exit t =           97
```

```
. save, replace
file c:\data\aids1.dta saved
```

stdes yields a brief description of how our survival-time data are structured. In this simple example we have only one record per subject, so some of this information is unneeded.

```
. stdes
```

```
        failure _d:  aids
  analysis time _t:  time
               id:  case
```

| Category | total | |-------------- per subject --------------| | | |
		mean	min	median	max
no. of subjects	51				
no. of records	51	1	1	1	1
(first) entry time		0	0	0	0
(final) exit time		62.03922	1	67	97
subjects with gap	0				
time on gap if gap	0	.	.	.	.
time at risk	3164	62.03922	1	67	97
failures	25	.4901961	0	0	1

The **stsum** command obtains summary statistics. We have 25 failures out of 3,164 person-months, giving an incidence rate of $25/3164 = .0079014$. The percentiles of survival time derive from a Kaplan–Meier survivor function (next section). This function estimates about a 25% chance of developing AIDS within 41 months after diagnosis, and 50% within 81 months. Over the observed range of the data (up to 97 months) the probability of AIDS does not reach 75%, so there is no 75th percentile given.

```
. stsum
```

```
        failure _d:  aids
   analysis time _t:  time
                id:  case


         |                   incidence    no. of   |------ Survival time -----|
         | time at risk        rate      subjects      25%       50%        75%
---------+------------------------------------------------------------------------
   total |        3164      .0079014          51        41        81         .
```

If the data happen to include a grouping or categorical variable such as *sex* (0 = male, 1 = female), we could obtain summary statistics on survival time separately for each group by a command of the following form:

```
. stsum, by(sex)
```

Later sections describe more formal methods for comparing survival times from two or more groups.

Count-Time Data

Survival-time (**st**) datasets like *aids1.dta* contain information on individual people or things, with variables indicating the time at which failure or censoring occurred for each individual. A different type of dataset called count-time (**ct**) contains aggregate data, with variables counting the number of individuals that failed or were censored at time *t*. For example, *diskdriv.dta* contains hypothetical test information on 25 disk drives. All but 5 drives failed before testing ended at 1,200 hours.

```
Contains data from C:\data\diskdriv.dta
  obs:            6                   Count-time data on disk drives
  vars:           3                   13 Jul 2001 10:32
  size:          48 (99.6% of memory free)
-------------------------------------------------------------------------------
              storage  display   value
variable name  type    format    label      variable label
-------------------------------------------------------------------------------
hours          int     %8.0g                Hours of continuous operation
failures       byte    %8.0g                Number of failures observed
censored       byte    %9.0g                Number still working
-------------------------------------------------------------------------------
Sorted by:
```

```
. list

        hours  failures   censored
  1.     200        2          0
  2.     400        3          0
  3.     600        4          0
  4.     800        8          0
  5.    1000        3          0
  6.    1200        0          5
```

To set up a count-time dataset, we specify the time variable, the number-of-failures variable, and the number-censored variable, in that order. After **ctset** , the **cttost** com-mand automatically converts our count-time data to survival-time format.

```
. ctset hours failures censored

    dataset name:  C:\data\diskdriv.dta
           time:  hours
       no. fail:  failures
       no. lost:  censored
      no. enter:  --                        (meaning all enter at time 0)

. cttost

(data is now st)

    failure event:  failures ~= 0 & failures ~= .
obs. time interval:  (0, hours]
exit on or before:  failure
           weight:  [fweight=w]

--------------------------------------------------------------------------------
        6  total obs.
        0  exclusions
--------------------------------------------------------------------------------
        6  physical obs. remaining, equal to
       25  weighted obs., representing
       20  failures in single record/single failure data
    19400  total analysis time at risk, at risk from t =         0
                          earliest observed entry t =            0
                            last observed exit t =            1200

. list

        hours   failures          w        _st         _d          _t        _t0
    1.   1200         0           5          1          0        1200          0
    2.    200         1           2          1          1         200          0
    3.    400         1           3          1          1         400          0
    4.    600         1           4          1          1         600          0
    5.    800         1           8          1          1         800          0
    6.   1000         1           3          1          1        1000          0

. stdes

        failure _d:  failures
   analysis time _t:  hours
            weight:  [fweight=w]
```

| | | \|-------------- per subject --------------\| | | | |
Category	unweighted total	unweighted mean	min	unweighted median	max
no. of subjects	6				
no. of records	6	1	1	1	1
(first) entry time		0	0	0	0
(final) exit time		.700	200	700	1200
subjects with gap	0				
time on gap if gap	0				
time at risk	4200	700	200	700	1200
failures	5	.8333333	0	1	1

The **cttost** command defines a set of frequency weights, *w*, in the resulting **st**-format dataset. **st***** commands automatically recognize and use these weights in any survival-time analysis, so the data now are viewed as containing 25 observations (25 disk drives) instead of the previous 6 (6 time periods):

```
. stsum

   failure time:  hours
failure/censor:  failures
        weight:  [fweight=w]

       |                 incidence    no. of    |---- Survival time ----|
       | time at risk      rate      subjects     25%      50%      75%
-------+-------------------------------------------------------------------
 total |        19400    .0010309        25       600      800     1000
```

Kaplan–Meier Survivor Functions

Let n_t represent the number of observations that have not failed, and are not censored, at the beginning of time period *t*. d_t represents the number of failures that occur to these observations during time period *t*. The Kaplan–Meier estimator of surviving beyond time *t* is the product of survival probabilities in *t* and the preceding periods:

$$S(t) = \prod_{j=t0}^{t} \{ (n_j - d_j) / n_j \} \qquad\qquad [11.1]$$

For example, in the AIDS data seen earlier, one of the 51 individuals developed symptoms only one month after diagnosis. No observations were censored this early, so the probability of "surviving" (meaning, not developing AIDS) beyond *time* = 1 is

$$S(1) = (51 - 1) / 51 = .9804$$

A second patient developed symptoms at *time* = 2, and a third at *time* = 9:

$$S(2) = .9804 \times (50 - 1) / 50 = .9608$$
$$S(9) = .9608 \times (49 - 1) / 49 = .9412$$

Graphing *S(t)* against *t* produces a Kaplan–Meier survivor curve, like the one seen in Figure 11.1. Stata draws such graphs automatically with the **sts graph** command. For example,

```
. use aids, clear
(AIDS (Selvin 1995:453))

. sts graph

        failure _d:  aids
  analysis time _t:  time
              id:  case
```

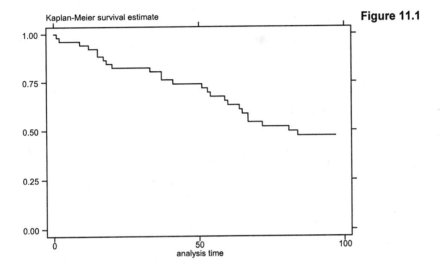

Figure 11.1

For a second example of survivor functions, we turn to data in *smoking1.dta*, adapted from Rosner (1995). The observations are 234 former smokers, attempting to quit. Most did not succeed. Variable *days* records how many days elapsed between quitting and starting up again. The study lasted one year, and variable *smoking* indicates whether an individual resumed smoking before the end of this study (*smoking* = 1, "failure") or not (*smoking* = 0, "censored"). With new data, we should begin by using **stset** to set the data up for survival-time analysis:

```
Contains data from C:\data\smoking1.dta
  obs:           234                       Smoking (Rosner 1995:607)
  vars:            8                       13 Jul 2001 11:03
  size:        3,744 (99.1% of memory free)
-------------------------------------------------------------------------
              storage  display   value
variable name   type    format   label    variable label
-------------------------------------------------------------------------
id              int     %9.0g              Case ID number
days            int     %9.0g              Days abstinent
smoking         byte    %9.0g              Resumed smoking
age             byte    %9.0g              Age in years
sex             byte    %9.0g     sex      Sex (female)
cigs            byte    %9.0g              Cigarettes per day
co              int     %9.0g              Carbon monoxide x 10
minutes         int     %9.0g              Minutes elapsed since last cig
-------------------------------------------------------------------------
Sorted by:
```

```
. stset days, failure(smoking)

        failure event:  smoking ~= 0 & smoking ~= .
   obs. time interval:  (0, days]
   exit on or before:  failure

------------------------------------------------------------------------
      234  total obs.
        0  exclusions
------------------------------------------------------------------------
      234  obs. remaining, representing
      201  failures in single record/single failure data
    18946  total analysis time at risk, at risk from t =         0
                                earliest observed entry t =         0
                                   last observed exit t =       366
```

The study involved 110 men and 124 women. Incidence rates for both sexes appear to be similar:

```
. stsum, by(sex)

        failure _d:  smoking
   analysis time _t:  days

          |                incidence    no. of   |------ Survival time -----|
sex       | time at risk     rate      subjects       25%       50%        75%
----------+-------------------------------------------------------------------
   Male   |      8813     .0105526       110          4        15         68
   Female |     10133     .0106582       124          4        15         91
----------+-------------------------------------------------------------------
   total  |     18946     .0106091       234          4        15         73
```

Figure 11.2 confirms this similarity, showing little difference between the survivor functions of men and women. That is, both sexes returned to smoking at about the same rate. The survival probabilities of nonsmokers decline very steeply during the first 30 days after quitting. For either sex, there is less than a 15% chance of surviving beyond a full year.

```
. sts graph, by(sex)

        failure _d:  smoking
   analysis time _t:  days
```

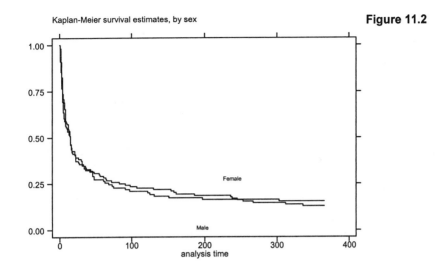

Kaplan-Meier survival estimates, by sex

Figure 11.2

We can also formally test for the equality of survivor functions using a log-rank test. Unsurprisingly, this test finds no significant difference ($P = .6772$) between the smoking recidivism of men and women.

```
. sts test sex

          failure _d:  smoking
    analysis time _t:  days


Log-rank test for equality of survivor functions

        |   Events          Events
sex     | observed        expected
--------+--------------------------
Male    |       93           95.88
Female  |      108          105.12
--------+--------------------------
Total   |      201          201.00

          chi2(1)  =      0.17
          Pr>chi2  =      0.6772
```

Cox Proportional Hazard Models

Regression methods allow us to take survival analysis further and examine the effects of multiple continuous or categorical predictors. One widely-used method known as Cox regression employs a proportional hazard model. The hazard rate for failure at time t is defined as

$$h(t) = \frac{\text{probability of failing between times } t \text{ and } t + \Delta t}{(\Delta t)\,(\text{probability of failing after time } t)} \qquad [11.2]$$

We model this hazard rate as a function of the baseline hazard (h_0) at time t, and the effects of one or more x variables,

$$h(t) \quad = \quad h_0(t) \exp(\beta_1 x_1 + \beta_2 x_2 + \ldots + \beta_k x_k) \qquad [11.3a]$$

or, equivalently,

$$\ln[h(t)] = \quad \ln[h_0(t)] + \beta_1 x_1 + \beta_2 x_2 + \ldots + \beta_k x_k \qquad [11.3b]$$

"Baseline hazard" means the hazard for an observation with all x variables equal to 0. Cox regression estimates this hazard nonparametrically and obtains maximum-likelihood estimates of the β parameters in [11.3]. Stata's **stcox** procedure ordinarily reports hazard ratios, which are estimates of $\exp(\beta)$. These indicate proportional changes relative to the baseline hazard rate.

Does age affect the onset of AIDS symptoms? Dataset *aids.dta* contains information that helps answer this question. Note that with **stcox**, unlike most other Stata model-fitting commands, we list only the independent variable(s). The survival-analysis dependent variable, a survivor or hazard function, is understood automatically with **stset** data.

```
. use aids
(AIDS (Selvin 1995:453))

. stcox age, nolog

         failure _d:  aids
   analysis time _t:  time
              id:  case

Cox regression -- Breslow method for ties

No. of subjects =          51            Number of obs   =          51
No. of failures =          25
Time at risk    =        3164
                                         LR chi2(1)      =        5.00
Log likelihood  =   -86.576295           Prob > chi2     =      0.0254

------------------------------------------------------------------------------
        _t |
        _d | Haz. Ratio   Std. Err.      z    P>|z|     [95% Conf. Interval]
-----------+------------------------------------------------------------------
       age |   1.084557    .0378623     2.33   0.020     1.01283    1.161363
------------------------------------------------------------------------------
```

We might interpret the estimated hazard ratio, 1.084557, with reference to two HIV-positive individuals whose ages are a and $a + 1$. The older person is 8.5% more likely to develop AIDS symptoms over a short period of time (that is, the ratio of their respective hazards is 1.084557). This ratio differs significantly ($P = .020$) from 1. If we wanted to state our findings for a five-year difference in age, we could raise the hazard ratio to the fifth power:

```
. display exp(_b[age])^5
1.5005865
```

Thus, the hazard of AIDS onset is about 50% higher when the second person is five years older than the first. Alternatively, we could learn the same thing (and obtain the new confidence interval) by repeating the regression after creating a new version of *age* measured in five-year units. The **nolog noshow** options below suppress display of the iteration log and the **st-**dataset description.

```
. generate age5 = age/5
. label variable age5 "age in 5-year units"
. stcox age5, nolog noshow
```

```
Cox regression -- Breslow method for ties

No. of subjects =          51            Number of obs   =          51
No. of failures =          25
Time at risk    =        3164
                                         LR chi2(1)      =        5.00
Log likelihood  =  -86.576295            Prob > chi2     =      0.0254

------------------------------------------------------------------------
      _t |
      _d | Haz. Ratio  Std. Err.       z    P>|z|   [95% Conf. Interval]
---------+--------------------------------------------------------------
    age5 |  1.500587   .2619305      2.33   0.020    1.065815    2.112711
------------------------------------------------------------------------
```

Like ordinary regression, Cox models can have more than one independent variable. Dataset *heart.dta* contains survival-time data from Selvin (1995) on 35 patients with very high cholesterol levels. Variable *time* gives the number of days each patient was under observation. *coronary* indicates whether a coronary event occurred during this time (*coronary* = 1) or not (*coronary* = 0). The data also include cholesterol levels and other factors thought to affect heart disease. File *heart.dta* was previously set up for survival-time analysis by an **stset time, failure(coronary)** command, so we can go directly to **st** analysis.

```
. describe patient - ab

              storage  display   value
variable name  type    format    label      variable label
------------------------------------------------------------------------
patient        byte    %9.0g                Patient ID number
time           int     %9.0g                Time in days
coronary       byte    %9.0g                Coronary event (1) or none (0)
weight         int     %9.0g                Weight in pounds
sbp            int     %9.0g                Systolic blood pressure
chol           int     %9.0g                Cholesterol level
cigs           byte    %9.0g                Cigarettes smoked per day
ab             byte    %9.0g                Type A (1) or B (0) personality

. stdes

         failure _d:  coronary
    analysis time _t:  time

                            |-------------- per subject --------------|
Category            total        mean       min     median        max
------------------------------------------------------------------------
no. of subjects       35
no. of records        35           1          1          1          1

(first) entry time                 0          0          0          0
(final) exit time           2580.629        773       2875       3141

subjects with gap      0
time on gap if gap     0
time at risk       90322    2580.629        773       2875       3141

failures               8    .2285714          0          0          1
------------------------------------------------------------------------
```

Cox regression finds that cholesterol level and cigarettes both significantly increase the hazard of a coronary event. Counterintuitively, weight appears to decrease the hazard. Systolic blood pressure and A/B personality do not have significant net effects.

```
. stcox weight sbp chol cigs ab, noshow nolog

Cox regression -- no ties

No. of subjects =            35          Number of obs   =         35
No. of failures =             8
Time at risk    =         90322
                                         LR chi2(5)      =      13.97
Log likelihood  =   -17.263231           Prob > chi2     =     0.0158

------------------------------------------------------------------------
  _t  |
  _d  | Haz. Ratio   Std. Err.      z    P>|z|     [95% Conf. Interval]
------+-----------------------------------------------------------------
weight |  .9349336    .0305184   -2.06   0.039     .8769919    .9967034
   sbp |  1.012947    .0338061    0.39   0.700     .9488087    1.081421
  chol |  1.032142    .0139984    2.33   0.020     1.005067    1.059947
  cigs |  1.203335    .1071031    2.08   0.038     1.010707    1.432676
    ab |   3.04969    2.985616    1.14   0.255     .4476492    20.77655
------------------------------------------------------------------------
```

After estimating the model, **stcox** can also generate new variables holding the esti-mated baseline hazard and survivor functions. Since "baseline" refers to a situation with all *x* variables equal to zero, however, we first need to recenter some variables so that 0 values make sense. A patient who weighs 0 pounds, or has 0 blood pressure, does not provide a useful comparison. Guided by the minimum values actually in our data, we might shift *weight* so that 0 indicates 120 pounds, *sbp* so that 0 indicates 100, and *chol* so that 0 indicates 340:

```
. summarize patient - ab

   Variable |     Obs        Mean    Std. Dev.      Min       Max
------------+--------------------------------------------------------
    patient |      35          18    10.24695         1        35
       time |      35    2580.629    616.0796       773      3141
   coronary |      35    .2285714     .426043         0         1
     weight |      35    170.0857    23.55516       120       225
        sbp |      35    129.7143    14.28403       104       154
       chol |      35    369.2857    51.32284       343       645
       cigs |      35    17.14286    13.07702         0        40
         ab |      35    .5142857    .5070926         0         1

. replace weight = weight - 120
(35 real changes made)

. replace sbp = sbp - 100
(35 real changes made)

. replace chol = chol - 340
(35 real changes made)

. summarize patient - ab

   Variable |     Obs        Mean    Std. Dev.      Min       Max
------------+--------------------------------------------------------
    patient |      35          18    10.24695         1        35
       time |      35    2580.629    616.0796       773      3141
   coronary |      35    .2285714     .426043         0         1
     weight |      35    50.08571    23.55516         0       105
        sbp |      35    29.71429    14.28403         4        54
```

```
chol |      35     29.28571   51.32284          3        305
cigs |      35     17.14286   13.07702          0         40
  ab |      35     .5142857   .5070926          0          1
```

Zero values for all the *x* variables now make more substantive sense. To create new variables holding the baseline survivor and hazard function estimates, we repeat the regression with **basesurv()** and **basechaz()** options:

```
. stcox weight sbp chol cigs ab, noshow nolog basesurv(survivor)
    basechaz(hazard)
```

```
Cox regression -- no ties

No. of subjects =          35            Number of obs   =          35
No. of failures =           8
Time at risk    =       90322
                                         LR chi2(5)      =       13.97
Log likelihood  =   -17.263231           Prob > chi2     =      0.0158

--------------------------------------------------------------------------
      _t |
      _d | Haz. Ratio   Std. Err.      z    P>|z|     [95% Conf. Interval]
---------+----------------------------------------------------------------
  weight |   .9349336    .0305184    -2.06   0.039     .8769919    .9967034
     sbp |   1.012947    .0338061     0.39   0.700     .9488087    1.081421
    chol |   1.032142    .0139984     2.33   0.020     1.005067    1.059947
    cigs |   1.203335    .1071031     2.08   0.038     1.010707    1.432676
      ab |    3.04969    2.985616     1.14   0.255     .4476492    20.77655
--------------------------------------------------------------------------
```

Note that recentering three *x* variables had no effect on the hazard ratios, standard errors, and so forth. The command created two new variables, arbitrarily named *survivor* and *hazard*. To graph the baseline survivor function, we plot *survivor* against *time* and connect data points with a step function (**connect(J)**), as seen in Figure 11.3:

```
. graph survivor time, connect(J) symbol(.) sort ylabel xlabel
```

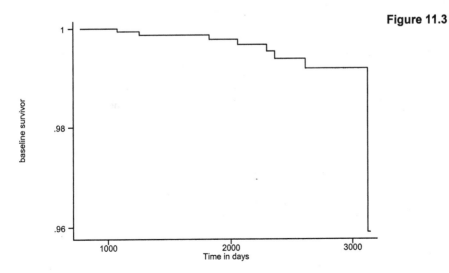

Figure 11.3

The baseline survivor function—which depicts survival probabilities for patients having "0" weight (120 pounds), "0" blood pressure (100), "0" cholesterol (340), 0 cigarettes per day, and a type B personality—declines with time. Although this decline looks precipitous at the right, notice that the probability really only falls from 1 to about .96. Given less favorable values of the predictor variables, the survival probabilities would fall much farther.

The same baseline survivor-function graph could have been obtained another way, without **stcox** . The alternative, shown in Figure 11.4, employs an **sts graph** command with **adjustfor()** option listing the predictor variables:

```
. sts graph, adjustfor(weight sbp chol cigs ab)
```

```
        failure _d:  coronary
 analysis time _t:  time
```

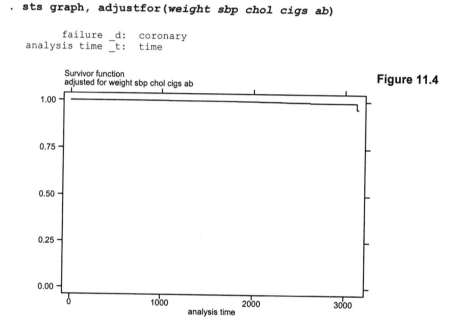

Figure 11.4

Figure 11.4, unlike Figure 11.3, follows the usual survivor-function convention of scaling the vertical axis from 0 to 1. Apart from this difference in scaling, Figures 11.3 and 11.4 depict the same curve.

Figure 11.5 graphs the estimated cumulative baseline hazard against time, using the variable (*hazard*) generated by our **stcox** command. This graph shows the baseline hazard increasing in 8 steps (because 8 patients "failed" or had coronary events), from near 0 to .033.

```
. graph hazard time, connect(J) ylabel xlabel sort
```

Figure 11.5

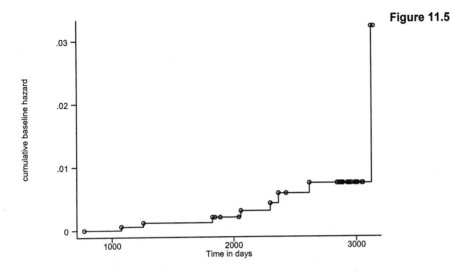

Exponential and Weibull Regression

Cox regression estimates the baseline survivor function empirically without reference to any theoretical distribution. Several alternative "parametric" approaches begin instead from assumptions that survival times do follow a known theoretical distribution. Possible distribution families include the exponential, Weibull, lognormal, log-logistic, Gompertz, or generalized gamma. Models based on any of these can be estimated through the **streg** command. Such models have the same general form as Cox regression (equations [11.2] and [11.3]), but define the baseline hazard $h_0(t)$ differently. Two examples appear in this section.

If failures occur randomly, with a constant hazard, then survival times follow an exponential distribution and could be analyzed by *exponential regression*. Constant hazard means that the individuals studied do not "age," in the sense that they are no more or less likely to fail late in the period of observation than they were at its start. Over the long term, this assumption seems unjustified for machines or living organisms, but it might approximately hold if the period of observation covers a relatively small fraction of their life spans. An exponential model implies that logarithms of the survivor function, $\ln(S(t))$, are linearly related to t.

A second common parametric approach, *Weibull regression*, is based on the more general Weibull distribution. This does not require failure rates to remain constant, but allows them to increase or decrease smoothly over time. The Weibull model implies that $\ln(-\ln(S(t)))$ is a linear function of $\ln(t)$.

Graphs provide a useful diagnostic for the appropriateness of exponential or Weibull models. For example, returning to *aids.dta*, we construct a graph (Figure 11.6) of $\ln(S(t))$ versus time, after first generating Kaplan–Meier estimates of the survivor function $S(t)$:

```
. use aids, clear
(AIDS (Selvin 1995:453))

. sts gen S = S

. generate logS = ln(S)
```

. **graph** *logS time*, **ylabel xlabel**

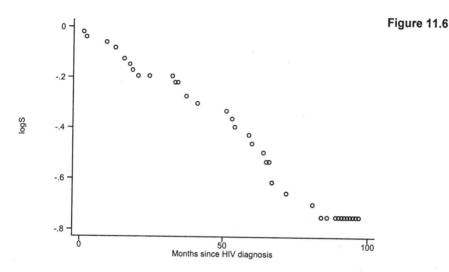

Figure 11.6

The pattern in Figure 11.6 appears somewhat linear, encouraging us to try an exponential regression:

. **streg** *age*, **dist(exponential) nolog noshow**

```
Exponential regression -- log relative-hazard form

No. of subjects =           51                  Number of obs   =          51
No. of failures =           25
Time at risk    =         3164
                                                LR chi2(1)      =        4.34
Log likelihood  =    -59.996976                 Prob > chi2     =      0.0372

------------------------------------------------------------------------------
        _t | Haz. Ratio   Std. Err.      z    P>|z|     [95% Conf. Interval]
-----------+------------------------------------------------------------------
       age |   1.074414    .0349626     2.21   0.027     1.008028    1.145172
------------------------------------------------------------------------------
```

The hazard ratio (1.074) and standard error (.035) estimated by this exponential regression do not greatly differ from their counterparts (1.085 and .038) in our earlier Cox regression. The similarity reflects the degree of correspondence between empirical and exponential hazard functions. According to this exponential model, the hazard of an HIV-positive individual developing AIDS increases about 7.4% with each year of age.

After **streg**, the **stcurve** command draws a graph of the models' cumulative hazard, survival, or hazard functions. By default, **stcurve** draws these curves holding all *x* variables in the model at their means. We can specify other *x* values by using the **at()** option. The individuals in *aids.dta* ranged from 26 to 50 years old. We could graph the survival function at *age* = 26 by issuing a command such as

. **stcurve, surviv at(***age***=26)**

A more informative graph uses the **at1()** and **at2()** options to show the survival curve at two different sets of *x* values, such as the low and high extremes of *age*:

```
. stcurve, survival at1(age=26) at2(age=50) connect(ss)
     ylabel(0,.2 to 1) xlabel(0,20 to 100) border
```

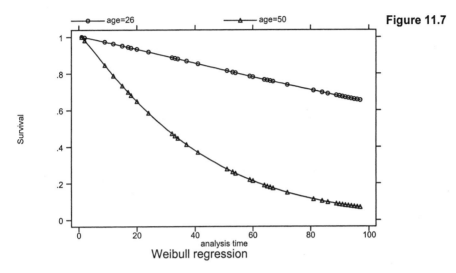

Weibull regression

Figure 11.7

Figure 11.7 shows the predicted survival curve (for transition from HIV diagnosis to AIDS) falling more steeply among older patients. The significant *age* hazard ratio greater than 1 in our exponential regression table implied the same thing, but using **stcurve** with **at1()** and **at2()** values gives a strong visual interpretation of this effect. These options work in a similar manner with all three types of **stcurve** graphs:

stcurve, survival	Survival function.
stcurve, hazard	Hazard function.
stcurve, cumhaz	Cumulative hazard function.

Instead of the exponential distribution, **streg** can also estimate survival models based on the Weibull distribution. A Weibull distribution might appear curvilinear in a plot of $\ln(S(t))$ versus *t*, but it should be linear in a plot of $\ln(-\ln(S(t)))$ versus $\ln(t)$, such as Figure 11.8. An exponential distribution, on the other hand, will appear linear in both plots and have a slope equal to 1 in the $\ln(-\ln(S(t)))$ versus $\ln(t)$ plot. In fact, the data points in Figure 11.8 are not far from a line with slope 1, suggesting that our previous exponential model is adequate.

```
. generate loglogS = ln(-ln(S))
```

```
. generate logtime = ln(time)
```

. **graph** *loglogS logtime*, **ylabel xlabel(0,1 to 5)**

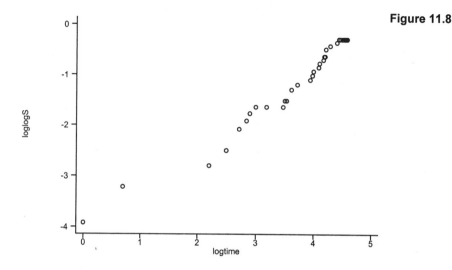

Figure 11.8

Although we do not need the additional complexity of a Weibull model with these data, results are given below for illustration.

. **streg** *age*, **dist(weibull) noshow nolog**

```
Weibull regression -- log relative-hazard form

No. of subjects =          51              Number of obs   =          51
No. of failures =          25
Time at risk    =        3164
                                           LR chi2(1)      =        4.68
Log likelihood  =    -59.778257            Prob > chi2     =      0.0306


        _t | Haz. Ratio   Std. Err.      z    P>|z|     [95% Conf. Interval]
-----------+----------------------------------------------------------------
       age |   1.079477    .0363509     2.27   0.023     1.010531    1.153127
-----------+----------------------------------------------------------------
     /ln_p |   .1232638    .1820858     0.68   0.498    -.2336179     .4801454
-----------+----------------------------------------------------------------
         p |   1.131183    .2059723                      .7916643    1.616309
       1/p |   .8840305    .1609694                      .6186934    1.263162
----------------------------------------------------------------------------
```

The Weibull regression obtains a hazard ratio estimate (1.079) intermediate between our previous Cox and exponential results. The most noticeable difference from those earlier models is the presence of three new lines at the bottom of the table. These refer to the Weibull distribution shape parameter p. A p value of 1 corresponds to an exponential model: the hazard does not change with time. $p > 1$ indicates that the hazard increases with time; $p < 1$ indicates that the hazard decreases. A 95% confidence interval for p ranges from .79 to 1.62, so we have no reason to reject an exponential ($p = 1$) model here. Different, but mathematically equivalent, parameterizations of the Weibull model focus on $\ln(p)$, p, or $1/p$, so Stata provides all three. **stcurve** draws survival, hazard, or cumulative hazard functions after **streg**,

dist(weibull) just as it does after **streg, dist(exponential)** or other **streg** models.

Exponential or Weibull regression is preferable to Cox regression when survival times actually follow an exponential or Weibull distribution. When they do not, these models are misspecified and can yield misleading results. Cox regression, which makes no *a priori* assumptions about distribution shape, remains useful in a wider variety of situations.

In addition to exponential and Weibull models, **streg** can fit models based on the Gompertz, lognormal, log-logistic, or generalized gamma distributions. Type **help streg**, or see **st streg** in the *Reference Manuals*, for syntax and a list of current options.

Poisson Regression

If events occur independently and with constant probability, then counts of events over a given period of time follow a Poisson distribution. Let r_j represent the incidence rate:

$$ r_j \;=\; \frac{\text{count of events}}{\text{number of times event could have occurred}} \qquad [11.4] $$

The denominator in [11.4] is termed the "exposure" and is often measured in units such as person-years. We model the logarithm of incidence rate as a linear function of one or more predictor (*x*) variables:

$$ \ln(r_t) \;=\; \beta_0 + \beta_1 x_1 + \beta_2 x_2 + \ldots + \beta_k x_k \qquad [11.5a] $$

Equivalently, the model describes logs of expected event counts:

$$ \ln(\textit{expected count}) \;=\; \ln(\textit{exposure}) + \beta_0 + \beta_1 x_1 + \beta_2 x_2 + \ldots \beta_k x_k \qquad [11.5b] $$

Assuming that a Poisson process underlies the events of interest, Poisson regression finds maximum-likelihood estimates of the β parameters.

Data on radiation exposure and cancer deaths among workers at Oak Ridge National Laboratory provide an example. The 56 observations in dataset *oakridge.dta* represent 56 age/radiation-exposure categories (7 categories of age × 8 categories of radiation). For each combination, we know the number of deaths and the number of person-years of exposure.

```
Contains data from C:\data\oakridge.dta
  obs:            56                        Radiation (Selvin 1995:474)
  vars:            4                        15 Jul 2001 15:14
  size:          616 (98.9% of memory free)
-------------------------------------------------------------------------
              storage  display    value
variable name  type    format     label    variable label
-------------------------------------------------------------------------
age            byte    %9.0g      ageg      Age group
rad            byte    %9.0g                Radiation exposure level
deaths         byte    %9.0g                Number of deaths
pyears         float   %9.0g                Person-years
-------------------------------------------------------------------------
Sorted by:
```

```
. summarize

    Variable |     Obs        Mean   Std. Dev.        Min        Max
-------------+--------------------------------------------------------
         age |      56           4      2.0181          1          7
         rad |      56         4.5    2.312024          1          8
      deaths |      56    1.839286    3.178203          0         16
      pyears |      56    3807.679    10455.91         23      71382

. list in 1/6

        age     rad     deaths     pyears
  1.   < 45       1          0      29901
  2.  45-49       1          1       6251
  3.  50-54       1          4       5251
  4.  55-59       1          3       4126
  5.  60-64       1          3       2778
  6.  65-69       1          1       1607
```

Does the death rate increase with exposure to radiation? Poisson regression finds a statistically significant effect:

```
. poisson deaths rad, nolog exposure(pyears) irr

Poisson regression                              Number of obs   =         56
                                                LR chi2(1)      =      14.87
                                                Prob > chi2     =     0.0001
Log likelihood =  -169.7364                     Pseudo R2       =     0.0420

------------------------------------------------------------------------------
      deaths |         IRR   Std. Err.       z    P>|z|     [95% Conf. Interval]
-------------+----------------------------------------------------------------
         rad |    1.236469    .0603551    4.35    0.000     1.123657    1.360606
      pyears |  (exposure)
------------------------------------------------------------------------------
```

For the regression above, we specified the event count (*deaths*) as the dependent variable and radiation (*rad*) as the independent variable. The Poisson "exposure" variable is *pyears*, or person-years in each category of *rad*. The `irr` option calls for incidence rate ratios rather than regression coefficients in the results table—that is, we get estimates of $\exp(\beta)$ instead of β, the default. According to this incidence rate ratio, the death rate becomes 1.236 times higher (increased by 23.6%) with each increase in radiation category. Although that ratio is statistically significant, the fit is not impressive; the pseudo R^2 (see equation [10.4]) is only .042.

To perform a goodness-of-fit test, comparing the Poisson model's predictions with the observed counts, use the follow-up command `poisgof`:

```
. poisgof

        Goodness-of-fit chi2  =   254.5475
        Prob > chi2(54)       =     0.0000
```

These goodness-of-fit test results ($\chi^2 = 254.5, P < .00005$) indicate that our model's predictions are significantly different from the actual counts—another sign that the model fits poorly.

We obtain better results when we include *age* as a second predictor. Pseudo R^2 then rises to .5966, and the goodness-of-fit test no longer leads us to reject our model.

```
. poisson deaths rad age, nolog exposure(pyears) irr
```

```
Poisson regression                          Number of obs   =          56
                                            LR chi2(2)      =      211.41
                                            Prob > chi2     =      0.0000
Log likelihood =   -71.4653                 Pseudo R2       =      0.5966
```

```
------------------------------------------------------------------------------
      deaths |      IRR   Std. Err.      z    P>|z|     [95% Conf. Interval]
-------------+----------------------------------------------------------------
         rad |  1.176673   .0593446     3.23   0.001     1.065924    1.298929
         age |  1.960034   .0997536    13.22   0.000     1.773955    2.165631
       pyears | (exposure)
------------------------------------------------------------------------------
```

```
. poisgof
```

```
        Goodness-of-fit chi2  =    58.00534
        Prob > chi2(53)       =     0.2960
```

For simplicity, to this point we have treated *rad* and *age* as if both were continuous variables, and we expect their effects on the log death rate to be linear. In fact, however, both independent variables are measured as ordered categories. *rad* = 1, for example, means 0 radiation exposure; *rad* = 2 means 0 to 19 milliseiverts; *rad* = 3 means 20 to 39 milliseiverts; and so forth. An alternative way to include radiation exposure categories in the regression, while watching for nonlinear effects, is as a set of dummy variables. Below we use the **gen()** option of **tabulate** to create 8 dummy variables, *r1* to *r8*, representing each of the 8 values of *rad*.

```
. tabulate rad, gen(r)
```

```
  Radiation |
   exposure |
      level |      Freq.     Percent        Cum.
------------+-----------------------------------
          1 |          7       12.50       12.50
          2 |          7       12.50       25.00
          3 |          7       12.50       37.50
          4 |          7       12.50       50.00
          5 |          7       12.50       62.50
          6 |          7       12.50       75.00
          7 |          7       12.50       87.50
          8 |          7       12.50      100.00
------------+-----------------------------------
      Total |         56      100.00
```

```
. describe
```

```
Contains data from C:\data\oakridge.dta
  obs:            56                          Radiation (Selvin 1995:474)
  vars:           12                          15 Jul 2001 15:14
  size:        1,064 (98.9% of memory free)
-------------------------------------------------------------------------------
              storage  display    value
variable name   type   format     label      variable label
-------------------------------------------------------------------------------
age           byte    %9.0g      ageg       Age group
rad           byte    %9.0g                 Radiation exposure level
deaths        byte    %9.0g                 Number of deaths
pyears        float   %9.0g                 Person-years
r1            byte    %8.0g                 rad==      1.0000
```

```
r2            byte    %8.0g              rad==    2.0000
r3            byte    %8.0g              rad==    3.0000
r4            byte    %8.0g              rad==    4.0000
r5            byte    %8.0g              rad==    5.0000
r6            byte    %8.0g              rad==    6.0000
r7            byte    %8.0g              rad==    7.0000
r8            byte    %8.0g              rad==    8.0000
-----------------------------------------------------------------------
Sorted by:
```

We now include seven of these dummies (omitting one to avoid multicollinearity) as regression predictors. The additional complexity of this dummy-variable model brings little improvement in fit. It does, however, add to our interpretation. The overall effect of radiation on death rate appears to come primarily from the two highest radiation levels (*r7* and *r8*, corresponding to 100 to 119 and 120 or more milliseiverts). At these levels, the incidence rates are about four times higher.

`. poisson deaths r2-r8 age, nolog exposure(pyears) irr`

```
Poisson regression                           Number of obs   =        56
                                             LR chi2(8)      =    215.44
                                             Prob > chi2     =    0.0000
Log likelihood = -69.451814                  Pseudo R2       =    0.6080


------------------------------------------------------------------------
     deaths |       IRR    Std. Err.      z    P>|z|    [95% Conf. Interval]
------------+-----------------------------------------------------------
         r2 |  1.473591     .426898     1.34   0.181    .8351884    2.599975
         r3 |  1.630688    .6659257     1.20   0.231     .732428    3.630587
         r4 |  2.375967    1.088835     1.89   0.059    .9677429    5.833389
         r5 |  .7278113    .7518255    -0.31   0.758    .0961018    5.511957
         r6 |  1.168477     1.20691     0.15   0.880    .1543195    8.847472
         r7 |  4.433729    3.337738     1.98   0.048    1.013863    19.38915
         r8 |   3.89188    1.640978     3.22   0.001    1.703168    8.893267
        age |  1.961907    .1000652    13.21   0.000    1.775267    2.168169
      pyears |  (exposure)
------------------------------------------------------------------------
```

Radiation levels 7 and 8 seem to have similar effects, so we might simplify the model by combining them. First, we test whether their coefficients are significantly different. They are not:

`. test r7 = r8`

```
 ( 1)   [deaths]r7 - [deaths]r8 = 0.0

           chi2( 1) =     0.03
         Prob > chi2 =    0.8676
```

Next, generate a new dummy variable *r78*, which equals 1 if either *r7* or *r8* equals 1:

`. generate r78 = (r7 | r8)`

Finally, substitute the new predictor for *r7* and *r8* in the regression:

```
. poisson deaths r2-r6 r78 age, irr ex(pyears) nolog
```

```
Poisson regression                          Number of obs   =         56
                                            LR chi2(7)      =     215.41
                                            Prob > chi2     =     0.0000
Log likelihood = -69.465332                 Pseudo R2       =     0.6079
```

deaths	IRR	Std. Err.	z	P>|z|	[95% Conf. Interval]	
r2	1.473602	.4269013	1.34	0.181	.8351949	2.599996
r3	1.630718	.6659381	1.20	0.231	.7324415	3.630655
r4	2.376065	1.08888	1.89	0.059	.9677823	5.833629
r5	.7278387	.7518538	-0.31	0.758	.0961055	5.512165
r6	1.168507	1.206942	0.15	0.880	.1543236	8.847704
r78	3.980326	1.580024	3.48	0.001	1.828214	8.665833
age	1.961722	.100043	13.21	0.000	1.775122	2.167937
pyears	(exposure)					

We could proceed to simplify the model further in this fashion. At each step, **test** helps to evaluate whether combining two dummy variables is justifiable.

Generalized Linear Models

Generalized linear models (GLM) have the form

$$g[E(y)] = \beta_0 + \beta_1 x_1 + \beta_2 x_2 + \ldots + \beta_k x_k, \qquad y \sim F \qquad [11.6]$$

where $g[\ ]$ is the *link function* and F the distribution family. This general formulation encompasses many specific models. For example, if $g[\ ]$ is the identity function and y follows a normal (Gaussian) distribution, we have a linear regression model:

$$E(y) = \beta_0 + \beta_1 x_1 + \beta_2 x_2 + \ldots + \beta_k x_k, \qquad y \sim \text{Normal} \qquad [11.7]$$

If $g[\ ]$ is the logit function and y follows a Bernoulli distribution, we have logit regression instead:

$$\text{logit}[E(y)] = \beta_0 + \beta_1 x_1 + \beta_2 x_2 + \ldots + \beta_k x_k, \qquad y \sim \text{Bernoulli} \qquad [11.8]$$

Because of its broad applications, GLM could have been introduced at several different points in this book. Its relevance to this chapter comes from the ability to fit event models. Poisson regression, for example, requires that $g[\ ]$ is the natural log function and that y follows a Poisson distribution:

$$\ln[E(y)] = \beta_0 + \beta_1 x_1 + \beta_2 x_2 + \ldots + \beta_k x_k, \qquad y \sim \text{Poisson} \qquad [11.9]$$

As might be expected with such a flexible method, Stata's **glm** command permits many different options. Users can specify not only the distribution family and link function, but also details of the variance estimation, fitting procedure, output, and offset. These options make **glm** a useful alternative even when applied to models for which a dedicated command (such as **regress**, **logistic**, or **poisson**) already exists.

We might represent a "generic" **glm** command as follows:

```
. glm y x1 x2 x3, family(familyname) link(linkname)
     lnoffset(exposure) eform jknife
```

where **family()** specifies the y distribution family, **link()** the link function, and **lnoffset()** an "exposure" variable such as that needed for Poisson regression. The **eform** option asks for regression coefficients in exponentiated form, $\exp(\beta)$ rather than β. Standard errors are estimated through jackknife (**jknife**) calculations.

Possible distribution families are

family(gaussian)	Gaussian or normal (default)
family(igaussian)	Inverse Gaussian
family(binomial)	Bernoulli binomial
family(poisson)	Poisson
family(nbinomial)	Negative binomial
family(gamma)	Gamma

We can also specify a number or variable indicating the binomial denominator N (number of trials), or a number indicating the negative binomial variance and deviance functions, by declaring them in the **family()** option:

family(binomial #)
family(binomial *varname***)**
family(nbinomial #)

Possible link functions are

link(identity)	Identity (default)
link(log)	Log
link(logit)	Logit
link(probit)	Probit
link(cloglog)	Complementary log-log
link(opower #)	Odds power
link(power #)	Power
link(nbinomial)	Negative binomial
link(loglog)	Log-log
link(logc)	Log-complement

Coefficient variances or standard errors can be estimated in a variety of ways. A partial list of **glm** variance-estimating options is given below:

opg	Berndt, Hall, Hall, and Hausman "B-H-cubed" variance estimator.
oim	Observed information matrix variance estimator.
robust	Huber/White/sandwich estimator of variance.
unbiased	Unbiased sandwich estimator of variance
nwest	Heteroskedasticity and autocorrelation-consistent variance estimator.
jknife	Jackknife estimate of variance.
jknife1	One-step jackknife estimate of variance.

bstrap	Bootstrap estimate of variance. The default is 199 repetitions; specify some other number by adding the **bsrep(#)** option.

For a full list of options with some technical details, look up **glm** in the *Reference Manuals*. A more in-depth treatment of GLM topics can be found in Hardin and Hilbe (2001).

Chapter 6 began with the simple regression of mean composite SAT scores (*csat*) on per-pupil expenditures (*expense*) of the 50 U.S. states and District of Columbia (*states.dta*):

```
. regress csat expense
```

We could fit the same model and obtain exactly the same estimates with the following command:

```
. glm csat expense, link(identity) family(gaussian)

Iteration 0:   log likelihood = -279.99869

Generalized linear models                No. of obs       =         51
Optimization      : ML: Newton-Raphson   Residual df      =         49
                                         Scale param      =   3577.678
Deviance       =     175306.2097         (1/df) Deviance  =   3577.678
Pearson        =     175306.2097         (1/df) Pearson   =   3577.678

Variance function: V(u) = 1              [Gaussian]
Link function    : g(u) = u              [Identity]
Standard errors  : OIM

Log likelihood  = -279.9986936           AIC              =   11.05877
BIC             =   175298.346

------------------------------------------------------------------------
      csat |    Coef.    Std. Err.      z    P>|z|    [95% Conf. Interval]
-----------+------------------------------------------------------------
   expense | -.0222756    .0060371    -3.69   0.000   -.0341082   -.0104431
     _cons |  1060.732    32.7009     32.44   0.000    996.6399    1124.825
------------------------------------------------------------------------
```

Because **link(identity)** and **family(gaussian)** are default options, we could actually have left them out of the previous **glm** command.

The **glm** command can do more than just duplicate our **regress** results, however. For example, we could fit the same OLS model but obtain bootstrap standard errors:

```
. glm csat expense, link(identity) family(gaussian) bstrap

Iteration 0:   log likelihood = -279.99869

Bootstrap iterations (199)
----+--- 1 ---+--- 2 ---+--- 3 ---+--- 4 ---+--- 5
.................................................. 50
.................................................. 100
.................................................. 150
..................................................

Generalized linear models                 No. of obs      =         51
Optimization      : ML: Newton-Raphson     Residual df     =         49
                                           Scale param     =   4124.656
Deviance       =   175306.2097             (1/df) Deviance =   3577.678
Pearson        =   175306.2097             (1/df) Pearson  =   3577.678

Variance function: V(u) = 1                [Gaussian]
Link function    : g(u) = u                [Identity]
Standard errors  : Bootstrap

Log likelihood   = -279.9986936            AIC             =   11.05877
BIC              =   175298.346

--------------------------------------------------------------------------
             |              Bootstrap
        csat |    Coef.    Std. Err.      z    P>|z|     [95% Conf. Interval]
-------------+------------------------------------------------------------
     expense |  -.0222756   .0039284    -5.67   0.000    -.0299751   -.0145762
       _cons |  1060.732    25.36566    41.82   0.000     1011.017    1110.448
--------------------------------------------------------------------------
```

The bootstrap standard errors reflect observed variation among coefficients estimated from 199 samples of $n = 51$ cases each, drawn by random sampling with replacement from the original $n = 51$ dataset. In this example, the bootstrap standard errors are less than the corresponding theoretical standard errors, and the resulting confidence intervals are narrower.

Similarly, we could use **glm** to repeat the first **logistic** regression of Chapter 10. In the following example, we ask for jackknife standard errors and odds ratio or exponential-form (**eform**) coefficients:

```
. glm any date, link(logit) family(bernoulli) eform jknife

Iteration 0:   log likelihood = -12.995268
Iteration 1:   log likelihood = -12.991098
Iteration 2:   log likelihood = -12.991096

Jackknife iterations (23)
----+--- 1 ---+--- 2 ---+--- 3 ---+--- 4 ---+--- 5
.....................
Generalized linear models                  No. of obs      =         23
Optimization      : ML: Newton-Raphson     Residual df     =         21
                                           Scale param     =          1
Deviance          =  25.98219269           (1/df) Deviance =   1.237247
Pearson           =   22.8885488           (1/df) Pearson  =   1.089931

Variance function: V(u) = u*(1-u)          [Bernoulli]
Link function    : g(u) = ln(u/(1-u))      [Logit]
Standard errors  : Jackknife

Log likelihood    = -12.99109634           AIC             =   1.303574
BIC               =  19.71120426
```

		Jackknife				
any	Odds Ratio	Std. Err.	z	P>|z|	[95% Conf. Interval]	
date	1.002093	.0015486	1.35	0.176	.9990623	1.005133

The final **poisson** regression of the present chapter corresponds to this **glm** model:

```
. glm deaths r2-r6 r78 age, link(log) family(poisson)
     lnoffset(pyears) eform
```

Although **glm** can replicate the models fit by many specialized commands, and adds some new capabilities, the specialized commands have their own advantages including speed and customized options. A particular attraction of **glm** is its ability to estimate models for which Stata has no specialized command.

12

Principal Components and Factor Analysis

Principal components and factor analysis provide methods for simplification, combining many correlated variables into a smaller number of underlying dimensions. Along the way to achieving simplification, the analyst must choose from a daunting variety of options. If the data really do reflect distinct underlying dimensions, different options might nonetheless converge on similar results. In the absence of distinct underlying dimensions, however, different options often lead to divergent results. Experimenting with these options can tell us how stable a particular finding is, or how much it depends on arbitrary choices about the specific analytical technique.

Stata accomplishes principal components and factor analysis with four basic commands:

factor   Extracts principal components or factors of several different types.

greigen   Constructs a scree graph (plot of the eigenvalues) from the recent **factor** .

rotate   Performs orthogonal (uncorrelated factors) or oblique (correlated factors) rotation, after **factor** .

score   Generates factor scores (composite variables) after **factor** or **rotate** .

The composite variables generated by **score** can subsequently be saved, listed, graphed, or analyzed like any other Stata variable.

Users who create composite variables by the older method of adding other variables together without doing factor analysis could assess their results by calculating an α reliability coefficient:

alpha   Cronbach's α reliability

Example Commands

. `factor x1-x20, pc mineigen(1)`

Obtains principal components of the variables *x1* through *x20*. Retains components having eigenvalues greater than 1.

. `factor x1-x20, ml factor(5)`

Performs maximum likelihood factor analysis of the variables *x1* through *x20*. Retains only the first five factors.

. **greigen**

Graphs eigenvalues versus factor or component number from the most recent **factor** command (also known as a "scree graph").

. **rotate, varimax factors(2)**

Performs orthogonal (varimax) rotation of the first two factors from the most recent **factor** command.

. **rotate, promax factors(3)**

Performs oblique (promax) rotation of the first three factors from the most recent **factor** command.

. **score** *f1 f2 f3*

Generates three new factor score variables named *f1*, *f2*, and *f3*, based upon the most recent **factor** and **rotate** commands.

. **alpha** *x1-x10*

Calculates Cronbach's α reliability coefficient for a composite variable defined as the sum of *x1-x10*. The sense of items entering negatively is ordinarily reversed. Options can override this default, or form a composite variable by adding together either the original variables or their standardized values.

Principal Components

To illustrate basic principal components and factor analysis commands, we will use a small dataset describing the nine major planets of this solar system (from Beatty et al. 1981). The data include several variables in both raw and natural logarithm form. Logarithms are employed here to reduce skew and linearize relations among the variables.

```
Contains data from C:\data\planets.dta
  obs:            9                      Solar system data
  vars:          12                      16 Jul 2001 08:59
  size:         441 (100.0% of memory free)
-------------------------------------------------------------------
              storage  display   value
variable name  type    format    label    variable label
-------------------------------------------------------------------
planet         str7    %9s                 Planet
dsun           float   %9.0g               Mean dist. sun, km*10^6
radius         float   %9.0g               Equatorial radius in km
rings          byte    %8.0g     ringlbl   Has rings?
moons          byte    %8.0g               Number of known moons
mass           float   %9.0g               Mass in kilograms
density        float   %9.0g               Mean density, g/cm^3
logdsun        float   %9.0g               natural log dsun
lograd         float   %9.0g               natural log radius
logmoons       float   %9.0g               natural log (moons + 1)
logmass        float   %9.0g               natural log mass
logdense       float   %9.0g               natural log dense
-------------------------------------------------------------------
Sorted by:  dsun
```

To extract initial factors or principal components, use the command **factor** followed by a variable list (variables in any order) and one of the following options:

pc	Principal components
pcf	Principal components factoring
pf	Principal factoring (default)
ipf	Principal factoring with iterated communalities
ml	Maximum-likelihood factoring

For example, to obtain principal components factors, type

```
. factor rings logdsun - logdense, pcf
```

(obs=9)

	(principal component factors; 2 factors retained)			
Factor	Eigenvalue	Difference	Proportion	Cumulative
1	4.62365	3.45469	0.7706	0.7706
2	1.16896	1.05664	0.1948	0.9654
3	0.11232	0.05395	0.0187	0.9842
4	0.05837	0.02174	0.0097	0.9939
5	0.03663	0.03657	0.0061	1.0000
6	0.00006	.	0.0000	1.0000

	Factor Loadings		
Variable	1	2	Uniqueness
rings	0.97917	0.07720	0.03526
logdsun	0.67105	-0.71093	0.04427
lograd	0.92287	0.37357	0.00875
logmoons	0.97647	0.00028	0.04651
logmass	0.83377	0.54463	0.00821
logdense	-0.84511	0.47053	0.06439

Only the first two components have eigenvalues greater than 1, and these two components explain over 96% of the six variables' combined variance. The unimportant 3rd through 6th principal components might safely be disregarded in subsequent analysis.

Two **factor** options provide control over the number of factors extracted:

factors(#)	where # specifies the number of factors
mineigen(#)	where # specifies the minimum eigenvalue for retained factors

The principal components factoring (**pcf**) procedure automatically drops factors with eigenvalues below 1, so

```
. factor rings logdsun - logdense, pcf
```

is equivalent to

```
. factor rings logdsun - logdense, pcf mineigen(1)
```

In this example, we would also have obtained the same results by typing

```
. factor rings logdsun - logdense, pcf factors(2)
```

To see a scree graph (plot of eigenvalues versus component or factor number) after any **factor**, simply type

```
. greigen
```

graph options can be added to **greigen** in order to draw a more presentable image (Figure 12.1):

```
. quietly factor rings logdsun - logdense, pcf
. greigen, yline(1) ylabel(0,1 to 5) xlabel(1,2 to 6)
     b2(Component Number)
```

Figure 12.1

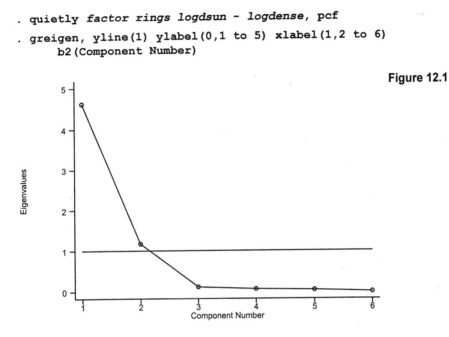

Figure 12.1 again emphasizes the unimportance of components 3 through 6.

Rotation

Rotation further simplifies factor structure. After factoring, type **rotate** followed by either

varimax   Varimax orthogonal rotation, for uncorrelated factors or components (default).

promax()   Promax oblique rotation, allowing correlated factors or components. Choose a number (promax power) ≤ 4; the higher the number, the greater the degree of interfactor correlation. **promax(3)** is the default.

   Two additional **rotate** options are

factors()   As it does with **factor**, this option specifies how many factors to retain.

horst   Horst modification to varimax and promax rotation.

   Rotation (and factor scoring) can be performed following any factor analysis, whether it employed the **pcf**, **pf**, **ipf**, or **ml** options. In this section, we will follow through on our **pcf** example. For orthogonal (default) rotation of the first two components found in the planetary data, we type

```
. rotate
```

```
                  (varimax rotation)
                   Rotated Factor Loadings
        Variable |      1          2      Uniqueness
     ------------+---------------------------------
           rings |   0.52848    0.82792    0.03526
         logdsun |   0.97173    0.10707    0.04427
          lograd |   0.25804    0.96159    0.00875
        logmoons |   0.58824    0.77940    0.04651
         logmass |   0.06784    0.99357    0.00821
        logdense |  -0.88479   -0.39085    0.06439
```

This example accepts all the defaults: varimax rotation and the same number of factors retained in the last **factor**. We could have asked for the same rotation explicitly, with the following command:

```
. rotate, varimax factors(2)
```

For oblique promax rotation (allowing correlated factors) of the most recent factoring, type

```
. rotate, promax
```

```
                  (promax rotation)
                   Rotated Factor Loadings
        Variable |      1          2      Uniqueness
     ------------+---------------------------------
           rings |   0.34664    0.76264    0.03526
         logdsun |   1.05196   -0.17270    0.04427
          lograd |   0.00599    0.99262    0.00875
        logmoons |   0.42747    0.69070    0.04651
         logmass |  -0.21543    1.08534    0.00821
        logdense |  -0.87190   -0.16922    0.06439
```

By default, this example used a promax power of 3. We could have asked for the same promax power and number of factors explicitly:

```
. rotate, promax(3) factors(2)
```

promax(4) would permit further simplification of the loading matrix, at the cost of stronger interfactor correlations and hence less simplification.

After promax rotation, *rings*, *lograd*, *logmoons*, and *logmass* load most heavily on factor 2. This appears to be a "large size/many satellites" dimension. *logdsun* and *logdense* load higher on factor 1, forming a "far out/low density" dimension. The next section shows how to create new variables representing these dimensions.

Factor Scores

Factor scores are linear composites, formed by standardizing each variable to zero mean and unit variance, and then weighting with factor score coefficients and summing for each factor. **score** performs these calculations automatically, using the most recent **rotate** or **factor** results. In the **score** command we supply names for the new variables, unimagi-natively called *f1* and *f2*.

```
. score f1 f2
```

```
            (based on rotated factors)
                Scoring Coefficients
    Variable |      1            2
-------------+------------------------
       rings |   0.12674     0.22099
     logdsun |   0.48769    -0.09689
      lograd |  -0.03840     0.30608
    logmoons |   0.16664     0.19543
     logmass |  -0.14338     0.34386
    logdense |  -0.39127    -0.01609
```

```
. label variable f1 "Far out/low density"
```

```
. label variable f2 "Large size/many satellites"
```

```
. list planet f1 f2
```

```
        planet         f1          f2
 1.    Mercury  -1.256881    -.9172388
 2.      Venus  -1.188757    -.5160229
 3.      Earth  -1.035242    -.3939372
 4.       Mars   -.5970106   -.6799535
 5.    Jupiter   .3841085    1.342658
 6.     Saturn   .9259058    1.184475
 7.     Uranus   .9347457     .7682409
 8.    Neptune   .8161058     .647119
 9.      Pluto   1.017025    -1.43534
```

Being standardized variables, the new factor scores *f1* and *f2* have means (approximately) equal to zero and standard deviations equal to one:

```
. summarize f1 f2
```

```
    Variable |     Obs        Mean   Std. Dev.        Min          Max
-------------+------------------------------------------------------------
          f1 |       9    9.93e-09           1   -1.256881     1.017025
          f2 |       9   -3.31e-09           1    -1.43534     1.342658
```

Thus, the factor scores are measured in units of standard deviations from their means. Mercury, for example, is about 1.26 standard deviations below average on the far out/low density (*f1*) dimension because it is actually close to the sun and high density. Mercury is .92 standard deviations below average on the large size/many satellites (*f2*) dimension because it is small and has no satellites. Saturn, in contrast, is .93 and 1.18 standard deviations above average on these two dimensions.

Promax rotation permits correlations between factor scores:

```
. correlate f1 f2
```

(obs=9)

```
             |       f1         f2
-------------+--------------------
          f1 |   1.0000
          f2 |   0.4974     1.0000
```

Scores on factor 1 have a moderate positive correlation with scores on factor 2: far out/low density planets are more likely also to be larger, with many satellites.

If we employ varimax instead of promax rotation, we get uncorrelated factor scores:

```
. quietly factor rings logdsun - logdense, pcf

. quietly rotate

. quietly score varimax1 varimax2

. correlate varimax1 varimax2
```

(obs=9)

```
             | varimax1 varimax2
-------------+------------------
    varimax1 |   1.0000
    varimax2 |   0.0000   1.0000
```

Once created by **score**, factor scores can be treated like any other Stata variable—listed, analyzed, graphed, and so forth. Graphs of principal component factors sometimes help to identify multivariate outliers or clusters of observations that stand apart from the rest. For example, Figure 12.2 reveals three distinct types of planets. The following **graph** command has no bottom or left axis (**noaxis**), but instead has lines at $y = 0$ and $x = 0$ (**yline** and **xline**). We use values of *planet*, the planets' names, as plotting symbols at 140% of the usual size (**psize**).

```
. graph f1 f2, noaxis yline(0) xline(0) ylabel xlabel(-2,-1 to 2)
       symbol([planet]) psize(140)
```

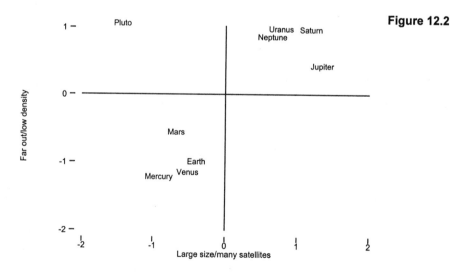

Figure 12.2

The inner, rocky planets (such as Mercury, low on "far out/low density" factor 1; low also on "large size/many satellites" factor 2) cluster together at the lower left. The outer gas giants have opposite characteristics, and cluster together at the upper right. Pluto, which physically resembles some outer-system moons, is unique among planets for being high on the "far out/low density" dimension, and at the same time low on the "large size/many satellites" dimension.

This example employed rotation. Factor scores obtained by principal components without rotation are often used to analyze large datasets in physical-science fields such as climatology and remote sensing. In these applications, principal components are termed "empirical orthogonal functions." The first empirical orthogonal function, or EOF1, equals the factor score for the first unrotated principal component. EOF2 is the score for the second principal component, and so forth.

Principal Factoring

Principal factoring extracts principal components from a modified correlation matrix, in which the main diagonal consists of communality estimates instead of 1's. The **factor** options **pf** and **ipf** both perform principal factoring. They differ in how communalities are estimated:

pf   Communality estimates equal R^2 from regressing each variable on all the others.

ipf   Iterative estimation of communalities.

Whereas principal components analysis focuses on explaining the variables' variance, principal factoring explains intervariable correlations.

We apply principal factoring with iterated communalities (**ipf**) to the planetary data:

```
. factor rings logdsun - logdense, ipf
```

(obs=9)

(iterated principal factors; 5 factors retained)

Factor	Eigenvalue	Difference	Proportion	Cumulative
1	4.59663	3.46817	0.7903	0.7903
2	1.12846	1.05107	0.1940	0.9843
3	0.07739	0.06438	0.0133	0.9976
4	0.01301	0.01176	0.0022	0.9998
5	0.00125	0.00137	0.0002	1.0000
6	-0.00012	.	-0.0000	1.0000

Factor Loadings

Variable	1	2	3	4	5	Uniqueness
rings	0.97599	0.06649	0.11374	-0.02065	-0.02234	0.02916
logdsun	0.65708	-0.67054	0.14114	0.04471	0.00816	0.09663
lograd	0.92670	0.37001	-0.04504	0.04865	0.01662	-0.00036
logmoons	0.96738	-0.01074	0.00781	-0.08593	0.01597	0.05636
logmass	0.83783	0.54576	0.00557	0.02824	-0.00714	-0.00069
logdense	-0.84602	0.48941	0.20594	-0.00610	0.00997	0.00217

Only the first two factors have eigenvalues above 1. With **pcf** or **pf** factoring, we can simply disregard minor factors. Using **ipf** , however, we must decide how many factors to retain, and then repeat the analysis asking for exactly that many factors. Here we will retain two factors:

```
. factor rings logdsun - logdense, ipf factor(2)
```

(obs=9)

```
                (iterated principal factors; 2 factors retained)
    Factor       Eigenvalue      Difference     Proportion      Cumulative
-------------------------------------------------------------------------------
      1           4.57495         3.47412         0.8061          0.8061
      2           1.10083         1.07631         0.1940          1.0000
      3           0.02452         0.02013         0.0043          1.0043
      4           0.00439         0.00795         0.0008          1.0051
      5          -0.00356         0.02182        -0.0006          1.0045
      6          -0.02537            .            -0.0045          1.0000
```

```
                       Factor Loadings
     Variable  |      1           2        Uniqueness
-------------+-------------------------------------
        rings |    0.97474      0.05374      0.04699
      logdsun |    0.65329     -0.67309      0.12016
       lograd |    0.92816      0.36047      0.00858
     logmoons |    0.96855     -0.02278      0.06139
      logmass |    0.84298      0.54616     -0.00890
     logdense |   -0.82938      0.46490      0.09599
```

After this final factor analysis, we can create composite variables by **rotate** and **score**. Rotation of the **ipf** factors produces results similar to those found earlier with **pcf** : a far out/low density dimension and a large size/many satellites dimension. When variables have a strong factor structure, as these do, the specific techniques we choose make less of a difference.

Maximum-Likelihood Factoring

Maximum-likelihood factoring, unlike Stata's other **factor** options, provides formal hypothesis tests that help in determining the appropriate number of factors. To obtain a single maximum-likelihood factor for the planetary data, type

```
. factor rings logdsun - logdense, ml nolog factor(1)
```

(obs=9)

```
                (maximum likelihood factors; 1 factor retained)
    Factor        Variance       Difference     Proportion      Cumulative
-------------------------------------------------------------------------------
      1           4.47258            .            1.0000          1.0000
```

```
Test:  1 vs. no    factors.  Chi2(   6) =   62.02, Prob > chi2 =  0.0000
Test:  1 vs. more factors.  Chi2(   9) =   51.73, Prob > chi2 =  0.0000
```

```
                       Factor Loadings
     Variable  |      1        Uniqueness
-------------+---------------------------
        rings |    0.98726      0.02535
      logdsun |    0.59219      0.64931
       lograd |    0.93654      0.12288
     logmoons |    0.95890      0.08052
      logmass |    0.86918      0.24451
     logdense |   -0.77145      0.40487
```

The **ml** output includes two χ^2 tests:

J vs. no factors
> This tests whether the current model, with *J* factors, fits the observed correlation matrix significantly better than a no-factor model. A low probability indicates that the current model is a significant improvement over no factors.

J vs. more factors
> This tests whether the current *J*-factor model fits significantly worse than a more complicated, perfect-fit model. A low *P*-value suggests that the current model does not have enough factors.

The previous 1-factor example yields these results:

1 vs. no factors
> χ^2 [6] = 62.02, *P* = 0.0000 (actually, meaning *P* < .00005). The 1-factor model significantly improves upon a no-factor model.

1 vs. more factors
> χ^2 [9] = 51.73, *P* = 0.0000 (*P* < .00005). The 1-factor model is significantly worse than a perfect-fit model.

Perhaps a 2-factor model will do better:

. **factor rings logdsun - logdense, ml nolog factor(2)**

```
(obs=9)
                (maximum likelihood factors; 2 factors retained)
    Factor      Variance      Difference    Proportion    Cumulative
    ----------------------------------------------------------------
       1         3.64200        1.67115        0.6489        0.6489
       2         1.97085           .           0.3511        1.0000

Test:  2 vs. no    factors.  Chi2( 12) = 134.14, Prob > chi2 = 0.0000
Test:  2 vs. more  factors.  Chi2(  4) =   6.72, Prob > chi2 = 0.1513

                  Factor Loadings
    Variable |      1          2      Uniqueness
    ---------+------------------------------------
       rings |   0.86551    -0.41545    0.07829
     logdsun |   0.20920    -0.85593    0.22361
      lograd |   0.98437    -0.17528    0.00028
    logmoons |   0.81560    -0.49982    0.08497
     logmass |   0.99965     0.02639    0.00000
    logdense |  -0.46434     0.88565    0.00000
```

Now we find the following:

2 vs. no factors
> χ^2 [12] = 134.14, *P* = 0.0000 (actually, *P* < .00005). The 2-factor model significantly improves upon a no-factor model.

2 vs. more factors
> χ^2 [4] = 6.72, *P* = 0.1513. The 2-factor model is *not* significantly worse than a perfect-fit model.

These tests suggest that two factors provide an adequate model.

Computational routines performing maximum-likelihood factor analysis often yield "improper solutions"—unrealistic results such as negative variance or zero uniqueness. When this happens (as it did in the 2-factor **m1** example), the χ^2 tests lack formal justification. Used descriptively, the tests can still provide informal guidance regarding the appropriate number of factors.

13

Time Series Analysis

Stata's evolving time series capabilities, already extensive, will eventually deserve a book of their own. This chapter begins by reviewing two elementary and useful analytical tools: time plots and smoothing. We then move on to illustrate the use of correlograms, ARIMA models, and tests for stationarity and white noise. Further applications, notably periodograms and the flexible ARCH family of models, are left to the reader's explorations.

A technical and thorough treatment of time series topics is found in Hamilton (1994). Other sources include Box, Jenkins, and Reinsel (1994), Chatfield (1996), Diggle (1990), Enders (1995), Johnston and DiNardo (1997), and Shumway (1988).

Example Commands

. **ac *fylltemp*, lags(8)**

Graphs autocorrelations with confidence intervals for lags 1 through 8.

. **arch D.*y*, arch(1/3) ar(1) ma(1)**

Estimates an ARCH (autoregressive conditional heteroskedasticity) model for first differences of *y*, including first- through third-order ARCH terms, and first-order AR and MA disturbances.

. **arima *y*, arima(3,1,2)**

Fits a simple ARIMA(3,1,2) model. Possible options include several estimation strategies, linear constraints, and robust estimates of variance.

. **arima D.*y* *x1* L1.*x1* *x2*, ar(1) ma(1 12)**

Regresses first differences of *y* on *x1*, lag-1 values of *x1*, and *x2*, including AR(1), MA(1), and MA(12) disturbances.

. **corrgram *y*, lags(8)**

Obtains autocorrelations, partial autocorrelations, and Q tests for lags 1 through 8.

. **dfuller *y***

Performs Dickey–Fuller unit root test for stationarity.

. **dwstat**

After **regress** , calculates a Durbin–Watson statistic testing first-order autocorrelation.

. **egen *newvar* = ma(*y*), nomiss t(7)**

Generates *newvar* equal to the span-7 moving average of *y*, replacing the start and end values with shorter, uncentered averages.

. `generate` *date* = `mdy`(*month,day,year*)
Creates variable *date*, equal to days since January 1, 1960, from the three variables *month*, *day*, and *year*.

. `generate` *date* = `date`(*str_date,* `"mdy"`)
Creates variable *date* from the string variable *str_date*, where *str_date* contains dates in month, day, year form such as "11/19/2001", "4/18/98", or "June 12, 1948". Type `help` `dates` for many other date functions and options.

. `generate` *newvar* = `L3.y`
Generates *newvar* equal to lag-3 values of *y*.

. `pac y, lags(8) needle`
Graphs partial autocorrelations with confidence intervals and residual variance for lags 1 through 8.

. `pergram y, generate(`*newvar*`)`
Draws the sample periodogram of variable *y* and creates *newvar* equal to the raw periodogram values.

. `prais y x1 x2`
Performs Prais–Winsten regression of *y* on *x1* and *x2*, correcting for first-order autoregressive errors. `prais y x1 x2,` `corc` does Cochrane–Orcutt instead.

. `smooth 73 y, generate(`*newvar*`)`
Generates *newvar* equal to span-7 running medians of *y*, re-smoothing by span-3 running medians. Compound smoothers such as "3RSSH" or "4253h,twice" are possible.

. `tsset` *date,* `format(%d)`
Defines the dataset as a time series. Time is indicated by variable *date*, which is formatted as daily. For "panel" data with parallel time series for a number of different units, such as cities, `tsset` `city year` identifies both panel and time variables. Most of the commands in this chapter require that the data be `tsset`.

. `wntestq y, lags(15)`
Box–Pierce portmanteau Q test for white noise (also provided by `corrgram`).

. `xcorr x y, lags(8) xline(0)`
Graphs cross-correlations between input (*x*) and output (*y*) variable for lags 1–8. `xcorr x y,` `table` gives a text version that includes the actual correlations.

Smoothing

Many time series exhibit rapid up-and-down fluctuations that make it difficult to discern underlying patterns. Smoothing such series breaks the data into two parts, one that varies gradually, and a second "rough" part containing the leftover rapid changes:

$$\text{data} = \text{smooth} + \text{rough}$$

Dataset *MILwater.dta* contains data on daily water consumption for the town of Milford, New Hampshire over seven months from January through July 1983 (Hamilton 1985b).

```
Contains data from MILwater.dta
    obs:           212              Milford daily water use, 1/1/83
                                     - 7/31/83
    vars:            4              25 Mar 2001 10:41
    size:     2,120 (100.0% of memory free)
-------------------------------------------------------------------------
              storage  display    value
variable name  type    format     label    variable label
-------------------------------------------------------------------------
month          byte    %9.0g               Month
day            byte    %9.0g               Date
year           int     %9.0g               Year
water          int     %9.0g               Water use in 1000 gallons
-------------------------------------------------------------------------
Sorted by:
```

Before further analysis, we need to convert the month, day, and year information into a single numerical index of time. Stata's **mdy()** function does this, creating an elapsed-date variable (named *date* here) indicating the number of days since January 1, 1960.

. **generate date = mdy(month,day,year)**

. **list in 1/5**

```
      month      day    year    water    date
1.      1         1     1983     520     8401
2.      1         2     1983     600     8402
3.      1         3     1983     610     8403
4.      1         4     1983     590     8404
5.      1         5     1983     620     8405
```

The January 1, 1960 reference date is an arbitrary default. We can provide more under-standable formatting for *date*, and also set up our data for later analyses, by using the **tsset** (**t**ime **s**eries **set**) command to identify *date* as the time index variable and to specify the **%d** (**d**aily) display option for this variable.

. **tsset date, format(%d)**
        time variable: date, 01jan1983 to 31jul1983

. **list in 1/5**

```
      month      day    year    water    date
1.      1         1     1983     520    01jan1983
2.      1         2     1983     600    02jan1983
3.      1         3     1983     610    03jan1983
4.      1         4     1983     590    04jan1983
5.      1         5     1983     620    05jan1983
```

Dates in the new *date* format, such as "05jan1983", are more readable than the underlying numerical values such as "8405" (days since January 1, 1960). If desired, we could use **%d** formatting to produce other formats, such as "05 Jan 1983" or "01/05/83". Stata offers a number of variable-definition, display-format, and dataset-format features that are important with time series. Many of these involve ways to input, convert, and display dates. Full descriptions of date functions are found in the *Reference Manual* and the *User's Guide*, or they can be explored within Stata by typing **help dates**.

The labeled values of *date* appear in a graph of *water* against *date*, which shows day-to-day variation, as well as an upward trend in water use as summer arrives (Figure 13.1):

. graph *water date*, connect(l) ylabel(300,400 to 900) xlabel symbol(.)

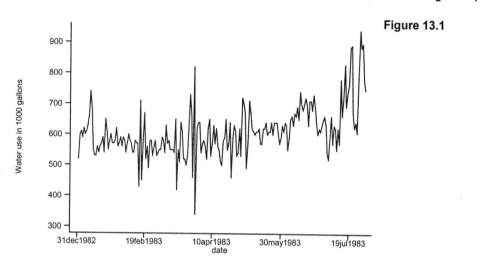

Figure 13.1

Visual inspection plays an important role in time series analysis. It often helps us to see underlying patterns in jagged series if we smooth the data by calculating a "moving average" at each point from its present, earlier, and later values. For example, a "moving average of span 3" refers to the mean of y_{t-1}, y_t, and y_{t+1}. We could use Stata's explicit subscripting to **generate** such a variable:

. generate *water3* = (*water*[_n-1] + *water*[_n] + *water*[_n+1])/3

Or, more conveniently, we could apply the **ma** (moving average) function of **egen** :

. egen *water3* = ma(*water*), nomiss t(3)

The **nomiss** option asks for shorter, uncentered moving averages in the tails; otherwise, the first and last values of *water3* would be missing. The **t(3)** option calls for moving averages of span 3. Any odd-number span ≥3 could be used; **t(5)** would yield a five-day moving average (from y_{t-2} to y_{t+2}), which is shown in Figure 13.2.

. graph *water water5 date*, connect(l[.]l) xlabel symbol(.i)
      ylabel(300,400 to 900)

The **con(l[.]l)** option above connected the *water* values in Figure 13.2 with dashed line segments **l[.]** to distinguish them from the solid line segments, **l**, connecting the smoothed variable, *water5*.

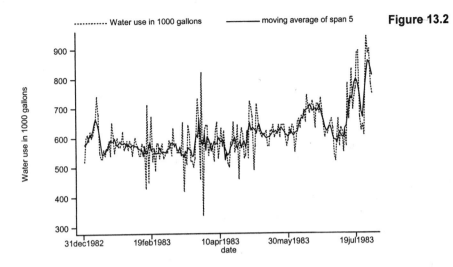

Figure 13.2

Moving averages share a drawback of other mean-based statistics: they have little resistance to outliers. Because outliers form prominent spikes in Figure 13.1, we might also try a different smoothing approach. The **smooth** command performs outlier-resistant nonlinear smoothing, employing methods and a terminology described in Velleman and Hoaglin (1981) and Velleman (1982). For example,

```
. smooth 5 water, gen(water5r)
```

creates a new variable named *water5r*, holding the values of *water* after smoothing by running medians of span 5. Compound smoothers using running medians of different spans, in combination with "hanning" (¼, ½, and ¼ -weighted moving averages of span 3) and other techniques, can be specified in Velleman's original notation. One compound smoother that seems particularly useful is called "4253h,twice." In Figure 13.3, we apply it to *water*:

```
. smooth 4253h,twice water, generate(water4r)
. graph water4r date, connect(1) symbol(.) xlabel
     ylabel(300,400 to 900)
```

Compare Figure 13.3 with 13.2 to see the subtle effects of 4253h,twice smoothing. Although both smoothers have similar spans, 4253h,twice does more to reduce the jagged variations.

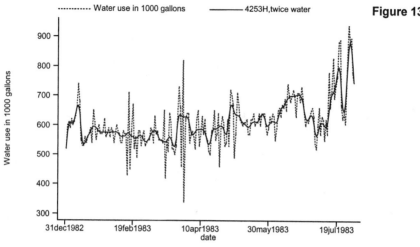

Figure 13.3

Sometimes our goal in smoothing is to look for patterns in smoothed plots. With these particular data, however, the "rough" or residuals after smoothing actually hold more interest. We can calculate the rough as the difference between data and smooth, and then graph the results in another time plot (Figure 13.4):

```
. generate rough = water - water4r
```

```
. label variable rough "Residuals from 4253h,twice"
```

```
. graph rough date, connect(1) symbol(.) xlabel
      ylabel(-200,-100,0,100,200)
```

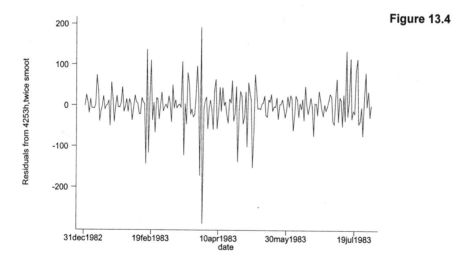

Figure 13.4

The wildest fluctuations in Figure 13.4 occur around March 27–29. Water use abruptly dropped, rose again, and then dropped even further before returning to more usual levels. On these days, local newspapers carried stories that hazardous chemical wastes had been discovered in one of the wells that supplied the town's water. Initial reports alarmed people, but they were reassured after the questionable well was taken offline.

The smoothing techniques described in this section require series with no missing values, and make the most sense when the observations are equally spaced in time. For series with missing values or uneven spacing, lowess regression (**ksm** command; see Chapter 8) provides a practical alternative.

Further Time Plot Examples

Dataset *atlantic.dta* contains time series of climate, ocean, and fisheries variables for the northern Atlantic from 1950–2000 (the original data sources include Buch 2000, and others cited in Hamilton, Brown, and Rasmussen 2001). The variables include sea temperatures on Fylla Bank off west Greenland; air temperatures in Nuuk, Greenland's capital city; two climate indexes called the North Atlantic Oscillation (NAO) and the Arctic Oscillation (AO); and catches of cod and shrimp in west Greenland waters.

```
Contains data from atlantic.dta
  obs:           51                      Greenland climate & fisheries
  vars:           8                      31 Mar 2001 12:25
  size:        1,734 (99.0% of memory free)
-----------------------------------------------------------------------
              storage  display   value
variable name  type    format    label    variable label
-----------------------------------------------------------------------
year           int     %ty                Year
fylltemp       float   %9.0g              Fylla Bank temp. at 0-40m
fyllsal        float   %9.0g              Fylla Bank salinity at 0-40m
nuuktemp       float   %9.0g              Nuuk air temperature
wNAO           float   %9.0g              Winter (Dec-Mar)
                                            Lisbon-Stykkisholmur NAO
wAO            float   %9.0g              Winter (Dec-Mar) AO index
tcod1          float   %9.0g              Division 1 cod catch, 1000t
tshrimp1       float   %9.0g              Division 1 shrimp catch, 1000t
-----------------------------------------------------------------------
Sorted by:  year
```

Before analyzing these time series, we **tsset** the dataset, which tells Stata that the variable *year* contains the time-sequence information.

```
. tsset year, yearly
        time variable:  year, 1950 to 2000
```

With a **tsset** dataset, two new qualifiers become available: **tin** (times **in**) and **twithin** (times **within**). To list Fylla temperatures and NAO values for the years 1950 through 1955, type

```
. list year fylltemp wNAO if tin(1950,1955)

        year   fylltemp    wNAO
  1.    1950       2.1      1.4
  2.    1951       1.9    -1.26
  3.    1952       1.6      .83
  4.    1953       2.1      .18
  5.    1954       2.3      .13
  6.    1955       1.2    -2.52
```

The **twithin** qualifier works similarly, but excludes the two endpoints:

. **list** *year fylltemp wNAO* **if twithin(1950,1955)**

	year	fylltemp	wNAO
2.	1951	1.9	-1.26
3.	1952	1.6	.83
4.	1953	2.1	.18
5.	1954	2.3	.13

Figure 13.5 shows a time plot of Fylla Bank temperatures (data from Buch 2000). To create this plot, we generated two new variables. *fyllmean* is simply a constant equal to the long-term mean (1.67):

. **egen** *fyllmean* = **mean(**fylltemp**)**

The second new variable, *fyll4*, contains 4253h,twice smoothed values of *fylltemp*:

. **smooth 4253h,twice** *fylltemp*, **gen(**fyll4**)**

The graph displays smoothed *fyll4* values and raw *fylltemp* values drawn as deviations from the mean (*fyllmean*). The **connect(||)** option links *fylltemp* to *fyllmean* by vertical bars.

. **graph** *fylltemp fyllmean fyllmean fyll4 year*, **connect(||ll)**
      ylabel(0,1,2,3) xlabel(1950,1960 to 2000) symbol(iiii)
      ll(Fylla Bank temperature, degrees C) t1(" ")

Figure 13.5

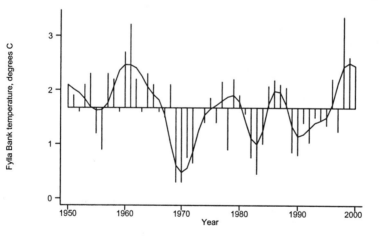

The smoothed values of Figure 13.5 exhibit irregular periods of generally warmer and cooler water. Of course, "warmer" is a relative term around Greenland; these summer sea temperatures rise no higher than 3.34 °C (37 °F).

Fylla Bank temperatures are influenced by a large-scale atmospheric pattern called the North Atlantic Oscillation, or NAO. Figure 13.6 graphs smoothed temperatures together with smoothed values of the NAO (a new variable named *wNAO4*). The **connect(ll[-])** option connects *wNAO4* values with dashed lines. **rescale** causes the vertical axes to be scaled separately for each variable; **rline(0)** draws a horizontal line at *wNAO4* = 0, and **rlabel()** specifies labels for the right-hand (*wNAO4*) axis.

```
. graph fyll4 wNAO4 year, xlabel(1950,1960 to 2000) connect(ll[-])
      symbol(Op) ylabel(0,1,2,3) ll(Fylla Bank temperature, degrees C)
      rescale rline(0) rlabel(-3,-2 to 3) rl(Winter NAO index)
```

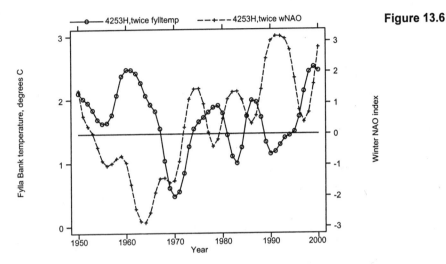

Figure 13.6

The **rescale** option provides a way to visually examine how several time series vary together. In Figure 13.6, we see evidence of a negative correlation: high-NAO periods correspond to low temperatures. The physical mechanism behind this correlation involves northerly winds that bring Arctic air and water to west Greenland during high-NAO phases. The negative Fylla–NAO correlation became stronger during the later part of this time series, roughly the years 1973 to 1997. We will return to this relationship in later sections.

Lags, Leads, and Differences

Time series analysts often work with lagged variables, or values from previous times. Lags can be specified by explicit subscripting. For example, the following command creates variable *wNAO_1*, equal to the previous year's NAO value:

```
. generate wNAO_1 = wNAO[_n-1]
(1 missing value generated)
```

An alternative way to achieve the same thing, using a **tsset** dataset, is with Stata's **L.** (lag) operator:

```
. generate wNAO_2 = L.wNAO
(1 missing values generated)
```

Lag operators are often simpler than an explicit-subscripting approach. More importantly, the lag operators also respect panel data. To generate lag 2 values, use

```
. generate wNAO_2 = L2.wNAO
(2 missing values generated)
. list year wNAO wNAO_1 wNAO_2 if tin(1950,1954)
```

	year	wNAO	wNAO_1	wNAO_2
1.	1950	1.4	.	.
2.	1951	-1.26	1.4	.
3.	1952	.83	-1.26	1.4
4.	1953	.18	.83	-1.26
5.	1954	.13	.18	.83

We could have obtained this same list without generating any new variables, by instead typing

```
. list year wNAO L.wNAO L2.wNAO if tin(1950,1954)
```

The **L.** operator is one of several that simplify the analysis of **tsset** datasets. Other time series operators are **F.** (lead), **D.** (difference), and **S.** (seasonal difference). These operators can be typed in upper or lowercase—for example, **F2.wNAO** or **f2.wNAO**.

Time Series Operators

L.	Lag y_{t-1} (**L1.** means the same thing)
L2.	2-period lag y_{t-2} (similarly, **L3.**, etc. **L(1/4).** means **L1.** through **L4.**)
F.	Lead y_{t+1} (**F1.** means the same thing)
F2.	2-period lead y_{t+2} (similarly, **F3.**, etc.)
D.	Difference $y_t - y_{t-1}$ (**D1.** means the same thing)
D2.	Second difference $(y_t - y_{t-1}) - (y_{t-1} - y_{t-2})$ (similarly, **D3.**, etc.)
S.	Seasonal difference $y_t - y_{t-1}$, (which is the same as **D.**)
S2.	Second seasonal difference $(y_t - y_{t-2})$ (similarly, **S3.**, etc.)

In the case of seasonal differences, **S12.** does not mean "12th difference," but rather a first difference at lag 12. For example, if we had monthly temperatures instead of yearly, we might want to calculate **S12.temp** , which would be the differences between December 2000 temperature and December 1999 temperature, November 2000 temperatures and November 1999 temperature, and so forth.

Lag operators can appear directly in most analytical commands. We could regress 1973–97 *fylltemp* on *wNAO*, including as additional predictors *wNAO* values from one, two, and three years previously, without first creating any new lagged variables.

```
. regress fylltemp wNAO L1.wNAO L2.wNAO L3.wNAO if tin(1973,1997)

      Source |       SS       df       MS              Number of obs =      25
-------------+------------------------------           F(  4,    20) =    4.57
       Model |  3.1884913        4  .797122826          Prob > F      =  0.0088
    Residual |  3.48929123      20  .174464562          R-squared     =  0.4775
-------------+------------------------------           Adj R-squared =  0.3730
       Total |  6.67778254      24  .278240939          Root MSE      =  .41769


------------------------------------------------------------------------------
    fylltemp |      Coef.   Std. Err.      t    P>|t|     [95% Conf. Interval]
-------------+----------------------------------------------------------------
wNAO         |
          -- |  -.1688424   .0412995    -4.09   0.001    -.2549917   -.0826931
          L1 |   .0043805   .0421436     0.10   0.918    -.0835294    .0922905
          L2 |  -.0472993    .050851    -0.93   0.363    -.1533725     .058774
          L3 |   .0264682   .0495416     0.53   0.599    -.0768738    .1298102
       _cons |   1.727913   .1213588    14.24   0.000     1.474763    1.981063
------------------------------------------------------------------------------
```

Equivalently, we could have typed

```
. regress fylltemp L(0/3).wNAO if tin(1973,1997)
```

The estimated model is

$$\text{predicted } fylltemp_t = 1.728 - .169 wNAO_t + .004 wNAO_{t-1} - .047 wNAO_{t-2} + .026 wNAO_{t-3}$$

Coefficients on the lagged terms are not statistically significant; it appears that current (unlagged) values of $wNAO_t$ provide the most parsimonious prediction. Indeed, if we re-estimate this model without the lagged terms, the adjusted R^2 rises from .37 to .43. Either model is very rough, however. A Durbin–Watson test for autocorrelated errors is inconclusive, but that is not reassuring given the small sample size.

```
. dwstat

Durbin-Watson d-statistic(  5,    25) =  1.423806
```

Autocorrelated errors are commonly encountered with time series, and they invalidate the usual OLS confidence intervals and tests. More suitable regression methods for time series are discussed later in this chapter.

Correlograms

Autocorrelation coefficients estimate the correlation between a variable and itself at particular lags. For example, first-order autocorrelation is the correlation between y_t and y_{t-1}; second order refers to $\text{Cor}[y_t, y_{t-2}]$; and so forth. A correlogram graphs correlation versus lags.

Stata's **corrgram** command provides a crude correlogram and related information. The maximum number of lags it shows can be limited by the data, by **matsize**, or to some arbitrary lower number that is set by specifying the **lags()** option:

```
. corrgram fylltemp, lags(9)
```

					-1	0	1 -1	0	1
LAG	AC	PAC	Q	Prob>Q	[Autocorrelation]		[Partial Autocor]		
1	0.4038	0.4141	8.8151	0.0030	|---		|---		
2	0.1996	0.0565	11.012	0.0041	|-		|		
3	0.0788	0.0045	11.361	0.0099	|		|		
4	0.0071	-0.0556	11.364	0.0228	|		|		
5	-0.1623	-0.2232	12.912	0.0242	-|		-|		
6	-0.0733	0.0880	13.234	0.0395	|		|		
7	0.0490	0.1367	13.382	0.0633	|		|-		
8	-0.1029	-0.2510	14.047	0.0805	|		--|		
9	-0.2228	-0.2779	17.243	0.0450	-|		--|		

Lags appear at the left side of the table, and are followed by columns for the autocorrelations (AC) and partial autocorrelations (PAC). For example, the correlation between *fylltemp*$_t$ and *fylltemp*$_{t-2}$ is .1996, and the partial autocorrelation (adjusted for lag 1) is .0565. The Q statistics (Box–Pierce portmanteau) test a series of null hypotheses that all autocorrelations up to and including each lag are zero. Because the probabilities seen here are mostly below .05, we can reject the null hypothesis, and conclude that *fylltemp* shows significant autocorrelation, across lags 1 through 9 at least. If none of the Q statistics had been below .05, we might conclude instead that the series was "white noise" with no significant autocorrelation.

At the right in this output are character-based plots of the autocorrelations and partial autocorrelations. Inspection of such plots plays a role in the specification of time series models. More refined graphical autocorrelation plots can be obtained through the **ac** command:

```
. ac fylltemp, lags(9)
```

The resulting correlogram, Figure 13.7, includes pointwise 95% confidence intervals. Correlations outside of these bounds are individually significant.

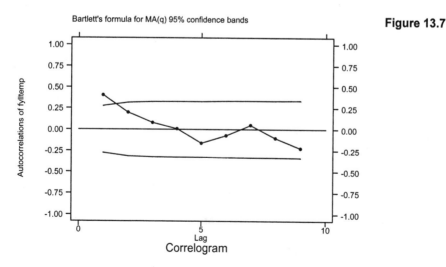

Bartlett's formula for MA(q) 95% confidence bands

Figure 13.7

A similar command, **pac**, produces the graph of partial autocorrelations seen in Figure 13.8. Approximate confidence intervals (estimating the standard error as $1/\sqrt{n}$) and residual variances also appear in Figure 13.8. The **needle** and **xlabel()** options could be used either with **ac** or **pac** (or with **xcorr**, described below).

```
. pac fylltemp, lags(9) needle xlabel(0,1 to 9)
```

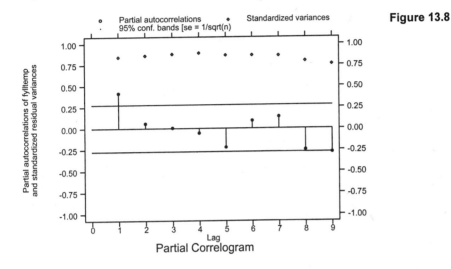

Figure 13.8

Partial Correlogram

Cross correlogram graphs help to explore relationships between two time series. Figure 13.9 shows the cross correlogram of *wNAO* and *fylltemp* over 1973–97. The cross-correlation is substantial and negative at 0 lag, but is closer to zero at other positive or negative lags. This suggests that the relationship between the two series is "instantaneous" (in yearly data) rather than delayed or distributed over several years. Recall the nonsignificance of lagged predictors from our earlier OLS regression.

```
. xcorr wNAO fylltemp if tin(1973,1997), lags(9) xline(0)
```

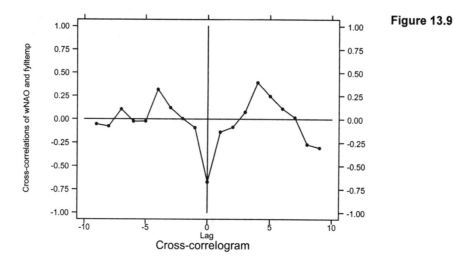

Figure 13.9

Cross-correlogram

If we list our input or independent variable first in the **xcorr** command, and the output or dependent variable second—as was done for Figure 13.9—then positive lags denote correlations between input at time *t* and output at time *t* +1, *t* +2, etc. Thus, we see a positive correlation of .394 between winter NAO index and Fylla temperature four years later.

The actual cross-correlation coefficients, and a text version of the cross correlogram, can be obtained with the **table** option:

```
. xcorr wNAO fylltemp if tin(1973,1997), lags(9) table
```

```
                         -1        0        1
     LAG       CORR     [Cross-correlation]
   -----------------------------------------
    -9      -0.0541              |
    -8      -0.0786              |
    -7       0.1040              |
    -6      -0.0261              |
    -5      -0.0230              |
    -4       0.3185              |--
    -3       0.1212              |
    -2       0.0053              |
    -1      -0.0909              |
     0      -0.6740         -----|
     1      -0.1386            -|
     2      -0.0865              |
     3       0.0757              |
     4       0.3940              |---
     5       0.2464              |-
     6       0.1100              |
     7       0.0183              |
     8      -0.2699           --|
     9      -0.3042           --|
```

ARIMA Models

Autoregressive integrated moving average (ARIMA) models for time series can be estimated through the **arima** command. This command encompasses simple autoregressive (AR), moving average (MA), or ARIMA models of any order. It also can estimate structural models that include one or more predictor variables with AR or MA errors. The general form of such structural models, in matrix notation, is

$$y_t = \mathbf{x}_t \boldsymbol{\beta} + \mu_t \qquad [13.1]$$

where y_t is the vector of dependent-variable values at time t, $\mathbf{x}_t$ is a matrix of predictor-variable values (usually including a constant), and μ_t is a vector of disturbances. Those disturbances can be autoregressive or moving-average, of any order. For example, ARMA(1,1) disturbances are

$$\mu_t = \rho\mu_{t-1} + \theta\epsilon_{t-1} + \epsilon_t \qquad [13.2]$$

where ρ is the first-order autocorrelation parameter, θ is the first-order moving average parameter, and ϵ is a white-noise (normal i.i.d.) disturbance. **arima** fits simple models as a special case of [13.1] and [13.2], with a constant (β_0) replacing the structural term $\mathbf{x}_t\boldsymbol{\beta}$. Therefore, a simple ARMA(1,1) model becomes

$$\begin{aligned} y_t &= \beta_0 + \mu_t \\ &= \beta_0 + \rho\mu_{t-1} + \theta\epsilon_{t-1} + \epsilon_t \end{aligned} \qquad [13.3]$$

Some sources present an alternative version. In the ARMA(1,1) case, they show y_t as a function of the previous y value (y_{t-1}) and the present (ϵ_t) and lagged (ϵ_{t-1}) disturbances:

$$y_t = \alpha + \rho y_{t-1} + \theta\epsilon_{t-1} + \epsilon_t \qquad [13.4]$$

Because in the simple structural model $y_t = \beta_0 + \mu_t$, equation [13.3] (Stata's version) is equivalent to [13.4], apart from rescaling the constant $\alpha = (1-\rho)\beta_0$.

Using **arima**, an ARMA(1,1) model (equation [13.3]) can be specified in either of two ways:

```
. arima y, ar(1) ma(1)
```

or,

```
. arima y, arima(1,0,1)
```

The **i** in **arima** stands for "integrated," referring to models that also involve differencing. To fit an ARIMA(2,1,1) model, use

```
. arima y, arima(2,1,1)
```

or, equivalently,

```
. arima D.y, ar(1 2) ma(1)
```

Either command specifies a model in which first differences of the dependent variable ($y_t - y_{t-1}$) are a function of first differences one and two lags previous ($y_{t-1} - y_{t-2}$ and $y_{t-2} - y_{t-3}$) and also of present and previous disturbances (ϵ_t and ϵ_{t-1}).

To estimate a structural model in which y_t depends on two predictor variables x (present and lagged values, x_t and x_{t-1}) and w (present values only, w_t), with ARIMA(1,0,1) errors, an appropriate command would be

```
. arima y x L.x w, arima(1,0,1)
```

Although seasonal differencing (e.g., **S12.y**) and/or seasonal lags (e.g., **L12.x**) can be included, as of this writing **arima** does not estimate multiplicative ARIMA(p,d,q)(P,D,Q)$_s$ seasonal models.

A time series *y* is considered "stationary" if its mean and variance do not change with time, and if the covariance between y_t and y_{t+u} depends only on the lag *u*, and not on the particular values of *t*. ARIMA modeling assumes that our series is stationary, or can be made stationary through appropriate differencing or transformation. We can check this assumption informally by inspecting time plots for trends in level or variance. Formal statistical tests for "unit roots" (a nonstationary AR(1) process in which $\rho_1 = 1$, also known as a "random walk") also help. Stata offers two unit root tests, **pperron** (Phillips–Perron) and **dfuller** (Dickey–Fuller). Applied to Fylla Bank temperatures, both tests reject the hypothesis of unit roots.

. **pperron** *fylltemp*

```
Phillips-Perron test for unit root              Number of obs   =        50
                                                Newey-West lags =         3

                   ---------- Interpolated Dickey-Fuller ---------
            Test       1% Critical      5% Critical     10% Critical
          Statistic       Value            Value            Value
-----------------------------------------------------------------------
Z(rho)     -29.871        -18.900         -13.300          -10.700
Z(t)        -4.440         -3.580          -2.930           -2.600
-----------------------------------------------------------------------
* MacKinnon approximate p-value for Z(t) = 0.0003
```

. **dfuller** *fylltemp*

```
Dickey-Fuller test for unit root                Number of obs   =        50

                   ---------- Interpolated Dickey-Fuller ---------
            Test       1% Critical      5% Critical     10% Critical
          Statistic       Value            Value            Value
-----------------------------------------------------------------------
Z(t)        -4.411         -3.580          -2.930           -2.600
-----------------------------------------------------------------------
* MacKinnon approximate p-value for Z(t) = 0.0003
```

In rejecting the unit root hypothesis, these tests confirm the visual impression of stationarity given by Figure 13.5.

For a stationary series, correlograms provide guidance about selecting a preliminary ARIMA model:

AR(*p*)   An autoregressive process of order *p* has autocorrelations that damp out gradually with increasing lag. Partial autocorrelations cut off after lag *p*.

MA(*q*)   A moving average process of order *q* has autocorrelations that cut off after lag *q*. Partial autocorrelations damp out gradually with increasing lag.

ARMA(*p,q*)   A mixed autoregressive–moving average process has autocorrelations and partial autocorrelations that damp out gradually with increasing lag.

Correlogram spikes at seasonal lags (for example, at 12, 24, 36 in monthly data) indicate a seasonal pattern. Identification of seasonal models follows similar guidelines, but applied to autocorrelations and partial autocorrelations at seasonal lags.

Figures 13.7–13.8 weakly suggest an AR(1) process, so we will try this as a simple model for *fylltemp*:

```
. arima fylltemp, arima(1,0,0) nolog
```

ARIMA regression

Sample: 1950 to 2000

Log likelihood = -48.66274

	Number of obs	=	51
	Wald chi2(1)	=	7.53
	Prob > chi2	=	0.0061

fylltemp	Coef.	OPG Std. Err.	z	P>|z|	[95% Conf. Interval]
fylltemp					
_cons	1.68923	.1513096	11.16	0.000	1.392669 1.985792
ARMA					
ar					
L1	.4095759	.1492491	2.74	0.006	.1170531 .7020987
/sigma	.627151	.0601859	10.42	0.000	.5091889 .7451131

After we fit an **arima** model, its coefficients and other results are saved temporarily in Stata's usual way. For example, to see the recent model's AR(1) coefficient and s.e., type

```
. display [ARMA]_b[L1.ar]
.4095759
```

```
. display [ARMA]_se[L1.ar]
.14924909
```

The AR(1) coefficient in this example is statistically distinguishable from zero ($t = 2.74$, $p = .006$), which gives one indication of model adequacy. A second indication is that the residuals should be uncorrelated "white noise." We can obtain residuals (also predicted values, and other case statistics) after **arima** through **predict**:

```
. predict fyllres, resid
```

```
. corrgram fyllres, lags(15)
```

LAG	AC	PAC	Q	Prob>Q	-1 0 1 [Autocorrelation]	-1 0 1 [Partial Autocor]
1	-0.0173	-0.0176	.0162	0.8987	|	|
2	0.0467	0.0465	.13631	0.9341	|	|
3	0.0386	0.0497	.22029	0.9742	|	|
4	0.0413	0.0496	.31851	0.9886	|	|
5	-0.1834	-0.2450	2.2955	0.8069	-|	-|
6	-0.0498	-0.0602	2.4442	0.8747	|	|
7	0.1532	0.2156	3.8852	0.7929	|-	|-
8	-0.0567	-0.0726	4.087	0.8492	|	|
9	-0.2055	-0.3232	6.8055	0.6574	-|	--|
10	-0.1156	-0.2418	7.6865	0.6594	|	-|
11	0.1397	0.2794	9.0051	0.6214	|-	|--
12	-0.0028	0.1606	9.0057	0.7024	|	|-
13	0.1091	0.0647	9.8519	0.7060	|	|
14	0.1014	-0.0547	10.603	0.7169	|	|
15	-0.0673	-0.2837	10.943	0.7566	|	--|

corrgram's Q test finds no significant autocorrelation among residuals out to lag 15. We could obtain exactly the same result by requesting a **wntestq** (white noise test Q statistic) for 15 lags:

```
. wntestq fyllres, lags(15)
```

```
Portmanteau test for white noise
-------------------------------------
  Portmanteau (Q) statistic =    10.9435
  Prob > chi2(15)           =     0.7566
```

By these criteria, our AR(1) or ARIMA(1,0,0) model appears adequate. More complicated versions, with MA or higher-order AR terms, do not offer much improvement in fit.

A similar AR(1) model fits *fylltemp* over just the years 1973–1997. During this period, however, information about the winter North Atlantic Oscillation (*wNAO*) significantly improves the predictions. For this model, we include *wNAO* as a predictor but keep an AR(1) term to account for autocorrelation of errors.

```
. arima fylltemp wNAO if tin(1973,1997), ar(1) nolog
```

ARIMA regression

```
Sample:  1973 to 1997                    Number of obs    =        25
                                         Wald chi2(2)     =     12.73
Log likelihood = -10.3481                Prob > chi2      =    0.0017
```

fylltemp		Coef.	OPG Std. Err.	z	P>|z|	[95% Conf. Interval]
fylltemp						
wNAO		-.1736227	.0531688	-3.27	0.001	-.2778317 -.0694138
_cons		1.703462	.1348599	12.63	0.000	1.439141 1.967782
ARMA						
ar						
L1		.2965222	.237438	1.25	0.212	-.1688478 .7618921
/sigma		.36536	.0654008	5.59	0.000	.2371767 .4935432

```
. predict fyllhat
(option xb assumed; predicted values)
```

```
. label variable fyllhat "predicted temperature"
```

```
. predict fyllres2, resid
```

```
. corrgram fyllres2, lags(9)
```

LAG	AC	PAC	Q	Prob>Q	-1 0 1 -1 0 1 [Autocorrelation] [Partial Autocor]
1	0.1485	0.1529	1.1929	0.2747	|- |-
2	-0.1028	-0.1320	1.7762	0.4114	| -|
3	0.0495	0.1182	1.9143	0.5904	| |
4	0.0887	0.0546	2.3672	0.6686	| |
5	-0.1690	-0.2334	4.0447	0.5430	-| -|
6	-0.0234	0.0722	4.0776	0.6662	| |
7	0.2658	0.3062	8.4168	0.2973	|-- |--
8	-0.0726	-0.2236	8.7484	0.3640	| -|
9	-0.1623	-0.0999	10.444	0.3157	-| |

wNAO has a significant, negative coefficient in this model. The AR(1) coefficient now is not statistically significant. If we dropped the AR term, however, our residuals would no longer pass **corrgram** 's test for white noise. Figure 13.10 graphs the predicted values, *fyllhat*, together with the observed temperature series *fylltemp*. The model does reasonably well in fitting the main warming/cooling episodes and a few of the minor variations. To have the *y*-axis labels displayed with the same number of decimal places (0.5, 1.0, 1.5,... instead of .5, 1, 1.5,...) in this graph, we specify a display format for the first-named *y* variable:

```
. format fylltemp %3.1f

. graph fylltemp fyllhat year if tin(1973,1997), connect(ll[_])
      symbol(Op) ll(Degrees C) ylabel(0.5,1 to 2.5)
      xlabel(1975,1980 to 1995)
```

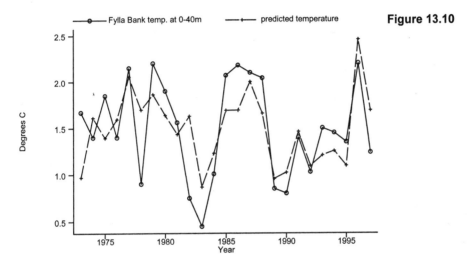

A technique called Prais–Winsten regression (**prais**), which corrects for first-order autoregressive errors, can also be illustrated with this example.

```
. prais fylltemp wNAO if tin(1973,1997), nolog
```

```
Prais-Winsten AR(1) regression -- iterated estimates
```

Source	SS	df	MS		
Model	3.35819258	1	3.35819258		
Residual	3.33743545	23	.145105889		
Total	6.69562803	24	.278984501		

Number of obs = 25
F(1, 23) = 23.14
Prob > F = 0.0001
R-squared = 0.5016
Adj R-squared = 0.4799
Root MSE = .38093

fylltemp	Coef.	Std. Err.	t	P>|t|	[95% Conf. Interval]
wNAO	-.17356	.037567	-4.62	0.000	-.2512733 -.0958468
_cons	1.703436	.1153695	14.77	0.000	1.464776 1.942096
rho	.2951576				

```
Durbin-Watson statistic (original)     1.344998
Durbin-Watson statistic (transformed)  1.789412
```

prais is an older method, more specialized than **arima**. Its regression-based standard errors assume that rho (ρ) is known rather than estimated. Because that assumption is untrue, the standard errors, tests, and confidence intervals given by **prais** tend to be anti-conservative, especially in small samples. **prais** provides a Durbin–Watson statistic ($d = 1.789$). In this example, the Durbin–Watson test agrees that after fitting the model, no significant first-order autocorrelation remains.

14

Introduction to Programming

As noted in Chapter 2, we can create a simple type of program by writing any sequence of Stata commands in an ASCII file, most conveniently by using Stata's Do-file Editor (click on Window – Do-file Editor or the icon ⬦). We then enter Stata and type a command with the form **do** *filename* that tells Stata to read *filename.do* and execute whatever commands it contains. More sophisticated programs are possible as well, making use of Stata's built-in programming language. Many of the commands used in previous chapters actually involve programs written in Stata. These programs might have originated either from Stata Corporation or from users, who wanted something beyond Stata's built-in features to accomplish a particular task.

Stata programs can access all the existing features of Stata, call other programs that call other programs in turn, and use model-fitting aids including matrix algebra and maximum likelihood estimation. Whether our purposes are broad, such as adding new statistical techniques, or narrowly specialized, such as managing a particular database, our ability to write programs in Stata greatly extends what we can do.

Substantial books (*Stata Programming Manual*; *Maximum Likelihood Estimation with Stata*) have been written about Stata programming. This engaging topic is also the focus of periodic NetCourses (see www.stata.com) and a section of the *User's Guide*. The present chapter has the modest aim of introducing a few basic tools and giving examples that show how these tools can be used.

Basic Concepts and Tools

Some elementary concepts and tools, combined with the Stata capabilities described in earlier chapters, suffice to get started.

Do-files

Do-files are ASCII (text) files, created by Stata's Do-file Editor, a word processor, or another text editor. They are typically saved with a *.do* extension. The file can contain any sequence of legitimate Stata commands. In Stata, typing the following command causes Stata to read *filename.do* and execute the commands it contains:

`. do filename`

Each command in *filename.do*, including the last, must end with a hard return—unless we have reset the delimiter to some other character, through a #delimit command. For example,

```
#delimit ;
```

This sets a semicolon as the end-of-line delimiter, so that Stata does not consider a line finished until it encounters a semicolon. Setting the semicolon as delimiter permits a single command to extend over more than one physical line. Later, we can reset "carriage return" as the usual end-of-line delimiter by typing the following command:

```
#delimit cr
```

Ado-files

Ado (automatic do) files are ASCII files containing sequences of Stata commands, much like do-files. The difference is that we need not type **do *filename*** in order to run an ado-file. Suppose we type the command

```
. clear
```

As with any command, Stata reads this and checks whether an intrinsic command by this name exists. If a **clear** command does not exist (and, in fact, it does not), then Stata next searches in its usual "ado" directories, trying to find a file named *clear.ado*. If Stata finds such a file (as it should), it then executes whatever commands the file contains. Ado-files have the extension *.ado*. User-written programs commonly go in a directory named C:\ado\personal, whereas the hundreds of official Stata ado-files get installed in C:\stata\ado. Type **sysdir** to see a list of the directories Stata currently uses. Type **help sysdir** or **help adopath** for advice on changing them.

The **which** command reveals whether a given command really is an intrinsic, hardcoded Stata command or one defined by an ado-file; and if it is an ado-file, where that resides. For example, **logit** is a command, but **logistic** is an ado-file:

```
. which logit
built-in command:  logit


. which logistic
C:\STATA\ado\base\l\logistic.ado
*! version 3.1.8  09jun2000
```

This distinction makes no difference to most users, because **logit** and **logistic** work with similar ease and syntax when called.

Programs

Both do-files and ado-files might be viewed as types of programs, but Stata uses the word "program" in a narrower sense, to mean a sequence of commands stored in memory and executed by typing a particular program name. Do-files, ado-files, or commands typed interactively can define such programs. The definition begins with a statement that names the program. For example, to create a program named *new*, we start with

```
program define new
```

Next should be the lines that actually define the program. Finally, we give an end command, followed by a hard return:

```
end
```

Once Stata has read the program-definition commands, it retains that definition of the program in memory and will run it any time we type the program's name as a command:

```
. new
```

Programs effectively make new commands available within Stata, so most users do not need to know whether a given command comes from Stata itself or from an ado-file-defined program.

Local Macros

Macros are names (up to 31 characters) that can stand for strings, program-defined results, or user-defined values. A *local macro* exists only within the program that defines it, and cannot be passed to another program. To create a local macro named `iterate`, standing for the number 0, type

```
local iterate = 0
```

To refer to the contents of a local macro (0 in this example), place the macro name *within left and right single quotes*. For example,

```
display `iterate'
```
0

Thus to increase the value of `iterate` by one, we write

```
local iterate = `iterate' + 1
```

Global Macros

Global macros are similar to local macros, but once defined, they remain in memory and can be used by other programs. To refer to a global macro's contents, we *preface the macro name with a dollar sign* (instead of enclosing the name in left and right single quotes as done with local macros):

```
global distance = 73
display $distance * 2
```
146

Version

Stata's capabilities and features have changed over the years. Consequently, programs written for an older version of Stata might not run directly under the current version. The `version` command works around this problem so that old programs remain usable. Once we tell Stata for what version the program was written, Stata makes the necessary adjustments, and the old program can run under a new version of Stata. For example, if we begin our program with the following statement, Stata interprets all the program's commands as it would have in Stata 3.1:

```
version 3.1
```

Comments

Stata does not attempt to execute any line that begins with an asterisk. Such lines can therefore be used to insert comments and explanation into a program. For example,

```
* This entire line is a comment.
```

Alternatively, we can include a comment within an executable line by bracketing it with /*
and */ . For example,

```
summarize income education      /* this part is the comment */
```

Programmers often use the /* */ comment notation as a trick for writing a command
more than one physical (typed) line long. At the end of the first typed line they place /* , and
at the beginning of the next */ . Then Stata skips over the line break and reads both typed
lines as a single command. For example, the following lines would be read as one table
command, even though they are separated by a hard return:

```
table gender kids school if contam==1, contents(mean lived /*
    */ median lived count lived)
```

Although this method is simple, if our program has more than a few long commands, the
#delimit ; approach (described earlier) might be easier to write and read.

Looping

A section of program with the following general form will repeatedly execute the commands
within curly { } brackets, so long as *expression* evaluates to "true":

```
while expression {
    command A
    command B
    . . . .
    }
command Z
```

When *expression* evaluates to "false," the looping stops and Stata goes on to execute
command Z. For example, here is simple program named count5 , that uses a loop to display
onscreen the iteration numbers from 1 through 5:

```
* Program that counts from one to five
program define count5
    version 7.0
    local iterate = 1
    while `iterate' <= 5 {
        display `iterate'
        local iterate = `iterate' + 1
    }
end
```

By typing these commands, we define program count5 . Alternatively, we could use the
Do-file Editor to save the same series of commands as an ASCII file named *count5.do.* Then,
typing the following causes Stata to read the file:

. **do count5**

Either way, by defining program count5 we make this available as a new command:

. **count5**
```
1
2
3
4
5
```

The *Programming Manual* describes other ways to create program loops.

If . . . else

The if and else commands tell a program to do one thing if an expression is true, and something else otherwise. They are set up as follows:

```
if expression {
    command A
    command B
    . . . .
}
else {
    command Z
}
```

For example, the following program segment checks whether the content of local macro span is an odd number, and informs the user of the result.

```
if int(`span'/2) != (`span' - 1)/2 {
    display "span is NOT an odd number"
}
else {
    display "span IS an odd number"
}
```

Arguments

Programs define new commands. In some instances (as with the earlier example, count5), we intend our command to do exactly the same thing each time it is used. Often, however, we need a command that is modified by arguments such as variable names or options. There are two ways we can tell Stata how to read and understand a command line that includes arguments. The simplest of these is the args command.

The following do-file (*listres1.do*) defines a program that performs a two-variable regression, and then lists the observations with the largest absolute residuals.

```
* Perform simple regression and list observations with #
* largest absolute residuals.
*    listres1 Yvariable Xvariable # IDvariable
program define listres1, sortpreserve
    version 7.0
    args Yvar Xvar number id
    quietly regress `Yvar' `Xvar'
    capture drop Yhat
    capture drop Resid
    capture drop Absres
    quietly predict Yhat
    quietly predict Resid, resid
    quietly gen Absres = abs(Resid)
    gsort -Absres
    drop Absres
    list `id' `Yvar' Yhat Resid in 1/`number'
end
```

The line args Yvar Xvar number id tells Stata that the command listresid should be followed by four arguments. These arguments could be numbers, variable names, or other strings separated by spaces. The first argument becomes the contents of a local macro named `Yvar`, the second a local macro named `Xvar`, and so forth. The program then uses the contents of these macros in other commands, such as the regression:

```
quietly regress `Yvar' `Xvar'
```

The program calculates absolute residuals (*Absres*), and then uses the `gsort` command (followed by a minus sign before the variable name) to sort the data in high-to-low order, with missing values last:

```
gsort -Absres
```

The option `sortpreserve` on the command line makes this program "sort-stable": it returns the data to their original order after the calculations are finished.

Dataset *nations.dta*, seen previously in Chapter 8, contains variables indicating life expectancy (*life*), per capita daily calories (*food*), and country name (*country*) for 109 countries. We can open this file, and use it to demonstrate our new program. A **do** command runs do-file *listres1.do*, thereby defining the program `listres1`:

```
. do listres1.do
```

Next, we use the newly-defined `listres1` command, followed by its four arguments. The first argument specifies the *y* variable, the second *x*, the third how many observations to list, and the fourth gives the case identifier. In this example, our command asks for a list of observations that have the five largest absolute residuals.

```
. listres1 life food 5 country
```

	country	life	Yhat	Resid
1.	Libya	60	76.6901	-16.69011
2.	Bhutan	44	60.49577	-16.49577
3.	Panama	72	58.13118	13.86882
4.	Malawi	45	58.58232	-13.58232
5.	Ecuador	66	52.45305	13.54695

Life expectancies are lower than predicted in Libya, Bhutan, and Malawi. Conversely, life expectancies in Panama and Ecuador are higher than predicted, based on food supplies.

Syntax

The `syntax` command provides a more complicated but also more powerful way to read a command line. The following do-file named *listres2.do* is similar to our previous example, but it uses `syntax` instead of `args`:

```
* Perform simple or multiple regression and list
* observations with # largest absolute residuals.
*   listres2 yvar xvarlist [if] [in], number(#) [id(varname)]
program define listres2, sortpreserve
version 7.0
syntax varlist(min=1) [if] [in], Number(integer) [Id(string)]
   marksample touse
   quietly regress `varlist' if `touse'
   capture drop Yhat
   capture drop Resid
   capture drop Absres
   quietly predict Yhat if `touse'
   quietly predict Resid if `touse', resid
```

```
      quietly gen Absres = abs(Resid)
      gsort -Absres
      drop Absres
      list `id' `1' Yhat Resid in 1/`number'
end
```

listres2 has the same purpose as the earlier listres1: it performs regression, then lists observations with the largest absolute residuals. This newer version contains several improvements, however, made possible by the syntax command. It is not restricted to two-variable regression, as was listres1. listres2 will work with any number of predictor variables, including none (in which case, predicted values equal the mean of *y*, and residuals are deviations from the mean). listres2 permits optional if and in qualifiers. A variable identifying the observations is optional with listres2, instead of being required as it was with listres1. For example, we could regress life expectancy on *food* and *energy*, while restricting our analysis to only those countries where per capita GNP is above 500 dollars:

. do *listres2.do*

. listres2 *life food energy if gnpcap* > 500, n(6) i(*country*)

```
        country     life      Yhat      Resid
  1.    YemenPDR      46    61.34964   -15.34964
  2.    YemenAR       45    59.85839   -14.85839
  3.      Libya       60    73.62516   -13.62516
  4.    S_Africa      55    67.9146    -12.9146
  5.    HongKong      76    64.64022    11.35978
  6.     Panama       72    61.77788    10.22212
```

The syntax line in this example illustrates some general features of the command:

```
      syntax varlist(min=1) [if] [in], Number(integer) [Id(string)]
```

The variable list for a listres2 command is required to contain at least one variable name (varlist(min=1)). Square brackets denote optional arguments, which here are the if and in qualifiers, and also the id() option. Capitalization of initial letters for the options indicates the minimum abbreviation that can be used. Since the syntax line in our example specified Number(integer) Id(string) , an actual command could be written:

. listres2 *life food*, number(6) id(*country*)

Or, equivalently,

. listres2 *life food*, n(6) i(*country*)

The contents of local macro number are required to be an integer, and id is a string (such as *country*, a variable's name).

This example also illustrates the marksample command, which marks the subsample (as qualified by if and in) to be used in subsequent analyses.

The syntax of syntax is outlined in the *Programming Manual*. Experimentation and studying other programs help in gaining fluency with this command.

Example Program: Moving Autocorrelation

The preceding sections presented basic ideas and example short programs. In this section, we apply those ideas to a slightly longer program that defines a new statistical procedure. The procedure obtains moving autocorrelations through a time series, as proposed for ocean-atmosphere data by Topliss (2000). The following do-file, *gossip.do*, defines a program that makes available a new command called `gossip`. Comments, in lines that begin with `*` or in phrases set off by `/* */` , explain what the program is doing. Indentation of lines has no effect on the program's execution, but makes it easier for the programmer to read.

```
capture program drop gossip    /* FOR WRITING & DEBUGGING; DELETE LATER */
program define gossip
version 7.0
* Syntax requires user to specify two variables (Yvar and TIMEvar), and
* the span of the moving window.  Optionally, the user can ask to generate
* a new variable holding autocorrelations, to draw a graph, or both.
syntax varlist(min=1 max=2 numeric), SPan(integer) [GENerate(string) GRaph]
if int(`span'/2) != (`span' - 1)/2 {
    display as error "Span must be an odd integer"
}
else {
* The first variable in `varlist' becomes Yvar, the second TIMEvar.
    tokenize `varlist'
        local Yvar `1'
        local TIMEvar `2'
    tempvar NEWVAR
    quietly gen `NEWVAR' = .
    local miss = 0
* spanlo and spanhi are local macros holding the observation number at the
* low and high ends of a particular window.  spanmid holds the observation
* number at the center of this window.
    local spanlo = 0
    local spanhi = `span'
    local spanmid = int(`span'/2)
    while `spanlo' <= _N-`span' {
        local spanhi = `span' + `spanlo'
        local spanlo = `spanlo' + 1
        local spanmid = `spanmid' + 1
* The next lines check whether missing values exist within the window.
* If they do exist, then no autocorrelation is calculated and we
* move on to the next window.  Users are informed that this occurred.
        quietly summ `Yvar' in `spanlo'/`spanhi'
        if r(N) != `span' {
            local miss = 1
        }
* The value of NEWVAR in observation `spanmid' is set equal to the first
* row, first column (1,1) element of the row vector of autocorrelations
* r(AC) saved by corrgram.
        else {
            quietly corrgram `Yvar' in `spanlo'/`spanhi', lag(1)
            quietly replace `NEWVAR' = el(r(AC),1,1) in `spanmid'
        }
    }
    if "`graph'" != "" {
        capture drop ZERO
        generate ZERO = 0
* The following "graph" command illustrates the use of comments to cause
* Stata to skip over line breaks, so it reads the next four lines as if
* they were one.
        graph `NEWVAR' ZERO `TIMEvar', connect(||) symbol(ii) yline(0) /*
```

```
        */ ylabel xlabel border /*
        */ ll("First-order autocorrelations of `Yvar' (span `span')")
        drop ZERO
    }
    if `miss' == 1 {
        display as error "Caution:  missing values exist"
    }
    if "`generate'" != "" {
        rename `NEWVAR' `generate'
        label variable `generate' /*
        */ "First-order autocorrelations of `Yvar' (span `span')"
    }
}
end
```

As the comments note, `gossip` requires time series (`tsset`) data. From an existing time series variable, `gossip` calculates a second time series consisting of lag-1 autocorrelation coefficients within a moving window of observations—for example, a moving 9-year span. Dataset *nao.dta* contains North Atlantic climate time series that can be used for illustration:

```
Contains data from C:\data\nao.dta
  obs:           159                    North Atlantic Oscillation &
                                          mean air temperature at
                                          Stykkisholmur, Iceland
  vars:            5                    23 Jul 2001 20:07
  size:        3,498 (99.6% of memory free)
-------------------------------------------------------------------------------
               storage  display    value
variable name   type    format     label    variable label
-------------------------------------------------------------------------------
year            int     %ty                 Year
wNAO            float   %9.0g               Winter NAO
wNAO4           float   %9.0g               Winter NAO smoothed
temp            float   %9.0g               Mean air temperature (C)
temp4           float   %9.0g               Mean air temperature smoothed
-------------------------------------------------------------------------------
Sorted by:  year
```

The variable *temp* records annual mean air temperatures at Stykkishólmur in west Iceland from 1841 to 1999. *temp4* contains 4253h, twice smoothed values of *temp* (see Chapter 13). Figure 14.1 graphs these two time series:

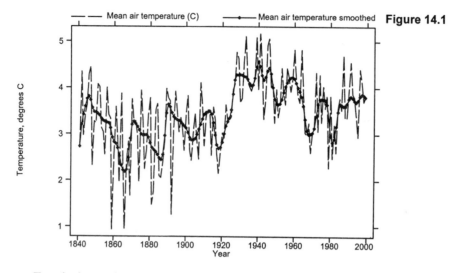

Figure 14.1

To calculate and graph a series of autocorrelations of *temp*, within a moving window of 9 years, we type the following commands. They produce the graph shown in Figure 14.2.

```
. do gossip.do

. gossip temp year, span(9) generate(autotemp) graph
```

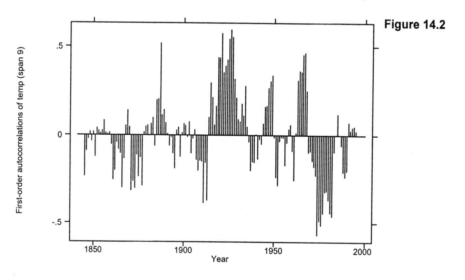

Figure 14.2

In addition to drawing Figure 14.2, gossip created a new variable named *autotemp*:

. describe

```
Contains data from C:\data\nao.dta
  obs:             159                    North Atlantic Oscillation &
                                          mean air temperature at
                                          Stykkisholmur, Iceland
  vars:              6                    23 Jul 2001 20:07
  size:          4,134 (99.8% of memory free)
---------------------------------------------------------------------------
                storage  display    value
variable name   type     format     label     variable label
---------------------------------------------------------------------------
year            int      %ty                  Year
wNAO            float    %9.0g                Winter NAO
wNAO4           float    %9.0g                Winter NAO smoothed
temp            float    %9.0g                Mean air temperature (C)
temp4           float    %9.0g                Mean air temperature smoothed
autotemp        float    %9.0g                First-order autocorrelations of
                                                temp (span 9)
---------------------------------------------------------------------------
Sorted by:  year
     Note:  dataset has changed since last saved
```

. list year temp autotemp in 1/10

```
         year      temp    autotemp
  1.     1841      2.73        .
  2.     1842      4.34        .
  3.     1843      2.97        .
  4.     1844      3.41        .
  5.     1845      3.62    -.2324837
  6.     1846      4.28    -.0883512
  7.     1847      4.45    -.0194607
  8.     1848      2.32     .0175247
  9.     1849      3.27     -.03303
 10.     1850      3.23     .0181154
```

autotemp values are missing for the first four years (1841 to 1844). In 1845, the autotemp value ($-.2324837$) equals the lag-1 autocorrelation of *temp* over the 9-year span from 1841 to 1849. This is the same coefficient we would obtain by typing the following command:

. corrgram temp in 1/9, lag(1)

```
                                          -1       0      1 -1      0       1
LAG      AC       PAC        Q     Prob>Q  [Autocorrelation]  [Partial Autocor]
---------------------------------------------------------------------------
1      -0.2325  -0.2398   .66885  0.4135        -|                  -|
```

In 1846, *autotemp* ($-.0883512$) equals the lag-1 autocorrelation of *temp* over the 9 years from 1842 to 1850, and so on through the data. *autotemp* values are missing for the last four years in the data (1996 to 1999), as they are for the first four.

The pronounced Arctic warming of the 1920s, visible in the temperatures of Figure 14.1, manifests in Figure 14.2 as a period of consistently positive autocorrelations. A briefer period of positive autocorrelations in the 1960s coincides with a cooling climate. Topliss (2000) suggests interpretation of such autocorrelations as indicators of changing feedbacks in ocean-atmosphere systems.

The do-file *gossip.do* was written incrementally, starting with input components such as the syntax statement and span macros, running the do-file to check how these work, and then adding other components. Not all of the trial runs produced satisfactory results. Typing the following command causes Stata to display programs line-by-line as they execute, so we can see exactly where an error occurs:

`. set trace on`

Later, we can turn this feature off by typing

`. set trace off`

gossip.do contains a first line, `capture program drop gossip`, that discards the program from memory before defining it again. This is helpful during the writing and debugging stage, when a previous version of our program might have been incomplete or incorrect. Such lines should be deleted once the program is mature, however. The next section describes further steps toward making `gossip` available as a regular Stata command.

Ado-File

Once we believe our do-file defines a useful program that we will want to use again, we can create an ado-file to make it available like any other Stata command. For the previous example, *gossip.do*, the change involves two steps:

1. With the Do-file Editor, delete the initial "DELETE LATER" line that had been inserted to streamline the program writing and debugging phase. We can also delete the comment lines. Doing so removes useful information, but it makes the program more compact and easier to read.

2. Save our modified file, renaming it to have an .ado extension (for example, *gossip.ado*), in a new directory. The recommended location is in C:\ado\personal; you might need to create this directory and subdirectory if they do not already exist. Other locations are possible, but review the *User's Manual* section on "Where does Stata look for ado-files?" before proceeding.

Once this is done, we can use `gossip` as a regular command within Stata. A listing of *gossip.ado* follows.

```
*! Version 1.0
*! L. Hamilton, Statistics with Stata (2002)
program define gossip
version 7.0
syntax varlist(min=1 max=2 numeric), SPan(integer) [GENerate(string) GRaph]
if int(`span'/2) != (`span' - 1)/2 {
    display as error "Span must be an odd integer"
}
else {
    tokenize `varlist'
        local Yvar `1'
        local TIMEvar `2'
    tempvar NEWVAR
    quietly gen `NEWVAR' = .
    local miss = 0
    local spanlo = 0
    local spanhi = `span'
    local spanmid = int(`span'/2)
    while `spanlo' <= _N -`span' {
        local spanhi = `span' + `spanlo'
        local spanlo = `spanlo' + 1
        local spanmid = `spanmid' + 1
```

```
               quietly summ `Yvar' in `spanlo'/`spanhi'
               if r(N) != `span' {
                   local miss = 1
               }
               else {
                   quietly corrgram `Yvar' in `spanlo'/`spanhi', lag(1)
                   quietly replace `NEWVAR' = el(r(AC),1,1) in `spanmid'
               }
          }
          if "`graph'" != "" {
               capture drop ZERO
               generate ZERO = 0
               graph `NEWVAR' ZERO `TIMEvar', connect(||) symbol(ii) yline(0) /*
               */ ylabel xlabel border /*
               */ l1("First-order autocorrelations of `Yvar' (span `span')")
               drop ZERO
          }
          if `miss' == 1 {
               display as error "Caution:  missing values exist"
          }
          if "`generate'" != "" {
               rename `NEWVAR' `generate'
               label variable `generate' /*
               */ "First-order autocorrelations of `Yvar' (span `span')"
          }
     }
}
end
```

The program could be refined further to make it more flexible, elegant, and user-friendly. Note the inclusion of comments stating the source and "version 1.0" in the first two lines, which both begin *!. The comment refers to version 1.0 of *gossip.ado*, not Stata. The Stata version suitable for this program is specified as 7.0 by the `version` command a few lines later. Although the *! comments do not affect how the program runs, they are visible to a **which** command:

```
. which gossip
c:\ado\personal\gossip.ado
*!  version 1.0
*!  L. Hamilton, Statistics with Stata (2002)
```

Once *gossip.ado* has been saved in the C:\ado\personal directory, the command `gossip` could be used at any time. If we are following the steps in this chapter, which previously defined a preliminary version of `gossip`, then before running the new ado-file version we should drop the old definition from memory by typing

```
. program drop gossip
```

We are now prepared to run the final, ado-file version. To see a graph of span-15 autocorrelations of variable *wNAO* from dataset *nao.dta*, for example, we would simply open *nao.dta* and type

```
. gossip wNAO year, span(15) graph
```

Help File

Help files are an integral aspect of using Stata. For a user-written program such as *gossip.ado*, they become even more important because no documentation exists in the printed manuals. We can write a help file for *gossip.ado* by using Stata's Do-file Editor to create a text file named *gossip.hlp*. This help file should be saved in the same ado-file directory (for example, C:\ado\personal) as *gossip.ado*.

Any text file, saved in one of Stata's recognized ado-file directories with a name of the form *filename.hlp*, will be displayed onscreen by Stata when we type **help** *filename*. For example, we might write the following in the Do-file Editor, and save it in directory C:\ado\personal as file *gossip1.hlp*. Typing **help gossip1** at any time would then cause Stata to display the text.

```
help for gossip                  L. Hamilton

Moving first-order autocorrelations

gossip yvar timevar, span(#) [ generate(newvar) graph ]

Description

calculates first-order autocorrelations of time series
yvar, within a moving window of span #.  For example, if we
specify span(7) gen(new), then the first
through 3rd values of new are missing.  The 4th value of new
equals the lag-1 autocorrelation of yvar across observations 1
through 7.  The 5th value of new equals the lag-1 autocorrelation
of yvar across observations 2 through 8, and so forth.  The last
3 values of new are missing.  See Topliss (2000) for a rationale
and applications of this statistic to atmosphere-ocean data.
Statistics with Stata (2002) discusses the gossip program itself.

gossip requires tsset data.  timevar is the time
variable to be used for graphing.

Options

span(#)    specifies the width of the window for
           calculating autocorrelations.  This option is required;
           # should be an odd integer.

gen(newvar)   creates a new variable holding the
           autocorrelation coefficients.

graph      requests a spike plot of lag-1 autocorrelations vs.
           timevar.

Examples

    . gossip water month, span(13) graph
    . gossip water month, span(9) gen(autowater)
    . gossip water month, span(17) gen(autowater) graph

References

Hamilton, Lawrence C.  2002.  Statistics with Stata.  Pacific Grove,
  CA:  Duxbury.
```

Topliss, Brenda J. 2000. "Climate variability I: A conceptual approach to ocean-atmosphere feedback." In Abstracts for AGU Chapman Conference, The North Atlantic Oscillation, Nov. 28 - Dec 1, 2000, Ourense, Spain.

Nicer help files containing links, text formatting, dialog boxes, and other features can be designed using Stata Markup and Control Language (SMCL). All official Stata help files, as well as log files and onscreen results, employ SMCL. The following is an SMCL version of the help file for gossip . Once this file has been saved in C:\ado\personal with the file name *gossip.hlp*, typing **help gossip** will produce a readable and official-looking display.

```
{smcl}
{* 27jul2001}{...}
{hline}
help for {hi:gossip}{right:(L. Hamilton)}
{hline}

{title:Moving first-order autocorrelations}

{p 8 12}{cmd:gossip} {it:yvar timevar} {cmd:,} {cmdab:sp:an}{cmd:(}
{it:#}{cmd:)} [ {cmdab:gen:erate}{cmd:(}{it:newvar}{cmd:)}
{cmdab:gr:aph} ]


{title:Description}

{p}{cmd:gossip} calculates first-order autocorrelations of time series
{it:yvar}, within a moving window of span {it:#}.  For example, if we
specify {cmd:span(}7{cmd:)} {cmd:gen(}{it:new}{cmd:)}, then the first
through 3rd values of {it:new} are missing.  The 4th value of {it:new}
equals the lag-1 autocorrelation of {it:yvar} across observations 1
through 7.  The 5th value of {it:new} equals the lag-1 autocorrelation
of {it:yvar} across observations 2 through 8, and so forth.  The last
3 values of {it:new} are missing.  See Topliss (2000) for a rationale
and applications of this statistic to atmosphere-ocean data.
{browse "http://www.stata.com/bookstore/sws.html":Statistics with Stata}
 (2002) discusses the {cmd:gossip} program itself.{p_end}

{p}{cmd:gossip} requires {cmd:tsset} data.  {it:timevar} is the time
variable to be used for graphing.{p_end}


{title:Options}

{p 0 4}{cmd:span(}{it:#}{cmd:)} specifies the width of the window for
calculating autocorrelations.  This option is required; {it:#} should be
 an odd integer.

{p 0 4}{cmd:gen(}{it:newvar}{cmd:)} creates a new variable holding the
autocorrelation coefficients.

{p 0 4}{cmd:graph} requests a spike plot of lag-1 autocorrelations vs.
{it:timevar}.


{title:Examples}

{p 8 12}{inp:. gossip water month, span(13) graph}{p_end}
{p 8 12}{inp:. gossip water month, span(9) gen(autowater)}{p_end}
{p 8 12}{inp:. gossip water month, span(17) gen(autowater) graph}{p_end}
```

```
{title:References}

{p 0 4}Hamilton, Lawrence C.  2002.
{browse "http://www.stata.com/bookstore/sws.html":Statistics with Stata}.
 Pacific Grove, CA:  Duxbury.{p_end}

{p 0 4}Topliss, Brenda J.  2000.  "Climate variability I:  A conceptual
approach to ocean-atmosphere feedback."  In Abstracts for AGU Chapman
Conference, The North Atlantic Oscillation, Nov. 28 - Dec 1, 2000, Ourense,
Spain. citation.{p_end}
```

The help file begins with {smcl}, which tells Stata to process the file as SMCL. Curly brackets {} enclose SMCL codes, many of which have the form {command:text} or {command arguments:text}. The following examples illustrate how these codes are interpreted.

{hline}	Draw a horizontal line.
{hi:gossip}	Highlight the text "gossip".
{title:Moving...}	Display the text "Moving . . ." as a title.
{right:L Hamilton}	Right-justify the text "L. Hamilton".
{p 8 12}	Format the following text as a paragraph, with the first line indented 8 columns and subsequent lines indented 12.
{cmd:gossip}	Display the text "gossip" as a command. That is, show "gossip" with whatever colors and font attributes are presently defined as appropriate for a command.
{it:yvar}	Display the text "yvar" in italics.
{cmdab:sp:an}	Display "span" as a command, with the letters "sp" marked as the minimum abbreviation.
{p}	Format the following text as a paragraph, until terminated by {p_end}.
{browse "http://www.stata.com/bookstore/sws.html":Statistics...}	Link the text "Statistics with Stata" to the web address (URL) http://www.stata.com/bookstore/sws.html. Clicking on the words "Statistics with Stata" should then launch your browser and connect it to this URL.

The *Programming Manual* supplies details about using these and many other SMCL commands.

Matrix Algebra

Matrix algebra provides essential tools for statistical modeling. Stata's matrix commands are too diverse to describe adequately here; the subject occupies more than 60 pages in the *Programming Manual* and *User's Guide*. A few examples help to impart the flavor of these commands.

The built-in Stata command **regress** performs ordinary least squares (OLS) regression, among other things. But as an exercise, we could write an OLS program ourselves. *ols1.do* (following) defines a primitive regression program that does nothing except calculate and display the vector of estimated regression coefficients according to the familiar OLS equation:

$$\mathbf{b} = (\mathbf{X'X})^{-1}\mathbf{X'y}$$

```
* A very simple program, "ols1" estimates linear regression
* coefficients using ordinary least squares (OLS).
program define ols1
* The syntax allows only for a variable list with one or more
* numeric variables.
    syntax varlist(min=1 numeric)
* "tempname..." assigns names to temporary matrices to be used in this
* program.  When ols1 has finished, these matrices will be dropped.
    tempname crossYX crossX crossY b
* "matrix accum..." forms a cross-product matrix.  The K variables in
* varlist, and the N observations with nonmissing values on all K variables,
* comprise an N row, K column data matrix we might call yX.
* The cross product matrix crossYX equals the transpose of yX times yX.
* Written algebraically:
*         crossYX = (yX)'yX
    quietly matrix accum `crossYX' = `varlist'
* Matrix crossX extracts rows 2 through K, and columns 2 through K,
* from crossYX:
*         crossX = X'X
    matrix `crossX' = `crossYX'[2...,2...]
* Column vector crossY extracts rows 2 through K, and column 1 from crossYX:
*         crossY = X'y
    matrix `crossY' = `crossYX'[2...,1]
* The column vector b contains OLS regression coefficients, obtained by
* the classic estimating equation:
*         b = inverse(X'X)X'y
    matrix `b' = syminv(`crossX') * `crossY'
* Finally, we list the coefficient estimates, which are the contents of b.
    matrix list `b'
end
```

Comments explain each command in *ols1.do*. A comment-free version named *ols2.do* (following) gives a clearer view of the matrix commands:

```
program define ols2
    syntax varlist(min=1 numeric)
    tempname crossYX crossX crossY b
    quietly matrix accum `crossYX' = `varlist'
    matrix `crossX' = `crossYX'[2...,2...]
    matrix `crossY' = `crossYX'[2...,1]
    matrix `b' = syminv(`crossX') * `crossY'
    matrix list `b'
end
```

Neither *ols1.do* nor *ols2.do* make any provision for in or if qualifiers, syntax errors, or options. They also do not calculate standard errors, confidence intervals, or the other ancillary statistics we usually want with regression. To see just what they do accomplish, we will analyze a small dataset on nuclear power plants (*reactor.dta*):

```
Contains data from c:\data\reactor.dta
  obs:            5                      Reactor decommissioning costs
                                            (from Brown et al. 1986)
  vars:           6                      28 Jul 2001 10:53
  size:         130 (100.0% of memory free)
-------------------------------------------------------------------------
              storage  display   value
variable name  type    format    label    variable label
-------------------------------------------------------------------------
site          str14    %14s               Reactor site
decom         byte     %8.0g              Decommissioning cost, millions
capacity      int      %8.0g              Generating capacity, megawatts
years         byte     %9.0g              Years in operation
start         int      %8.0g              Year operations started
close         int      %8.0g              Year operations closed
-------------------------------------------------------------------------
Sorted by:  start
```

The cost of decommissioning a reactor increases with its generating capacity and with the number of years in operation, as can be seen by using **regress** :

`. regress decom capacity years`

```
      Source |       SS       df       MS                Number of obs =      5
-------------+------------------------------             F(  2,    2) =  189.42
       Model | 4666.16571      2  2333.08286             Prob > F      = 0.0053
    Residual | 24.6342883      2  12.3171442             R-squared     = 0.9947
-------------+------------------------------             Adj R-squared = 0.9895
       Total |    4690.80      4    1172.70              Root MSE      = 3.5096


       decom |      Coef.   Std. Err.       t    P>|t|     [95% Conf. Interval]
-------------+----------------------------------------------------------------
    capacity |   .1758739   .0247774      7.10   0.019     .0692653    .2824825
       years |   3.899314   .2643087     14.75   0.005     2.762085    5.036543
       _cons |  -11.39963   4.330311     -2.63   0.119    -30.03146     7.23219
```

Our home-brewed program *ols2.do* yields exactly the same regression coefficients:

`. do ols2.do`

`. ols2 decom capacity years`

```
__000003[3,1]
                 decom
capacity     .1758739
   years    3.8993139
   _cons  -11.399633
```

Although its results are correct, the minimalist ols2 program lacks many features we would want in a useful modeling command. The following ado-file, *ols3.ado*, defines an improved program named ols3. This program permits in and if qualifiers, and optionally allows specification of the level for confidence intervals. It calculates and neatly displays regression coefficients in a table with their standard errors, *t* tests, and confidence intervals.

```
*! version 1.0 27jul2001
*! Matrix demonstration:  more complete OLS regression program.
program define ols3, eclass
    syntax varlist(min=1 numeric) [in] [if] [, Level(integer $S_level)]
    marksample touse
    tokenize "`varlist'"
    tempname crossYX crossX crossY b hat V
    quietly matrix accum `crossYX' = `varlist' if `touse'
    local nobs = r(N)
    local df = `nobs' - (rowsof(`crossYX') - 1)
    matrix `crossX' = `crossYX'[2...,2...]
    matrix `crossY' = `crossYX'[2...,1]
    matrix `b' = (syminv(`crossX') * `crossY')'
    matrix `hat' = `b' * `crossY'
    matrix `V' = syminv(`crossX') * (`crossYX'[1,1] - `hat'[1,1])/`df'
    estimates post `b' `V', dof(`df') obs(`nobs') depname(`1') /*
        */ esample(`touse')
    estimates local depvar "`1'"
    estimates local cmd "ols3"
    if `level' < 10 | `level' > 99 {
        display as error "level( ) must be between 10 and 99 inclusive."
        exit 198
    }
    estimates display, level(`level')
end
```

Because *ols3.ado* is an ado-file, we can simply type `ols3` as a command:

. ols3 *decom capacity years*

```
------------------------------------------------------------------------------
    decom |     Coef.    Std. Err.       t     P>|t|     [95% Conf. Interval]
----------+-------------------------------------------------------------------
 capacity |   .1758739    .0247774     7.10    0.019     .0692653    .2824825
    years |   3.899314    .2643087    14.75    0.005     2.762085    5.036543
    _cons |  -11.39963    4.330311    -2.63    0.119    -30.03146     7.23219
------------------------------------------------------------------------------
```

ols3.ado contains familiar elements including `syntax` and `marksample` commands, as well as `matrix` operations built upon those seen earlier in *ols1.do* and *ols2.do*. Note the use of a right single quote (') as the "matrix transpose" operator. We write the transpose of the coefficients vector (syminv(`crossX') * `crossY') as follows:

            (syminv(`crossX') * `crossY')'

The `ols3` program is defined as e-class, indicating that this is a statistical model-estimation command:

            program define ols3, eclass

E-class programs store their results with `e()` designations. After the previous `ols3` command, these have the following contents:

. estimates list

```
scalars:
                e(N) =   5
             e(df_r) =   2

macros:
              e(cmd) : "ols3"
           e(depvar) : "decom"
```

```
matrices:
                      e(b)  :   1 x 3
                      e(V)  :   3 x 3

functions:
            e(sample)

. display e(N)
5

. matrix list e(b)

e(b)[1,3]
          capacity          years          _cons
y1     .1758739      3.8993139     -11.399633

. matrix list e(V)

symmetric e(V)[3,3]
               capacity          years          _cons
capacity     .00061392
   years    -.00216732       .0698591
   _cons    -.01492755      -.942626      18.751591
```

The e() results from e-class programs remain in memory until the next e-class command. In contrast, r-class programs such as **summarize** store their results with r() designations, and these remain in memory only until the next e- or r-class command.

Several estimates lines in *ols3.ado* save the e() results, then use these in the output display :

```
estimates post `b' `V', dof(`df') obs(`nobs') depname(`1') /*
    */ esample(`touse')
```

The above command sets the contents of e() results, including the coefficient vector (b) and the variance-covariance matrix (V). This makes all the post-estimation features detailed in **help est** and **help postest** available. Options specify the residual degrees of freedom (df), the number of observations used in estimation (nobs), the dependent variable name (`1' , meaning the contents of the first macro obtained when we tokenize varlist), and the estimation sample marker (touse).

```
estimates local depvar "`1'"
```

This command sets the name of the dependent variable, macro 1 after tokenize varlist , to be the contents of macro e(depvar) .

```
estimates local cmd "ols3"
```

This sets the name of the command, ols3 , as the contents of macro e(cmd) .

```
estimates display, level(`level')
```

The estimates display command displays the coefficient table based on our previous estimates post . This table follows a standard Stata format: its first two columns contain coefficient estimates (from b) and their standard errors (square roots of diagonal elements from V). Further columns are *t* statistics (first column divided by second), two-tail *t* probabilities, and confidence intervals based on the level specified in the ols3 command line (or defaulting to 95%).

Bootstrapping

Bootstrapping refers to a process of repeatedly drawing random samples, with replacement, from the data at hand. Instead of trusting theory to describe the sampling distribution of an estimator, we approximate that distribution empirically. Drawing k bootstrap samples of size n (from an original sample also size n) yields k new estimates. The distribution of these bootstrap estimates provides an empirical basis for estimating standard errors or confidence intervals (Efron and Tibshirani 1986; for an introduction, see Stine in Fox and Long 1990). Bootstrapping seems most attractive in situations where the statistic of interest is theoretically intractable, or where the usual theory regarding that statistic rests on untenable assumptions.

Unlike Monte Carlo simulations, which fabricate their data, bootstrapping typically works from real data. For illustration, we turn to *islands.dta*, containing area and biodiversity measures for eight Pacific Island groups (from Cox and Moore 1993).

```
Contains data from c:\data\islands.dta
  obs:            8                        Pacific island biodiversity
                                           (Cox & Moore 1993)
  vars:           4                        28 Jul 2001 15:51
  size:         208 (99.8% of memory free)
-------------------------------------------------------------------------------
              storage  display    value
variable name   type   format     label    variable label
-------------------------------------------------------------------------------
island        str15    %15s                Island group
area          float    %9.0g               Land area, km^2
birds         byte     %8.0g               Number of bird genera
plants        int      %8.0g               Number flowering plant genera
-------------------------------------------------------------------------------
Sorted by:
```

Suppose we wish to form a confidence interval for the mean number of bird genera. The usual confidence interval for a means derives from a normality assumption. We might hesitate to make this assumption, however, given the skewed distribution that even in this tiny sample ($n = 8$) almost leads us to reject normality:

```
. sktest birds
```

```
                    Skewness/Kurtosis tests for Normality
                                               ------- joint ------
    Variable |  Pr(Skewness)   Pr(Kurtosis)   adj chi2(2)    Prob>chi2
-------------+-------------------------------------------------------------
       birds |     0.079          0.181           4.75          0.0928
```

Bootstrapping provides a more empirical approach to forming confidence intervals. An r-class command, **summarize, detail** unobtrusively stores its results as a series of macros. Some of these macros are:

r(N)	Number of observations
r(mean)	Mean
r(skewness)	Skewness
r(min)	Minimum
r(max)	Maximum
r(p50)	50th percentile or median

r(Var)	Variance
r(sum)	Sum
r(sd)	Standard deviation

Stored results simplify the job of bootstrapping any statistic. To obtain bootstrap confidence intervals for the mean of *birds*, based on 1,000 resamplings, and save the results in new file *boot1.dta*, type the following command:

```
. bs "summarize birds, detail"  "r(mean)", rep(1000) saving(boot1)

command:      summarize birds, detail
statistic:    r(mean)
(obs=8)

Bootstrap statistics

Variable |   Reps   Observed       Bias    Std. Err.    [95% Conf. Interval]
---------+-------------------------------------------------------------------
     bs1 |   1000    47.625     -.34575    12.59005    22.91902  72.33098   (N)
         |                                             25.3125       74.5   (P)
         |                                                  27    76.625  (BC)
---------+-------------------------------------------------------------------
                         N = normal, P = percentile, BC = bias-corrected
```

The **bs** command states in double quotes what analysis is to be bootstrapped (**"summ birds, detail"**). Following this comes the statistic to be bootstrapped, likewise in its own double quotes (**"r(mean) "**). More than one statistic could be listed, each separated by a space. The example above specifies two options:

rep(1000)   Calls for 1,000 repetitions, or drawing 1,000 bootstrap samples.

saving(boot1)   Saves the 1,000 bootstrap means in a new dataset named *boot1.dta*.

The **bs** results table shows the number of repetitions performed and the "observed" (original-sample) value of the statistic being bootstrapped—in this case, the mean *birds* value 47.625. The table also shows estimates of bias, standard error, and three types of confidence intervals. "Bias" here refers to the mean of the k bootstrap values of our statistic (for example, the mean of the 1,000 bootstrap means of *birds*), minus the observed statistic. The estimated standard error equals the standard deviation of the k bootstrap statistic values (for example, the standard deviation of the 1,000 bootstrap means of *birds*). This bootstrap standard error (12.59) is less than the conventional standard error (13.38) calculated by **ci** :

```
. ci birds

    Variable |     Obs       Mean    Std. Err.      [95% Conf. Interval]
-------------+-------------------------------------------------------------------
       birds |       8     47.625    13.38034      15.98552   79.26448
```

Normal-approximation (N) confidence intervals in the **bs** table are obtained as follows:

observed sample statistic $\pm t \times$ bootstrap standard error

where t is chosen from the theoretical t distribution with $k - 1$ degrees of freedom. Their use is recommended when the bootstrap distribution appears unbiased and approximately normal.

Percentile (P) confidence intervals simply use percentiles of the bootstrap distribution (for a 95% interval, the 2.5th and 97.5th percentiles) as lower and upper bounds. These might be appropriate when the bootstrap distribution appears unbiased but nonnormal.

The bias-corrected (BC) interval also employs percentiles of the bootstrap distribution, but chooses these percentiles following a normal-theory adjustment for the proportion of bootstrap values less than or equal to the observed statistic. When substantial bias exists (by one guideline, when bias exceeds 25% of one standard error), these intervals might be preferred.

Since we saved the bootstrap results in a file named *boot1.dta*, we can retrieve this and examine the bootstrap distribution more closely if desired. The **saving(boot1)** option created a dataset with 1,000 observations and a variable named *bs1*, holding the mean of each bootstrap sample.

```
Contains data from c:\data\boot1.dta
  obs:         1,000                        bs: summarize birds, detail
  vars:            1                        28 Jul 2001 17:30
  size:        8,000  (99.6% of memory free)
-------------------------------------------------------------------------
              storage  display   value
variable name   type   format    label    variable label
-------------------------------------------------------------------------
bs1             float   %9.0g              r(mean)
-------------------------------------------------------------------------
Sorted by:

. summarize

    Variable |     Obs        Mean    Std. Dev.       Min        Max
-------------+----------------------------------------------------------
        bs1 |    1000    47.27925    12.59005      13.625    103.375
```

Note that the standard deviation of these 1,000 bootstrap means equals the standard error (11.59) shown earlier in the **bs** results table. The mean of the 1,000 means, minus the observed (original sample) mean, equals the bias:

$$47.27925 - 47.625 = -.34575$$

Figure 14.3 shows the distribution of these 1,000 sample means. The distribution exhibits slight positive skew, but is not far from a theoretical normal curve.

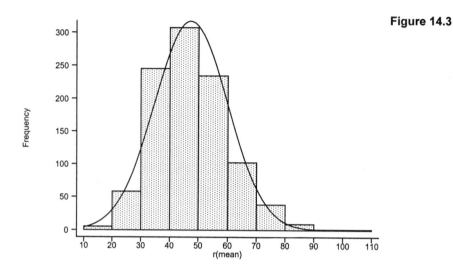

Figure 14.3

Biologists have observed that biodiversity, or the number of different kinds of plants and animals, tends to increase with island size. In *islands.dta*, we have data to test this proposition with respect to birds and flowering plants. As expected, a strong linear relation exists between *birds* and *area*:

```
. regress birds area
```

Source	SS	df	MS		
Model	9669.83255	1	9669.83255		
Residual	356.042449	6	59.3404082		
Total	10025.875	7	1432.26786		

				Number of obs =	8
				F(1, 6) =	162.96
				Prob > F =	0.0000
				R-squared =	0.9645
				Adj R-squared =	0.9586
				Root MSE =	7.7033

birds	Coef.	Std. Err.	t	P>|t|	[95% Conf. Interval]
area	.0026512	.0002077	12.77	0.000	.002143 .0031594
_cons	13.97169	3.79046	3.69	0.010	4.696773 23.24662

An e-class command, **regress** saves a set of e() results as noted earlier in this chapter. It also creates or updates a set of system variables containing the model coefficients (_b[*varname*]) and standard errors (_se[*varname*]). To bootstrap the slope and *y* intercept from the previous regression, saving the results in file *boot2.dta*, type

```
. bs "regress birds area" "_b[area] _b[_cons]", rep(1000)
      saving(boot2)
```

```
command:      regress birds area
statistics:   _b[area] _b[_cons]
(obs=8)
```

Bootstrap statistics

Variable	Reps	Observed	Bias	Std. Err.	[95% Conf. Interval]		
bs1	1000	.0026512	-.0000687	.0003649	.0019352	.0033672	(N)
					.0019569	.0029645	(P)
					.0019558	.0029384	(BC)
bs2	1000	13.97169	.5449429	3.617534	6.872857	21.07053	(N)
					7.702535	21.85249	(P)
					5.837525	21.00322	(BC)

N = normal, P = percentile, BC = bias-corrected

The bootstrap distribution of coefficients on *area* is severely skewed (skewness = 6.34). Whereas the bootstrap distribution of means (Figure 14.3) appeared approximately normal, and produced bootstrap confidence intervals narrower than the theoretical confidence interval, in this regression example bootstrapping obtains larger standard errors and wider confidence intervals.

In a regression context, **bs** ordinarily performs what is called "data resampling," or re-sampling intact observations. An alternative procedure called "residual resampling" (resampling only the residuals) requires a bit more programming work. Two additional commands make such do-it-yourself bootstrapping easier:

bsample   Draws a sample with replacement from the existing data, replacing the data in memory.

bstrap   Runs a user-defined program reps() times on bootstrap samples of size size().

The *Reference Manual* gives examples of programs for use with **bstrap** .

Monte Carlo Simulation

Monte Carlo simulations generate and analyze many samples of artificial data, allowing researchers to investigate the long-run behavior of their statistical techniques. The **simul** command makes designing a simulation straightforward so that it only requires a small amount of additional programming. This section gives two examples.

Simulation commands commonly take the form

. **simul** *progname*, **reps(500) dots**

This **simul** command asks Stata to find a program named *progname* and run that program 500 times. At the start of each repetition, a dot (optional) appears onscreen.

To begin a simulation, we need to define a program that generates one sample of random data, analyzes it, and stores the results of interest in memory. Here is a file defining a program named central. This program randomly generates 100 values of variable *x* from a standard normal distribution. It next generates 100 values of variable *w* from a "contaminated normal" distribution: N(0,1) with probability .95, and N(0,10) with probability .05. Contaminated normal distributions have often been used in robustness studies to simulate variables that contain occasional wild errors. After generating *x* and *w*, program `central` calculates two measures of central tendency, median and mean, for each variable.

```
* Creates samples containing n=100 observations of variables x and w.
* x~N(0,1)                                      x is standard normal
* w~N(0,1) with p=.95, w~N(0,10) with p=.05    w is contaminated normal
* For each sample, central calculates the mean and median of x and w.
program define central
    version 7.0
    if "`1'" == "?" {
        global S_1 "xmean xmedian wmean wmedian"
        exit
    }
    drop _all
    set obs 100
    generate x = invnorm(uniform())
    generate w = invnorm(uniform())
    replace w = 10*w if uniform() < .05
    summarize x, detail
    local XMEAN = r(mean)
    local XMEDIAN = r(p50)
    summarize w, detail
    post `1' (`XMEAN') (`XMEDIAN') (r(mean)) (r(p50))
* Macro names should ordinarily be in parentheses within a post command,
* unless they are part of an expression.
end
```

Once `central` has been defined, whether through a do-file, ado-file, or typing commands interactively, we can call this program with a **simul** command. To create a new dataset containing means and medians of *x* and *w* from 5,000 random samples, type

. **simul central, reps(5000)**

. **describe**

```
Contains data
  obs:          5,000
  vars:             4                     28 Jul 2001 19:54
  size:       100,000 (98.9% of memory free)
-------------------------------------------------------------------------------
              storage   display    value
variable name   type    format     label      variable label
-------------------------------------------------------------------------------
xmean          float    %9.0g
xmedian        float    %9.0g
wmean          float    %9.0g
wmedian        float    %9.0g
-------------------------------------------------------------------------------
Sorted by:
```

. **summarize**

```
    Variable |     Obs        Mean    Std. Dev.        Min        Max
-------------+----------------------------------------------------------
       xmean |    5000    .0025218    .1001628   -.3434822   .3699467
     xmedian |    5000    .0036662    .1240424   -.4210546   .4740642
       wmean |    5000    .0041213    .2444045   -1.002158   .9599783
     wmedian |    5000    .0030959    .1316304    -.457815   .4383048
```

The means of these means and medians, across 5,000 samples, are all close to 0—consistent with our expectation that the sample mean and median should both provide unbiased estimates of the true population means (0) for *x* and *w*. Also as theory predicts, the mean exhibits less sample-to-sample variation than the median when applied to a normally distributed variable. The standard deviation of *xmedian* is .124, noticeably larger than the standard deviation of *xmean* (.100). When applied to the outlier-prone variable *w*, on the other hand, the opposite holds true: the standard deviation of *wmedian* is much lower than the standard deviation of *wmean* (.132 vs. .244). This Monte Carlo experiment demonstrates that the median remains a relatively stable measure of center despite wild outliers in the contaminated distribution, whereas the mean breaks down and varies much more from sample to sample. Figure 14.4 draws the comparison graphically, with boxplots.

Figure 14.4

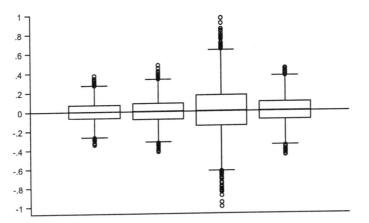

Our final example extends the inquiry to robust methods, bringing together several themes from this book. Program regsim generates 100 observations of *x* (standard normal) and two *y* variables. *y1* is a linear function of *x* plus standard normal errors. *y2* is also a linear function of *x*, but adding contaminated normal errors. These variables permit us to explore how various regression methods behave in the presence of normal and nonnormal errors. Four methods are employed: ordinary least squares (**regress**), robust regression (**rreg**), quantile regression (**qreg**), and quantile regression with bootstrapped standard errors (**bsqreg** , with 500 repetitions). Differences among these methods were discussed in Chapter 9. Program regsim applies each method to the regression of *y1* on *x* and then to the regression of *y2* on *x*. For this exercise, the program is defined by an ado-file, *regsim.ado*, saved in the C:\ado\personal directory.

```
program define regsim
* Performs one iteration of a Monte Carlo simulation comparing
* OLS regression (regress) with robust (rreg) and quantile
* (qreg and bsqreg) regression.  Generates one n = 100 sample
* with x ~ N(0,1) and y variables defined by the models:
*
*    MODEL 1:      y1 = 2x + e1       e1 ~ N(0,1)
*
*    MODEL 2:      y2 = 2x + e2       e2 ~ N(0,1) with p = .95
*                                     e2 ~ N(0,10) with p = .05
*
```

```
*
* Bootstrap standard errors for qreg involve 500 repetitions.
*
    version 7.0
    if "`1'" == "?" {
        #delimit ;
        global S_1 "b1 b1r se1r b1q se1q se1qb
                    b2 b2r se2r b2q se2q se2qb";
        #delimit cr
        exit
    }
    drop _all
    set obs 100
    generate x = invnorm(uniform())
    generate e = invnorm(uniform())
    generate y1 = 2*x + e
    reg y1 x
    local B1 = _b[x]
    rreg y1 x, iterate(25)
    local B1R = _b[x]
    local SE1R = _se[x]
    qreg y1 x
    local B1Q = _b[x]
    local SE1Q = _se[x]
    bsqreg y1 x, reps(500)
    local SE1QB = _se[x]
    replace e = 10 * e if uniform() < .05
    generate y2 = 2*x + e
    reg y2 x
    local B2 = _b[x]
    rreg y2 x, iterate(25)
    local B2R = _b[x]
    local SE2R = _se[x]
    qreg y2 x
    local B2Q = _b[x]
    local SE2Q = _se[x]
    bsqreg y2 x, reps(500)
    local SE2QB = _se[x]
    #delimit ;
    post `1' (`B1') (`B1R') (`SE1R') (`B1Q') (`SE1Q') (`SE1QB')
             (`B2') (`B2R') (`SE2R') (`B2Q') (`SE2Q') (`SE2QB');
    #delimit cr
end
```

The program posts coefficient or standard error estimates from eight regression analyses, storing them as variables *b1* (coefficient from OLS regression of *y1* on *x*), *b1r* (coefficient from robust regression of *y1* on *x*), *se1r* (standard error of robust coefficient from model 1), and so forth. All the robust and quantile regressions involve multiple iterations: typically 5 to 10 iterations for **rreg**, about 5 for **qreg**, and several thousand for **bsqreg** with its 500 bootstrap re-estimations of about 5 iterations each, *per sample*. Thus, a single execution of *regsim* demands more than two thousand regressions. The following command calls for ten repetitions:

```
. simul regsim, reps(10) dots
```

You might want to run a small simulation like this as a trial to get a sense of the time required on your computer. For research purposes, however, we would need a much larger experiment. Dataset *regsim.dta* contains results from an overnight experiment involving 5,000 repetitions of regsim —more than 10 million regressions. The regression coefficients and standard error estimates are summarized below. Figure 14.5 draws the distributions of coefficients as boxplots.

```
. describe

Contains data from c:\data\regsim.dta
   obs:         5,000                          Monte Carlo estimates of b;
                                               5000 samples of n=100
                                               (Statistics w/Stata, Ch. 14)
   vars:          12                           29 Jul 2001 13:13
   size:      260,000 (97.5% of memory free)
------------------------------------------------------------------------
              storage  display   value
variable name  type    format    label        variable label
------------------------------------------------------------------------
b1            float    %9.0g                   OLS b (normal errors)
b1r           float    %9.0g                   Robust b (normal errors)
se1r          float    %9.0g                   Robust SE[b] (normal errors)
b1q           float    %9.0g                   Quantile b (normal errors)
se1q          float    %9.0g                   Quantile SE[b] (normal errors)
se1qb         float    %9.0g                   Quantile bootstrap SE[b]
                                                 (normal errors)
b2            float    %9.0g                   OLS b (contaminated errors)
b2r           float    %9.0g                   Robust b (contaminated errors)
se2r          float    %9.0g                   Robust SE[b] (contaminated
                                                 errors)
b2q           float    %9.0g                   Quantile b (contaminated errors)
se2q          float    %9.0g                   Quantile SE[b] (contaminated
                                                 errors)
se2qb         float    %9.0g                   Quantile bootstrap SE[b]
                                                 (contaminated errors)
------------------------------------------------------------------------
Sorted by:


. summarize

    Variable |     Obs        Mean    Std. Dev.        Min         Max
-------------+----------------------------------------------------------
          b1 |    5000    2.000538    .1016268    1.592447    2.368651
         b1r |    5000    2.000541    .1046721    1.601189    2.363003
        se1r |    5000    .1042421    .0108813    .0619586     .147383
         b1q |    5000    2.000675    .1282029    1.492559     2.43919
        se1q |    5000    .1264252    .0290164    .0480243    .2975625
       se1qb |    5000    .1354706    .0320264    .0537904    .2733456
          b2 |    5000    1.999094    .2485003    .8323185    3.217123
         b2r |    5000    2.000961    .1091681    1.620961    2.413531
        se2r |    5000    .1081962    .0117687    .0656153    .1676369
         b2q |    5000    2.000405    .1341527    1.492559     2.47863
        se2q |    5000    .1327056    .0306925    .0480243    .3080628
       se2qb |    5000    .1430813    .0341174    .0570831    .3005342
```

Figure 14.5

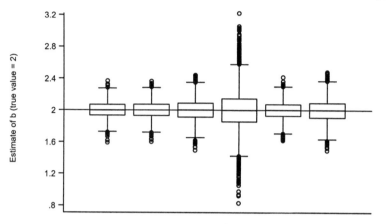

All three regression methods (OLS, robust, and quantile) produced mean coefficient estimates for both models that are not significantly different from the true value, $\beta = 2$. This can be confirmed through t tests such as

```
. ttest b2r = 2
```

One-sample t test

```
---------------------------------------------------------------------------
Variable |      Obs       Mean     Std. Err.    Std. Dev.   [95% Conf. Interval]
---------+-----------------------------------------------------------------
     b2r |     5000     2.000961    .0015439     .1091681    1.997934    2.003987
---------------------------------------------------------------------------
Degrees of freedom: 4999
```

Ho: mean(b2r) = 2

```
    Ha: mean < 2                   Ha: mean ~= 2                  Ha: mean > 2
      t =   0.6223                   t =   0.6223                   t =   0.6223
  P < t =   0.7331              P > |t| =   0.5338              P > t =   0.2669
```

All the regression methods thus yield unbiased estimates of β, but they differ in their sample-to-sample variation or efficiency. Applied to the normal-errors model 1, OLS proves the most efficient, as the famous Gauss–Markov theorem would lead us to expect. The observed standard deviation of OLS coefficients is .1016, compared with .1047 for robust regression and .1282 for quantile regression. Relative efficiency, expressing the OLS coefficient's observed variance as a percentage of another estimator's observed variance, provides a standard way to compare such statistics:

```
. quietly summarize b1
. global Varb1 = r(Var)
. quietly summarize b1r
. display 100*($Varb1/r(Var))
94.26595
. quietly summarize b1q
```

```
. display 100*($Varb1/r(Var))
62.837802
```

The calculations above use the `r(Var)` variance result from **summarize**. We first obtain the variance of the OLS estimates *b1*, and place this into global macro `Varb1`. Next the variances of the robust estimates *b1r*, and the quantile estimates *b1q*, are obtained and each compared with `Varb1`. This reveals that robust regression was about 94% as efficient as OLS when applied to the normal-errors model—close the large-sample efficiency of 95% that this robust method theoretically should have (Hamilton 1992a). Quantile regression, in contrast, achieves a relative efficiency of only 63% with the normal-errors model.

Similar calculations for the contaminated-errors model tell a different story. OLS was the best (most efficient) estimator with normal errors, but with contaminated errors it becomes the worst:

```
. quietly summarize b2

. global Varb2 = r(Var)

. quietly summarize b2r

. display 100*($Varb2/r(Var))
518.15827

. quietly summarize b2q

. display 100*($Varb2/r(Var))
343.12699
```

Outliers in the contaminated-errors model cause OLS coefficient estimates to vary wildly from sample to sample, as can be seen in the fourth boxplot of Figure 14.5. The variance of these OLS coefficients is more than five times greater than the variance of the corresponding robust coefficients, and more than three times greater than that of quantile coefficients. Put another way, both robust and quantile regression prove to be much more stable than OLS in the presence of outliers, yielding correspondingly lower standard errors and narrower confidence intervals. Robust regression outperforms quantile regression with both the normal-errors and the contaminated-errors models. In a scatterplot showing 5,000 pairs of regression coefficients (Figure 14.6), the **qreg** coefficients (vertical axis) vary more widely around the true value, 2.0, than **rreg** coefficients (horizontal axis) do.

The experiment also provides information about the estimated standard errors under each method and model. Mean estimated standard errors differ from the observed standard deviations of coefficients, but most of these discrepancies are small—less than 1% for the robust standard errors, and less than 2% for the quantile standard errors. The least satisfactory estimates appear to be the bootstrap standard errors obtained by **bsqreg**. Means of the bootstrap standard errors exceed the observed standard deviation of *b1q* and *b2q* by 6% or 7%. Bootstrapping apparently overestimated the sample-to-sample variation.

Monte Carlo simulation has become a key method in modern statistical research, and it plays a growing role in statistical teaching as well. These examples demonstrate how readily Stata supports Monte Carlo work.

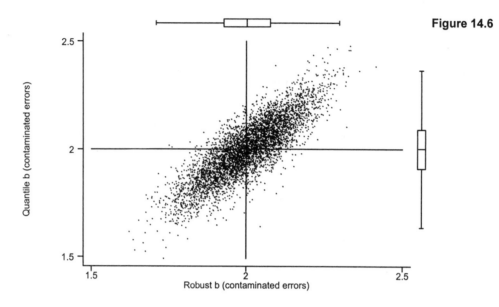

Figure 14.6

References

Barron's Educational Series. 1992. *Barron's Compact Guide to Colleges*, 8th ed. New York: Barron's Educational Series.

Beatty, J. Kelly, Brian O'Leary, and Andrew Chaikin (eds.). 1981. *The New Solar System*. Cambridge, MA: Sky.

Belsley, D. A., E. Kuh, and R. E. Welsch. 1980. *Regression Diagnostics: Identifying Influential Data and Sources of Collinearity*. New York: Wiley.

Box, G. E. P., G. M. Jenkins and G. C. Reinsel. 1994. *Time Series Analysis: Forecasting and Control*. 3rd ed. Englewood Cliffs, NJ: Prentice–Hall.

Brown, Lester R., William U. Chandler, Christopher Flavin, Cynthia Pollock, Sandra Postel, Linda Starke and Edward C. Wolf. 1986. *State of the World 1986*. New York: W. W. Norton.

Buch, E. 2000. *Oceanographic Investigations off West Greenland 1999*. Copenhagen: Danish Meteorological Institute.

Chambers, John M., William S. Cleveland, Beat Kleiner, and Paul A. Tukey (eds.). 1983. *Graphical Methods for Data Analysis*. Belmont, CA: Wadsworth.

Chatfield, C. 1996. *The Analysis of Time Series: An Introduction*. 5th ed. London: Chapman & Hall.

Chatterjee, S., A. S. Hadi, and B. Price. 2000. Regression Analysis by Example. 3rd edition. New York: John Wiley and Sons.

Cleveland, William S. 1994. *The Elements of Graphing Data*. Monterey, CA: Wadsworth.

Cook, R. Dennis and Sanford Weisberg. 1982. *Residuals and Influence in Regression*. New York: Chapman & Hall.

Cook, R. Dennis and Sanford Weisberg. 1994. *An Introduction to Regression Graphics*. New York: Wiley.

Council on Environmental Quality. 1988. *Environmental Quality 1987–1988*. Washington, DC: Council on Environmental Quality.

Cox, C. Barry and Peter D. Moore. 1993. *Biogeography: An Ecological and Evolutionary Approach*. London: Blackwell.

Cryer, Jonathan B. and Robert B. Miller. 1994. *Statistics for Business: Data Analysis and Modeling*, 2nd edition. Belmont, CA: Duxbury.

Davis, Duane. 2000. *Business Research for Decision Making*, 5th edition. Belmont, CA: Duxbury.

Diggle, P. J. 1990. *Time Series Analysis: A Biostatistical Introduction*. Oxford: Oxford University Press.

Efron, Bradley and R. Tibshirani. 1986. "Bootstrap methods for standard errors, confidence intervals, and other measures of statistical accuracy." *Statistical Science* 1(1):54–77.

Enders, W. 1995. *Applied Econometric Time Series*. New York: Wiley.

Federal, Provincial, and Territorial Advisory Commission on Population Health. 1996. *Report on the Health of Canadians.* Ottowa: Health Canada Communications.

Fox, John. 1991. *Regression Diagnostics.* Newbury Park, CA: Sage.

Fox, John and J. Scott Long (eds.). 1990. *Modern Methods of Data Analysis.* Beverly Hills: Sage.

Frigge, Michael, David C. Hoaglin, and Boris Iglewicz. 1989. "Some implementations of the boxplot." *The American Statistician* 43(1):50–54.

Gould, William and William Sribney. 1999. *Maximum Likelihood Estimation with Stata.* College Station, TX: Stata Press.

Hamilton, James D. 1994. *Time Series Analysis.* Princeton, NJ: Princeton University Press.

Hamilton, Lawrence C. 1985a. "Concern about toxic wastes: Three demographic predictors." *Sociological Perspectives* 28(4):463–486.

Hamilton, Lawrence C. 1985b. "Who cares about water pollution? Opinions in a small-town crisis." *Sociological Inquiry* 55(2):170–181.

Hamilton, Lawrence C. 1990. *Modern Data Analysis: A First Course in Applied Statistics.* Pacific Grove, CA: Brooks/Cole.

Hamilton, Lawrence C. 1992a. *Regression with Graphics: A Second Course in Applied Statistics.* Pacific Grove, CA: Brooks/Cole.

Hamilton, Lawrence C. 1992b. "Quartiles, outliers and normality: Some Monte Carlo results." Pp. 92–95 in Joseph Hilbe (ed.) *Stata Technical Bulletin Reprints, Volume 1.* College Station, TX: Stata.

Hamilton, Lawrence C. 1996. *Data Analysis for Social Scientists.* Belmont, CA: Duxbury.

Hamilton, Lawrence C. 1996. *Data Analysis for Social Scientists with StataQuest.* Belmont, CA: Duxbury.

Hamilton, Lawrence C., Benjamin C. Brown, and Rasmus Ole Rasmussen. 2001. "Local dimensions of climatic change: West Greenland's cod-to-shrimp transition." University of New Hampshire: NAArc Working Paper 01–3.

Hamilton, Lawrence C. and Carole L. Seyfrit. 1993. "Town-village contrasts in Alaskan youth aspirations." *Arctic* 46(3):255–263.

Hardin, James and Joseph Hilbe. 2001. *Generalized Linear Models and Extensions.* College Station, TX: Stata Press.

Hoaglin, David C., Frederick Mosteller and John W. Tukey (eds.). 1983. *Understanding Robust and Exploratory Data Analysis.* New York: Wiley.

Hoaglin, David C., Frederick Mosteller and John W. Tukey (eds.). 1985. *Exploring Data Tables, Trends and Shape.* New York: Wiley.

Hosmer, David W., Jr. and Stanley Lemeshow. 1999. *Applied Survival Analysis.* New York: Wiley.

Hosmer, David W., Jr. and Stanley Lemeshow. 2000. *Applied Logistic Regression*, 2nd edition. New York: Wiley.

Howell, David C. 1999. *Fundamental Statistics for the Behavioral Sciences*, 4th edition. Belmont, CA: Duxbury.

Howell, David C. 2002. *Statistical Methods for Psychology*, 5th edition. Belmont, CA: Duxbury.

Iman, Ronald L. 1994. *A Data-Based Approach to Statistics.* Belmont, CA: Duxbury.

Jentoft, Svein and Trond Kristoffersen. 1989. "Fishermen's co-management: The case of the Lofoten fishery." *Human Organization* 48(4):355–365.

Johnson, Anne M., Jane Wadsworth, Kaye Wellings, Sally Bradshaw and Julia Field. 1992. "Sexual lifestyles and HIV risk." *Nature* 360(3 December):410–412.

Johnston, Jack and John DiNardo. 1997. *Econometric Methods*, 4th edition. New York: McGraw-Hill.

Keller, Gerald, Brian Warrack and Henry Bartel. 2000. *Statistics for Management and Economics*, 5th edition. Belmont, CA: Duxbury.

League of Conservation Voters. 1990. *The 1990 National Environmental Scorecard*. Washington DC: League of Conservation Voters.

Lee, Elisa T. 1992. *Statistical Methods for Survival Data Analysis* (second edition). New York: Wiley.

Li, Guoying. 1985. "Robust regression." Pp. 281–343 in D. C. Hoaglin, F. Mosteller and J. W. Tukey (eds.) *Exploring Data Tables, Trends and Shape*. New York: Wiley.

Long, J. Scott. 1997. *Regression Models for Categorical and Limited Dependent Variables*. Thousand Oaks, CA: Sage.

Long, J. Scott and Jeremy Freese. 2001. *Regression Models for Categorical Outcomes Using Stata*. College Station, TX: Stata Press.

MacKenzie, Donald. 1990. *Inventing Accuracy: A Historical Sociology of Nuclear Missile Guidance*. Cambridge, MA: MIT.

Mallows, C. L.. 1986. "Augmented partial residuals." *Technometrics* 28:313–319.

McCullagh, D. W. Jr., and J. A. Nelder. 1989. *Generalized Linear Models*, 2nd edition. London: Chapman and Hall.

Mayewski, P. A., G. Holdsworth, M. J. Spencer, S. Whitlow, M. Twickler, M. C. Morrison, K. K. Ferland and L. D. Meeker. 1993. "Ice-core sulfate from three northern hemisphere sites: Source and temperature forcing implications." *Atmospheric Environment* 27A(17/18):2915–2919.

Mayewski, P. A., L. D. Meeker, S. Whitlow, M. S. Twickler, M. C. Morrison, P. Bloomfield, G. C. Bond, R. B. Alley, A. J. Gow, P. M. Grootes, D. A. Meese, M. Ram, K. C. Taylor and W. Wumkes. 1994. "Changes in atmospheric circulation and ocean ice cover over the North Atlantic during the last 41,000 years." *Science* 263:1747–1751.

Nash, James and Lawrence Schwartz . 1987. "Computers and the writing process." *Collegiate Microcomputer* 5(1):45–48.

National Center for Education Statistics. 1992. *Digest of Education Statistics 1992*. Washington, DC: U.S. Government Printing Office.

National Center for Education Statistics. 1993. *Digest of Education Statistics 1993*. Washington, DC: U.S. Government Printing Office.

Newton, H. Joseph and Jane L. Harvill. 1997. *StatConcepts: A Visual Tour of Statistical Ideas*. Pacific Grove,, CA: Duxbury.

Pagano, Marcello and Kim Gauvreau. 2000. *Principles of Biostatistics*, 2nd edition. Belmont, CA: Duxbury.

Rabe–Hesketh, Sophia and Brian Everitt. 2000. *A Handbook of Statistical Analysis Using Stata*, 2nd edition. Boca Raton, FL: Chapman & Hall.

Report of the Presidential Commission on the Space Shuttle Challenger Accident. 1986. Washington, DC.

Rosner, Bernard. 1995. *Fundamentals of Biostatistics*, 4th edition. Belmont, CA: Duxbury.

Selvin, Steve. 1995. *Practical Biostatistical Methods*. Belmont, CA: Duxbury.

Selvin, Steve. 1996. *Statistical Analysis of Epidemiologic Data*, 2nd edition. New York: Oxford University.

Seyfrit, Carole L.. 1993. *Hibernia's Generation: Social Impacts of Oil Development on Adolescents in Newfoundland*. St. John's: Institute of Social and Economic Research, Memorial University of Newfoundland.

Shumway, R. H. 1988. *Applied Statistical Time Series Analysis*. Upper Saddle River, NJ: Prentice–Hall.

Stata Corporation. 2001. *Getting Started with Stata for Macintosh*. College Station, TX: Stata Press.

Stata Corporation. 2001. *Getting Started with Stata for Unix*. College Station, TX: Stata Press.

Stata Corporation. 2001. *Getting Started with Stata for Windows*. College Station, TX: Stata Press.

Stata Corporation. 2001. *Stata Graphics Manual, Release 7*. College Station, TX: Stata Press.

Stata Corporation. 2001. *Stata Programming Manual, Release 7*. College Station, TX: Stata Press.

Stata Corporation. 2001. *Stata Reference Manual, Release 7*, Volumes 1–4. College Station, TX: Stata Press.

Stata Corporation. 2001. *Stata User's Guide, Release 7*. College Station, TX: Stata Press.

Stine, Robert and John Fox (eds.). 1997. *Statistical Computing Environments for Social Research*. Thousand Oaks, CA: Sage.

Topliss, Brenda J. 2000. "Climate variability I: A conceptual approach to ocean-atmosphere feedback." In Abstracts for AGU Chapman Conference, The North Atlantic Oscillation, Nov. 28 - Dec 1, 2000, Ourense, Spain.

Tufte, Edward R. 1997. *Visual Explanations: Images and Quantities, Evidence and Narratives*. Cheshire, CT: Graphics Press.

Tukey, John W. 1977. *Exploratory Data Analysis*. Reading, MA: Addison Wesley.

Velleman, Paul F. 1982. "Applied Nonlinear Smoothing," pp.141–177 in Samuel Leinhardt (ed.) *Sociological Methodology 1982*. San Francisco: Jossey-Bass.

Velleman, Paul F. and David C. Hoaglin. 1981. *Applications, Basics and Computing of Exploratory Data Analysis*. Boston: Wadsworth.

Ward, Sally and Susan Ault. 1990. "AIDS knowledge, fear, and safe sex practices on campus." *Sociology and Social Research* 74(3):158–161.

Werner, Al. 1990. "Lichen growth rates for the northwest coast of Spitsbergen, Svalbard." *Arctic and Alpine Research* 22(2):129–140.

World Bank. 1987. *World Development Report 1987*. New York: Oxford University.

World Resources Institute. 1993. *The 1993 Information Please Environmental Almanac*. Boston: Houghton Mifflin.

Index

A

absorbing categorical variables, 143–144
ac 277, 288
acprplot, 153, 158–159
added-variable plot, 154, 157–158
ado-file (automatic do), 298, 308–309, 314–316, 323–324
alpha reliability, 266–267
analysis of covariance (ANCOVA), 108–109, 119–121
analysis of variance (ANOVA), 108–109, 115–123, 203–208
anova, 108–109, 118–123
append, 12, 39–40
arch, 277
ARCH model. *See* **arch**
areg, 143–144
arguments, 301–303
args, 301–302
arima, 277, 290–296
ARIMA model. *See* **arima**
ASCII (text) file
    input data, 13, 36–38
    output data, 38
    output results. *See* log files
augmented component-plus-residual plot, 153, 158–159
autocorrelation
    cross, 289–290
    Durbin–Watson test, 123, 152–153, 277, 287, 295–296
    graphs, 287–290
    moving, 304–312
    partial, 288–289
    Prais–Winsten regression, 278, 295–296
autoregressive model. *See* **arima**
avplot, 154, 157–158
axis labels in graphs, 61–62

B

band regression, 171–174
bar chart, 58–59, 76–80
Bartlett's test for unequal variances, 116
batch-mode program, 55–56
bcskew0, 97
beta weights (standardized regression coefficients), 125
Bonferroni multiple-comparison adjustment
    correlation matrix, 136
    one-way ANOVA, 117
books about or using Stata, 8, 10–11
bootstrap, 194, 199–200, 263–264, 317–321, 323–327
Box–Cox transformation
    regression models, 171, 180–181
    symmetrize one variable, 97
Box–Pierce portmanteau white noise test, 277–278, 288, 293–294
boxcox, 171, 180–181
boxplot, 58–59, 66, 71–73
browse, 12
Browser, 12
bs, 317–319
bsqreg, 194, 199–200, 323–327
by, 89, 100–101

C

c chart (quality control), 84
case identification numbers, 35
case sensitive, 2
categorical variables, 24–25, 32–35, 213–214
cchart, 84
chi-squared
    deviance, 221, 225–227
    likelihood-ratio, 98, 218, 222–223, 228–229, 232–233
    Pearson, 98, 221, 224–226
    probability plot, 84

chi-squared, *continued*
>   quantile plot, 84
>   test of fit in classification table, 214
>   test of independence in cross-tabulation, 50, 98–101
Chronbach's α (alpha) reliability, 266–267
classification table (logit, etc.), 214, 220–222
collapse, 47–48
combine datasets. *See* **append** or **merge**
command line length in programs, 56
comments in programs, 299–300, 304–305, 309, 313
component-plus-residual plot, 154, 158–159
compress, 13, 36–37, 55, 240
conditional effect plot, 183–185, 223–224
confidence interval
>   bootstrap, 317–319
>   mean (normal), binomial, poisson, 92, 318
>   regression coefficient, 127
>   regression prediction, 133–135
>   robust, 208–209
constraint, 213
convert string to numeric variables, 31–32
Cook's *D*, 131, 153, 161–165
correlate, 4, 18, 125, 135–137
correlation
>   hypothesis test, 125, 135–136
>   Kendall's τ (tau-a and tau-b), 99, 138–139
>   Pearson product-moment, 4, 18, 135–139
>   regression coefficient estimates, 137, 169–170
>   Spearman rank, 138–139
correlograms, 277, 287–290
corrgram, 277, 287–288
count-time data, 242–244
covariance
>   regression coefficient estimates, 137, 170
>   variables, 136–137
COVRATIO, 131, 153, 161–165
Cox proportional hazard model, 238, 247–253
cprplot, 154, 158–159
Cramer's *V*, 99
create new dataset, 14–18
create new variables. *See* generating new variables
cross correlations, 289–290
cross-tabulation, 50, 97–103
ctset, 238, 242–243
cttost, 238, 242–244
cubic spline curve, 59

D
data dictionary, 37–38
database files. *See* importing data from other programs
dates, 27–28, 216–217, 278–280
decode, 30–31
define new variables. *See* generating new variables
delete observations or variables. *See* **drop** or **keep**
delimiter (end-of-line), 56
describe, 3, 16–17
dfbeta, 153, 161–165
DFBETAS, 131, 153, 161–165
DFITS, 131, 153, 161–164
dfuller, 277, 291–292
diagnostic plots
>   distribution shape, 80–84
>   logit, 224–228
>   regression, 153–154, 156–160
diagnostic statistics. *See* **predict**
Dickey–Fuller unit root test, 277, 291–292
differencing (time series), 35, 285–286
display, 28–29, 35, 133, 162, 166–167, 219, 248, 293, 316, 326–327
display formats. *See* **format**
do-file, 55–56, 297–298, 301–308, 312–314
Do-file Editor, 55–56, 297, 300, 308, 310
documentation, 6, 10–11
drawnorm, 13
drop, 21–22
dummy variable, 32–33, 98, 140–146
Durbin–Watson test, 123, 152–153, 277, 287, 295–296
dwstat, 153, 277, 287, 295–296

E
e-class command, 315–316, 320
EDA (exploratory data analysis), 71–73, 92–94
edit, 13
Editor, 13, 15–16, 18, 21
egen, 29, 277, 280–281, 284
eigenvalues, 266–269
empirical orthogonal function (EOF). *See* principal components
encode, 13, 30–31
error-bar chart, 109, 121–123
estimates, 315–316
event count models, 237, 239, 257–261
event history analysis. *See* survival analysis
Excel files. *See* importing data from other programs

exploratory data analysis (EDA), 71–73, 92–94
exponential growth model, 186–189

F

factor, 266–276
factor analysis, 266–276
factor rotation, 266–267, 274
factor scores, 266–267, 270–273
factorial ANOVA, 108–109, 118–119
FAQs (frequently asked questions), 7–8
fences, 59, 71–73, 93–94
Fisher's exact test, 98
format, 13, 23–24
format
    display. *See* **format**
    input data (fixed-column). *See* **infix**
frequency table, 88–89, 97–100, 230
frequency weights, 49–50, 105–107
functions
    date functions, 27–28
    mathematical functions, 25–26
    special functions, 28
    statistical functions, 26–27

G

generalized linear model, 213, 237, 239, 261–265
generate, 13, 22–29, 33–35
generate new variables 13, 22–35
gladder, 96
glm, 213, 215, 239, 261–265
Gompertz growth model, 187–192
Goodman and Kruskal's γ (gamma), 98
graph, 4–5, 58–81, 172–174, 224–228, 280–285
Graph window, 5
graphs
    copy to document, 5
    .gph format (Stata), 5
    printing, 5
    saving, 5
    .wmf format (Windows metafile), 5
greigen, 266–269
growth models, 172, 185–192

H

hat matrix diagonals, 131, 159–162
hazard function, 238–239, 247–257
help, 6–7
help file (writing for program), 310–312

heteroskedasticity-robust variance estimates. *See* robust, standard errors
heteroskedasticity test, 153, 155–156
hettest, 153, 155–156
histogram, 58, 60–64
Huber/White estimates of variance. *See* robust, standard errors

I

if qualifier, 18–22, 23, 160
importing data from other programs, 36–39
in qualifier, 18–22, 23, 130
incidence rate, 238, 239, 241–244, **246**, 257–261
infile, 13, 36–37
infix, 37–38
influence statistics
    logit, 214–215, 221, 224–228
    regression, 131
insert graph into document, 5
insert table into document, 4–5
insheet, 37
instrumental variables (2SLS), 125
interaction effects
    ANOVA, 118–120
    dummy variable regression, 144–**146**
interquartile range, 29, 48, 59, 91, 101–103

J

jackknife
    residuals (studentized residuals), 131
    standard errors, 262–265

K

Kaplan–Meier survivor function, 238–239, 241, 244–247, 253–254
keep, 22
keep observations or variables. *See* **drop** or **keep**
Kendall's τ (tau-a and tau-b), 99, 138–139
Kruskall–Wallis test, 109, 117–118
ksm, 172, 174–177
ktau, 138–139
kurtosis, 88, 90–91, 94–95
kwallis, 109, 117–118

L

L1-estimators, 197
label values, 24–25, 30
label variable, 15, 17, 22
ladder, 96

ladder of powers, 95–97
lags, 278, 285–294
letter-value display, 93–94
leverage (hat matrix diagonals), 131, 159–160, 200–202
leverage plot. *See* added-variable plot
leverage-vs.-squared-residuals plot, 159–160
lfit, 214
licensing, 7–8
likelihood ratio chi-squared, 98, 218, 222–223, 228–229, 232–233
link function. *See* generalized linear model
list, 3, 13, 16, 19–20, 31
listserv (Statalist), 8–9
log files, 2–3, 5–6
logical operators, 20
logistic, 213–214, 219–228
logistic growth model, 187–188
logit, 213–224
loop in program, 300–301
lowess regression (smoothing), 153, 171–172, 174–177, 189, 283
lrtest, 222–223, 228–229, 232–233
lstat, 214, 220–222
lv, 92–93
lvr2plot, 159–160

M
M-estimators, 197
macro
   global, 188–189, 299, 322, 324, 326–327
   local, 188–189, 299, 301–304, 308–309, 315–316, 322, 324,
Mann–Whitney *U* test, 109, 115
marksample, 302–303, 315
matrix algebra, 312–316
maximum, 3–4, 29, 48, 88, 90–91, 101–103
maximum likelihood factor analysis, 266, 268, 274–276
mean, 3–4, 48, 88–92, 101–104, 321–323
median, 48, 59, 66, 71–73, 82, 88–92, 101–104, 321–323
median ANOVA (quantile regression), 207–208
median regression (quantile regression), 193–194, 197, 199–202, 207–210
median tests (quantile regression), 208–209
memory, 14, 56–57
merge, 13, 40–43
minimum, 3–4, 29, 48, 88, 90–91, 101–103
missing values, 15, 20–21, 135, 137–138
mlogit, 230–236

Monte Carlo simulation, 94, 193, 199, 201, 321–328
moving average (smoothing), 277, 280–281
moving average (time series model). *See* **arima**
multicollinearity, 166–170
multiple-comparison tests
   correlation matrix, 136
   one-way ANOVA, 109, 116–117

N
negative exponential growth model, 187
net search, 7
nl, 172
nonlinear regression, 171–172, 185–192
normal (Gaussian) distributions
   curve with histogram, 58, 62–64, 81
   normal probability plot, 60, 82–84
   quantile-normal plot, 60, 82–84
   random variables from, 13
   test for normality, 94–97
numerical variables, 15, 29–32

O
observation number (_n), 35
odds ratio. *See* regression, logit
ologit, 228–230
one-sample *t* tests, 110–112
oneway, 109, 115–118
open file (dataset), 3
operators
   logical, 20
   relational, 19
order, 18
ordering Stata, 7–8
ordinal variables, 32–35
outfile, 38
outliers, 63, 66, 71–72, 93–94, 112, 118, 138, 154, 165, 193–209, 227–228, 272, 281, 321–328
ovtest, 153, 155

P
p chart (quality control), 84–85
pac, 278, 288–289
paired-difference tests, 110–113
panel data, 126, 214
parentheses, 20
partial autocorrelation, 278, 287–289
partial regression plot. *See* added-variable plot
pchart, 84–85
pchi, 84
Pearson correlations. *See* **correlate**

percentages in cross-tabulation, 50
percentiles, 48, 88, 90–92, 101–103
periodogram, 278
pie chart, 58–59, 73–75
plotting symbols in scatterplot, 67–70, 132–135, 225–227
pnorm, 84
poisson, 239, 257–261
polynomial regression, 149–152
power transformations, 95–97
predict
    after **anova**, 108, 121–123, 129–135
    after **logit** or **logistic**, 219, 221, 224–228
    after **regress**, 124, 129–135, 152–153, 160–165
predicted values. *See* **predict**
principal components, 266–273
print
    graph, 5
    results tables, 4
programming (user written), 55–56, 172, 187–189, 297–328
promax (oblique) rotation, 267, 269–273
pseudo R squared, 218
pseudo standard deviation, 93–94
pseudosigma, 93–94
pwcorr, 125, 135–139

Q

qchi, 84
qladder, 96–97
qnorm, 82–83
qqplot, 83
qreg, 193–194, 197, 199–202, 207–210, 323–328
qualifiers. *See* **in** qualifier, **if** qualifier
quality control graphs 60, 84–87
quantiles
    defined, 81–82
    quantile-normal plot, 60, 82–84, 96–97
    quantile plot, 82
    quantile-quantile plot, 83–84
    quantile regression, 193–194, 197, 199–202, 207–210
quartiles, 29, 71–73, 82, 90, 92–94
quietly, 139, 161, 166–167, 188, 222–224, 228–229, 301–304

R

r chart (quality control), 84, 86
r-class program, 316–318

Ramsey regression specification error test (RESET), 153, 155
random numbers, 51–52
random sampling. *See* **sample**
random variables, 13, 51–55
range, 190–191
range, 29
rangefinder box (also *see* boxplot), 59
rank, 29
ranksum, 109, 115, 118
rchart, 84, 86
relative efficiency of estimators, 326–327
regress, 124–134, 139–146, 171–172, 177–185, 193–212
regression
    beta weights (standardized regression coefficients), 125
    binomial, 213
    censored normal, 215
    coefficients (stored results), 130
    confidence bands, 163–135
    constant (or no constant), 127–128
    Cox, 238–239, 248–253
    curvilinear, 149–151, 171–172, 177–185
    diagnostics, 152–170
    dummy variables, 140–146
    exponential (survival time), 239, 253–257
    hypothesis tests, 125, 139–140
    influence analysis, 160–165
    logit, 213–236
    multicollinearity, 166–170
    multinomial logit, 230–236
    multiple, 128–129
    negative binomial, 148, 237, 239, 262
    nonlinear, 171–172, 185–192
    ordered logit, 228–230
    outliers, 197–203
    panel data, 126
    Poisson, 237, 239, 257–262, 265
    polynomial, 149–152
    predictions. *See* **predict**
    probit, 213–214
    residuals. *See* **predict**
    robust, 193–210
    simple, 124–128
    stepwise, 125
    survey data, 126
    tobit, 214
    transformed variables. *See* regression, curvilinear
    Weibull, 239, 253, 255–257
    weighted least squares (WLS), 125

relational operators, 19
relative risk ratios, 215, 231–233
rename, 15, 17
replace, 13, 22–29, 34
reshape, 45–47
residuals. *See* **predict**
residuals vs. predicted values (residual vs. fitted) plot, 124, 132–133, 154, 156, 163
residuals vs. predictor plot, 154
restructure dataset, 43–48
results tables, 4–5
Results window, 4
robust
    ANOVA, 203–208
    confidence intervals, 208–209
    regression, 193–210, 323–328
    standard errors, 124, 209–212
ROC (receiver operating characteristic) curve, 214
rotate, 266–267, 269–274
rreg, 193–210, 323–328
rvfplot, 154, 156
rvpplot, 154, 156

S

sample, 14, 55
sandwich estimator of variance. *See* robust, standard errors
SAS files. *See* importing data from other programs
save, 14, 16–17, 22
save dataset. *See* **save**
scatterplot
    connect points (lines or curves), 59, 68–70, 132–135, 159
    matrix, 58, 66–67, 125, 137–138
    oneway, 58–59, 66, 72–73
    plotting symbols, 67–70, 132–135, 225–227
    regression line, 124, 132–135, 144–146
    twoway, 58–59, 64–70
    weighting, 65, 163, 227
Schefffé multiple-comparison test, 109, 116–117
score, 266–267, 270–274
scree graph. *See* **greigen**
scrolling output, 56
search, 7
seasonal differencing (time series), 286, 291
serrbar, 109, 121–123
set memory, 14
sfrancia, 95

Shapiro–Francia *W'* test for normality, 95
Shapiro–Wilk *W* test for normality, 95
Šidák multiple-comparison adjustment
    correlation matrix, 136
    one-way ANOVA, 117
signrank, 109, 112–113
signtest, 110–111
simul, 321–325
simulation. *See* **simul**, Monte Carlo
skewness, 88, 91–92, 94–97
skewness–kurtosis test for normality, 94–95
sktest, 94–95
smcl (Stata markup and control language), 2–3
smooth, 278, 281–283
smoothing
    lowess, 153, 171–172, 174–177, 189, 283
    moving averages, 277, 280–281
    purpose, 278–282
    resistant nonlinear, 281–283
sort (also *see* **qsort**), 14, 16, 19–20, 35
spearman, 138–139
spikeplot, 64
spreadsheet files. *See* importing data from other programs
SPSS files. *See* importing data from other programs
stack bar chart, 77
standard deviation, 3–4, 29, 88–92, 101–103
standard error
    ANOVA, 121–123
    mean, 91–92
    regression predictions, 131–135
    robust, 124, 209–212
standardized regression coefficients, 125
standardized residuals, 131, 161–162
standardized value, 29
star chart, 58–59, 75–76
Stat/Transfer, 38–39
Stata Corporation, 7–8
Stata Journal, 8–10
Stata Technical Bulletin, 6, 9–10
Stata web site (www.stata.com), 6
Statalist listserv, 8–9
STB FTP, 8
stcox, 223, 238–239, 248–252
stcurve, 239, 254–256
stdes, 238, 241, 243, 249
stem, 92–93
stem-and-leaf display, 92–93
stepwise regression, logit, probit, etc., 125, 146–148
streg, 239, 253–257